PROTEIN TARGETING WITH SMALL MOLECULES

PROTEIN TARGETING WITH SMALL MOLECULES

Chemical Biology Techniques and Applications

Edited by

HIROYUKI OSADA
RIKEN Advanced Science Institute
Wako, Saitama, Japan

WILEY

A JOHN WILEY & SONS, INC., PUBLICATION

For general information on our other products and services or for technical support, please contact our Customer Care Department within the United States at (800) 762-2974, outside the United States at (317) 572-3993 or fax (317) 572-4002.

Wiley also publishes its books in a variety of electronic formats. Some content that appears in print may not be available in electronic formats. For more information about Wiley products, visit our web site at www.wiley.com.

Library of Congress Cataloging-in-Publication Data:

Protein targeting with small molecules : chemical biology techniques and applications / [edited by] Hiroyuki Osada.
    p. ; cm.
  Includes bibliographical references and index.
  ISBN 978-0-470-12053-8 (cloth)
  1. Protein binding. 2. Molecular probes. I. Osada, H. (Hiroyuki), 1954–
  [DNLM: 1. Protein Transport–physiology. 2. Proteins–metabolism. 3. Small Molecule Libraries–metabolism. QU 55 P9690124 2009]
  QP517.P76P76 2009
  572′.633–dc22

                                                                    2009007430

Printed in the United States of America

10 9 8 7 6 5 4 3 2 1

CONTENTS

CONTRIBUTORS

Udayanath Aich, The Johns Hopkins University, Baltimore, Maryland

Prabhani U. Atukorale, The Johns Hopkins University, Baltimore, Maryland

Christopher T. Campbell, The Johns Hopkins University, Baltimore, Maryland

Leonid L. Chepelev, Carleton University, Ottawa, Ontario, Canada

Nikolai L. Chepelev, Carleton University, Ottawa, Ontario, Canada

Yoon Sun Cho, Yonsei University, Seoul, Republic of Korea

Sean S. Choi, The Johns Hopkins University, Baltimore, Maryland

Michel Dumontier, Carleton University, Ottawa, Ontario, Canada

Hiroshi Handa, Tokyo Institute of Technology, Yokohama, Kanagawa, Japan

Yuichi Hashimoto, The University of Tokyo, Tokyo, Japan

Mamoru Hatakeyama, Tokyo Institute of Technology, Yokohama, Kanagawa, Japan

Masaya Imoto, Keio University, Yokohama, Kanagawa, Japan

Yasuaki Kabe, Tokyo Institute of Technology, Yokohama, Kanagawa, Japan

Naoki Kanoh, Tohoku University, Sendai, Miyagi, Japan

Tatsuro Kawamura, Keio University, Yokohama, Kanagawa, Japan

Mitsuhiro Kitagawa, Keio University, Yokohama, Kanagawa, Japan

Ho Jeong Kwon, Yonsei University, Seoul, Republic of Korea

M. Adam Meledeo, The Johns Hopkins University, Baltimore, Maryland

Adrian P. Neal, European Molecular Biology Laboratory, Heidelberg, Germany

Shinichi Nishimura, RIKEN Advanced Science Institute, Wako, Saitama, Japan

Kosuke Nishio, Tokyo Institute of Technology, Yokohama, Kanagawa, Japan

Hiroyuki Osada, RIKEN Advanced Science Institute, Wako, Saitama, Japan

Akiko Saito, RIKEN Advanced Science Institute, Wako, Saitama, Japan

Satoshi Sakamoto, Tokyo Institute of Technology, Yokohama, Kanagawa, Japan

Carsten Schultz, European Molecular Biology Laboratory, Heidelberg, Germany

Hooman Shadnia, Carleton University, Ottawa, Ontario, Canada

Takeo Usui, University of Tsukuba, Tsukuba, Ibaraki, Japan

William G. Willmore, Carleton University, Ottawa, Ontario, Canada

James S. Wright, Carleton University, Ottawa, Ontario, Canada

Kevin J. Yarema, The Johns Hopkins University, Baltimore, Maryland

Yoko Yashiroda, RIKEN Advanced Science Institute, Wako, Saitama, Japan

Minoru Yoshida, RIKEN Advanced Science Institute, Wako, Saitama, Japan

PREFACE

Chemical biology is recognized as a new frontier research area between chemistry and biology. One of the goals of this research is to understand the complex biological systems inspired by chemistry or chemical tools. Nowadays, research in this field is highlighted because chemical biology is useful not only for basic research but also for drug discovery. The term "bandwagon effect" implies that people often associate with the majority without deep consideration. The current "buzz" surrounding chemical biology may be like the bandwagon. If one does not possess deep insight and a good grasp of the techniques, one will be discarded after the boom is finished. It is very important to have one's own opinion and technique in chemical biology.

This book offers you the contemporary knowledge and techniques necessary to understand the entire research field of chemical biology. New technologies to dissect the interactions between small molecules and proteins are introduced with some examples of the identification of binding proteins of small molecules. The final chapter will be useful to get a bird's-eye view of recent progress on small molecules targeting proteins.

By offering an overview of chemical biology clarified by detailed examples and descriptions of important techniques, it is the aim of this book to stimulate young chemical biologists and inform them of the opportunities to apply the power of chemistry to important problems in biology.

Finally, I thank the members of my laboratory, especially Akiko Saito, for their assistance in the preparation of the book.

Director of Chemical Biology Department   HIROYUKI OSADA
RIKEN Advanced Science Institute
February 2009

1

CHEMICAL BIOLOGY BASED ON SMALL MOLECULE–PROTEIN INTERACTION

HIROYUKI OSADA

Chemical Biology Department, RIKEN Advanced Science Institute, Wako, Saitama, Japan

1.1 INTRODUCTION

Chemical biology is a new research field that applies chemistry to the investigation of biology and drug discovery. Bioactive organic compounds called *bioprobes* have been effective tools for elucidating the mechanism of many important cellular processes (Fig. 1-1). Such bioprobes are considered potential leads for the development of new therapeutics. One of the most important purposes of chemical biology is the search for small molecules that can control specific functions of proteins that will clarify currently unknown biological phenomena.

In this book new technologies are introduced for dissecting the interactions between small molecules and proteins and examples are provided of the identification of binding proteins of small molecules. Recent applications of small molecules as bioprobes to investigate biological systems are also described as

Protein Targeting with Small Molecules: Chemical Biology Techniques and Applications,
Edited by Hiroyuki Osada
Copyright © 2009 John Wiley & Sons, Inc.

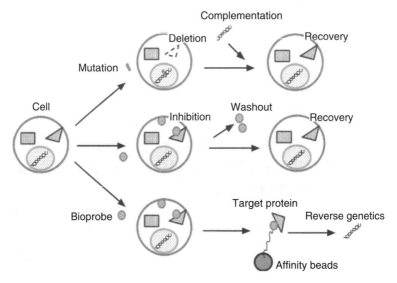

Figure 1-1   Bioprobes for chemical biology.

a typical topic of chemical biology. Chapter 12 provides a bird's-eye view of recent progress on small molecule–targeting proteins.

1.2   BIOPROBES AS A TOOL OF CHEMICAL BIOLOGY

The term *bioprobe* has been used by our group from the middle of 1990s to represent small organic compounds that are useful for investigating biological functions [1,2]. Until now, many bioactive small molecules that have shown antitumor, immunosuppressive, and other activities have been discovered from natural products or chemical libraries. Some of them are already in use as therapeutic medicines, but most of them have not yet been put into medicinal use. If the mechanism of action of a certain compound is revealed, the compound can be used for the investigation of biological functioning as a bioprobe. In general, when we are investigating a certain function of a protein, techniques of molecular biology, such as siRNA, are very useful. In the same way, an approach that uses a bioprobe to inhibit a certain function of a protein is also useful for dissecting a complex biological system. The approach that uses bioprobes instead of genetic tools, called *chemical genetics*, is a part of chemical biology. Much of the literature has reported that identification of a molecular target of small molecules is a successful strategy to use to reveal the biological functioning of a protein (see Chapters 2 to 11). Moreover, it is possible to identify the gene responsible for the protein by sequencing the protein captured by using affinity beads linked to bioprobes.

1.3   FROM CHEMICAL BIOLOGY TO CHEMICAL GENOMICS

Originally, chemical biology was thought to dissect complex biological systems from the standpoint of organic chemistry. The nature of this field can be exemplified by the research that has been performed on the mode of action of two immunosuppressive agents (FK506 and cyclosporine); this research pioneered the field of chemical biology [3]. The immunosuppressive agents bind to the first target protein, called *immunophilin* (FK506 binding protein or cyclophilin A) [4]. At the beginning, inhibition of the enzymatic activity, peptidyl–prolyl *cis-trans* isomerase activity, which is carried by the immunophilin, is thought to be important. However, a later intensive study revealed that the complex of the immunosuppressive agent and immunophilin inhibited the final target, calcineurin [5]. In addition to the technique of molecular biology, the chemicals FK506 and cyclosporine A were used as bioprobes in these studies. The chemicals were used like an antibody to co-precipitate target proteins, which inspired the field of chemical biology [6].

Biochemistry reductively elucidates biological phenomena and describes biological phenomena at the chemical level. On the contrary, chemical biology starts at the chemical level and describes the big picture of biological phenomena. Molecular biologists often use the gene knockout technique and siRNA to investigate a particular protein function, and chemical biologists use chemicals or bioprobes instead. Simplicity, speed, and reversibility are the benefits of using bioprobes to induce the alteration of cellular phenotypes. There might be a problem if the specificity of the bioprobe is low, but it is very convenient to use chemicals instead of genetic tools. In contrast to genetic methods, the use of bioprobes does not require a complicated experimental procedure. When a bioprobe is added to a cell culture, the morphological alteration can be observed within a relatively short period. It is also possible to observe the reversal of the phenotype after washing out the chemical if the effect of the compound is reversible.

Whereas chemical biology is considered to be a basic science that focuses on a specific biological phenomenon, chemical genomics deals with genome-wide phenomena that are associated with informatics. Chemical genomics will enable rapid identification of novel drugs and drug targets. The goal of chemical genomics is to discover ligands (chemical probes) for all existing proteins. These ligands must be a very powerful tool for the study of proteins, cell functions, and biological processes that are relevant to physiology and disease.

1.4   HOW TO EXPLORE SMALL MOLECULES

Recent advancements in combinatorial chemistry have increased the number of organic compounds (small molecules) and given us new opportunities to use them in screening to develop new agents. Thus, pharmaceutical companies explore drug candidates from chemically synthesized compounds. It might be a rational approach to develop medicines that are applicable for oral administration.

Because the molecular weight of natural products is usually larger than that of synthetic compounds, natural products are not considered to be suitable for medicines taken orally. However, it is not necessary to limit the screening source for chemical compounds when bioprobes are explored. Natural products vary widely in chemical structure as well as biological activity. Numerous compounds have been isolated from microbes, such as bacteria, actinomycetes, and fungi, and some are potential drug candidates.

Development of bioprobes starts from bioassay-based screening to explore the biological activities of a particular compound in extracts of microbial fermentation broth. After detecting sufficient activity, the active components are isolated by bioassay-guided purification. Physicochemical properties of the active components should be determined at the earliest possible stage and compared with those of known compounds to eliminate duplicate isolation. For novel compounds, sufficiently pure material should be obtained by large-scale fermentation of microorganisms to continue elucidation of the chemical structure and the evaluation of biological activity. Because the bioassay to detect desired compounds is very important, unique bioassays have been devised using intact cells or genetically engineered cells to express specific proteins. In the following sections, in addition to the conventional cell-based screening system, target-oriented high-throughput screening systems are described.

1.4.1 Screening System Based on the Conventional Paper Disk Agar Method

The paper disk agar plate method (halo assay) is the most conventional and commonly used bioassay system for antibiotics screening. Nowadays, this bioassay is applicable to genetically engineered microorganisms that express specific proteins or receptors as indicators. As an example, I would like to introduce the screening system that is used to detect antiviral compounds that are active against human immunodeficiency virus type I (HIV-1). HIV-1 is well known to be the causal virus of AIDS. However, it has not yet been clarified how HIV-1 infection induces CD4$^+$ T-cell depletion and eventual immunodeficiency in vivo. Cumulative evidence suggests that HIV-1 viral protein R (Vpr), one of the HIV-1-encoded proteins, may have an important role in its pathogenesis [7,8]. Until now, no clinical drug that targets Vpr has been developed; therefore, we have started a screen to isolate inhibitors of Vpr activity by using genetically engineered yeast. The growth of *Saccharomyces cerevisiae*, which carries the *vpr* gene, is suppressed when Vpr protein is overexpressed. Using this indicator microorganism, we carried out a screen for a Vpr inhibitory compound from fungal metabolites. After many actinomycetes and fungal extracts were applied to this bioassay, we found inhibitory activity against Vpr in one fungal extract [9] and identified it as fumagillin, which already had been known as a potent inhibitor of angiogenesis [10]. Fumagillin not only suppressed the growth inhibitory activity of Vpr in yeast and human cells but also inhibited Vpr-dependent viral gene expression and HIV-1 virion production on viral infection of human macrophages. Our results demonstrate that fumagillin can be a lead compound for the development of a novel anti-AIDS drug that targets Vpr activity.

1.4.2 Screening Systems Based on Cellular Phenotypes

During embryonic development, precursor cells differentiate into functional mature cells under the control of cytokines and hormones. For example, tumor cells, including leukemia and neuroblastoma, are thought to forfeit differentiation programs. Osteoporosis is explained simply as the hyperactivation of osteoclasts rather than osteoblasts. Osteoclasts are bone-resorptive multinucleated cells that are derived from hematopoietic stem cells of the monocyte/macrophage lineage. Osteoclasts are involved in remodeling bone matrix in concert with osteoblasts/stromal cells. Osteoclasts are differentiated from precursor cells by the aid of two critical factors: receptor activator of NF-κB ligand (RANKL) and macrophage colony-stimulating factor (M-CSF), which are supplied by osteoblasts [11].

Because the abnormal enhancement of osteoclasts is implicated in a variety of human diseases, including osteoporosis, rheumatoid arthritis, and cancer bone metastasis, it is worthwhile controlling osteoclast differentiation by small molecules [12]. Osteoclastogenesis inhibitors not only would be useful as tools in osteoclast biology but can also be developed as therapeutic drugs for these bone-related diseases. To identify small molecules that inhibit osteoclast differentiation, we have performed cellular phenotype-based assays. In the course of screening we found gerfelin and methylgerfelin [13] as active compounds against osteoclastogenesis. Gerfelin and methylgerfelin were revealed to be inhibitors of osteoclastogenesis through the inhibition of glyoxalase 1 activity [13].

1.4.3 Screening System Based on Physical Interaction

Microbial metabolites contain a wide variety of compounds that regulate cell proliferation through the specific inhibition of key proteins that are required for cell proliferation. Although it is possible to find such a cell growth inhibitor by the direct observation of cell growth, it is also possible to find cell-cycle inhibitors by a screening method that is based on protein–protein interactions, which is a method similar to the receptor binding assay [14]. A fluorescently labeled ligand is synthesized and applied to the target receptor/protein, which is immobilized on the surface of a 96-well plate. The binding between the fluorescently labeled ligand and the receptor is measured by a fluorometer. This assay system is suitable for high-throughput screening and is applicable to various types of screening for protein–protein interaction inhibitors.

As an example, I will describe the screening of polo-like kinase 1 (Plk1) inhibitors. Plk1 is one of the key regulators of mitotic cell division [15]. Localization of Plk1 changes dramatically in the M phase. During late G2 and prophase, Plk1 localizes to the centrosomes. At metaphase, Plk1 localizes to the kinetochores, where it is involved in the regulation of spindle assembly and chromosome segregation. In anaphase, Plk1 is found at the center of the spindle, and in telophase, Plk1 is found on the contractile ring. The C-terminal polo-box domain (PBD) is involved in the localization of Plk1. PBD was recently revealed to be a phosphopeptide-binding domain in various important proteins that are required for cell-cycle regulation. PBD-dependent binding is not only important for the subcellular localization of Plk1 but is also necessary for targeting of the protein to specific substrates. Therefore, we have established a screening system to find inhibitors of PBD-dependent binding in a high-throughput manner [16]. To quantify binding, fluorescent fusion proteins (which contain monomeric Venus and PBD) were expressed in *Escherichia coli*. Phosphopeptides of a PBD-binding sequence that was derived from human Wee1A were chemically synthesized and bound covalently to 96-well plates. The fusion protein was mixed with candidate compounds and placed into each well. The unbound PBD was washed out, and bound PBD was quantified by a spectrofluorometer. When a binding inhibitor of PBD is present in the assay system, it is expected that the fluorescence level decreases. Among about 2500 compounds that were tested, purpurogallin (PPG), a benzotropolone derivative originally isolated from nutgalls [17], was identified as the most potent inhibitor of PBD-dependent binding. Interestingly, however, compounds similar to PPG that lacked a hydroxyl group did not have inhibitory activity, indicating that the hydroxyl group of PPG is essential for the inhibitory activity. PPG had previously been reported as an inhibitor of the interaction of the Bcl-XL and BAD proteins [18]. However, the inhibitory concentration of PPG against Bcl-XL and BAD was higher than that against Plk1. Therefore, it is thought that the action of PPG is rather specific to the PBD–Wee1 interaction. We then examined the role of PBD-dependent binding on the progression of mitosis using purpurogallin. Previous studies showed that inhibition of Plk1 kinase activity induces prometaphase arrest with abnormal spindles, leading to the observation of monopolar cells. However, monopolar spindle cells were not observed in PPG-treated cells. Mitotic spindles in PPG-treated cells apparently were normal, but careful examination after staining with anti α- and γ-tubulin antibodies revealed that the spindle poles in PPG-treated cells were diffuse and more distanced than those of the control cells. The majority of chromosomes in the cells aligned at the metaphase plate, but some chromosomes were unaligned and located close to the spindle poles. This phenotype is similar to that of PBD overexpression in the cell.

1.5   PROFILING: A POWERFUL STRATEGY FOR TARGET SPECULATION

When bioactive small molecules are discovered by cell-based screening, its molecular target must be elucidated. If a compound has the possibility of becoming a drug, potential targets must be defined carefully to predict the side effects. To know the molecular target, we have to examine a lot of possible targets. In some cases, there may be unknown molecular targets. For these reasons, the identification of molecular targets for small molecules is usually a difficult and time-consuming process. When there are many possible target candidates for a small molecule, we have to narrow the range of possible targets. There have been several reports on the identification of binding proteins, such as a direct method that uses affinity matrices that arm small molecules to pull down binding proteins [19] and profiling of the phenotypes of small molecules [20]. Comparison of cell growth inhibitory concentrations of small molecules against various cell lines is informative in predicting the target of a small molecule. Using their profiling method, which is based on the drug sensitivity of cancer cells, Yaguchi et al. identified the molecular target of ZSTK474 to be phosphatidylinositol 3-kinase [21]. Comprehensive analyses that integrate bioinformatics are suitable and may be more informative for this purpose.

Biologically active small molecules affect cellular processes, induce modification of their target proteins, and change protein expression levels. When cells are treated with small molecules that have the same target in the cell, the alteration of target proteins should be similar in the cell. Proteome analysis by two-dimensional fluorescence differential gel electrophoresis (2D-DIGE) is an effective method for detecting the alternation of proteins in small molecule–treated cells. Comparison of protein expression patterns between cells that are treated with a small molecule and cells that are treated with known inhibitors will lead to the identification of molecular targets of new small molecules. The proteome database of cells that are treated with authentic inhibitors will be useful for this analysis. To establish the basis of this database, we performed proteome analysis of HeLa cells treated with 18 well-known inhibitors by 2D-DIGE. As a result, inhibitors that shared the same target were categorized in the same cluster. Our results indicate that proteome profiling on small molecule–treated cells is a useful methodology for target prediction of small molecules.

1.6 DETECTION OF BINDING BETWEEN PROTEINS AND SMALL MOLECULES

Affinity matrices are very useful for the identification of ligand-binding proteins from cell lysates. There have been many reports on matrices, such as agarose, polyacrylamide, and so on. However, nonspecific binding of proteins to solid supports has often caused trouble in identifying specific target proteins. To solve the problem, Handa et al. devised new matrices to reduce nonspecific interactions with proteins (Chapter 2). The newly developed matrices are very effective in identifying target proteins. Kanoh et al. describe the conjugation technique of small molecules to matrices by using photoaffinity linkers [22]. Their approach depends on the reactivity of carbene species that are generated from trifluoromethylaryldiazirine on ultraviolet irradiation. It was demonstrated in model experiments that photo-generated carbenes were able to react with every small molecule that was tested and to produce multiple conjugates in most cases. It was also found in on-array immobilization experiments that various small molecules were immobilized, retaining their ability to interact with their target proteins. With this approach, photo-cross-linked chemical arrays of about 15,000 compounds were constructed. These chemical arrays are very useful not only for ligand screening but also for the study of structure–activity relationships of ligands and their target proteins. Previously, if one wanted to make affinity matrices, he or she had to investigate the structure–activity relationship (SAR). Now, an SAR study can be done on a chemical array that is immobilized with a series of derivatives by combining them with a fluorescently labeled protein [22]. When the target protein is applied onto the chemical array, we can detect binding between the ligand and the protein as a fluorescent spot. Because this immobilization method does not require any specific chemical structures, the compounds that are immobilized retain their binding ability to the protein targets. Chemical arrays have enabled the rapid detection of such ligand–protein interactions in a high-throughput manner, although a label on the protein is needed to observe these interactions. By combining surface plasmon resonance (SPR) imaging technology with the chemical array platform, we developed a novel platform that allows in situ observation of interactions between photo-linked chemicals on gold surfaces and unlabeled proteins in solution [23]. In synthesizing suitable photo-cross-linkers for SPR imaging, the interactions between small molecules and unlabeled proteins can be observed with a high S/N ratio.

1.7   TRENDS IN CHEMICAL BIOLOGY

There have been many reports on bioactive natural products. Because some of them are promising candidates for medicinal purposes, it is necessary to identify their target molecules in cells. Otherwise, the small molecules may show severe side effects. On the contrary, even if a small molecule has no worth for medicinal use, it can potentially be useful as bioprobe when it has a specific molecular target in the cell. For example, lactacystin was isolated from a *Streptomyces* sp. as a differentiation inducer of neuroblastoma cells [24]. Its target molecule has been identified to be a subunit of the 20S proteasome [25]. Now, lactacystin is used in a wide range of studies to reveal the biological roles of proteasome function in mammalian cells. Moreover, velcade, another proteasome inhibitor, was developed as an effective medicine for multiple myeloma [26]. Investigation on the properties of bioactive small molecules is a starting point in chemical biology research. New bioprobes will be characterized by our efforts, and some of them may be golden eggs of medicine to be developed by pharmaceutical industries.

REFERENCES

1. Osada, H. (1998). Bioprobes for investigating mammalian cell cycle control. *J. Antibiot.*, *51*, 973–981.
2. Osada, H. (2000). Trends in bioprobe research. In *Bioprobes*, H. Osada, ed. Springer-Verlag, Tokyo, pp. 1–14.
3. Schreiber, S. L. (1991). Chemistry and biology of immunophilins and their immuno-suppressive ligands. *Science*, *251*, 283–287.

4. Schreiber, S. L., Crabtree, G. R. (1992). The mechanism of action of cyclosporin A and FK506. *Immunol. Today*, *13*, 136–142.

5. Cliptone, N. A., Crabtree, R. C. (1992). Identification of calcineurin as a key signalling enzyme in T-lymphocyte activation. *Nature*, *357*, 695–697.

6. Schreiber, S. L., Liu, J., Albers, M. W., et al. (1992). Molecular recognition of immunophilins and immunophilin–ligand complexes. *Tetrahedron*, *48*, 2545–2558.

7. Dedera, D., Hu, W., Vander Heyden, N., et al. (1989). Viral protein R of human immunodeficiency virus types 1 and 2 is dispensable for replication and cytopathogenicity in lymphoid cells. *J. Virol.*, *63*, 3205–3208.

8. Cohen, E. A., Dehni, G., Sodroski, J. G., et al. (1990). Human immunodeficiency virus Vpr product is a virion-associated regulatory protein. *J. Virol.*, *64*, 3097–3099.

9. Watanabe, N., Nishihara, Y., Yamaguchi, T., et al. (2006). Fumagillin suppresses HIV-1 infection of macrophages through the inhibition of Vpr activity. *FEBS Lett.*, *580*, 2598–2602.

10. Ingber, D., Fujita, T., Kishimoto, S., et al. (1990). Synthetic analogues of fumagillin that inhibit angiogenesis and suppress tumour growth. *Nature*, *348*.

11. Teitelbaum, S. L. (2000). Bone resorption by osteoclasts. *Science*, *289*, 1504–1508.

12. Osada, H. (2000). Differentiation. In *Bioprobes*, H. Osada, ed. Springer-Verlag, Tokyo, pp. 43–65.

13. Kawatani, M., Okumura, H., Honda, K., et al. (2008). The identification of an inhibition through the osteoclastogenesis via inhibition of glyoxalase I. *Proc. Natl. Acad. Sci. USA*, *105*, 11691–11696.

14. Hovius, R., Schmid, E. L., Tairi, A. P., et al. (1999). Fluorescence techniques for fundamental and applied studies of membrane protein receptors: the 5-HT3 serotonin receptor. *J. Recept. Signal Transduct. Res.*, *19*, 533–545.

15. Simizu, S., Osada, H. (2000). Mutations in the Plk gene lead to instability of Plk protein in human tumor cell lines. *Nat. Cell Biol.*, *2*, 852–854.

16. Watanabe, N., Sekine, T., Takagi, M., et al. (2009). Deficiency in chromosome congression by the inhibition of Plk1 polo box domain-dependent recognition. *J. Biol. Chem.*, *284*, 2344–2353.

17. Inamori, Y., Muro, C., Sajima, E., et al. (1997). Biological activity of purpurogallin. *Biosci. Biotechnol. Biochem.*, *61*, 890–892.

18. Kitada, S., Leone, M., Sareth, S., et al. (2003). Discovery, characterization, and structure–activity relationships studies of proapoptotic polyphenols targeting B-cell lymphocyte/leukemia-2 proteins. *J. Med. Chem.*, *46*, 4259–4264.

19. Stoughton, R. B., Friend, S. H. (2005). How molecular profiling could revolutionize drug discovery. *Nat. Rev. Drug Discov.*, *4*, 345–350.

20. Terstappen, G. C., Schlupen, C., Raggiaschi, R., et al. (2007). Target deconvolution strategies in drug discovery. *Nat. Rev. Drug Discov.*, *6*, 891–903.

21. Yaguchi, S., Fukui, Y., Koshimizu, I., et al. (2006). Antitumor activity of ZSTK474, a new phosphatidylinositol 3-kinase inhibitor. *J. Natl. Cancer Inst.*, *98*, 545–556.

22. Kanoh, N., Nakamura, T., Honda, K., et al. (2008). Distribution of photo-cross-linked products from 3-aryl-3-trifluoromethyldiazirines and alcohols. *Tetrahedron*, *64*, 5692–5698.

23. Saito, A., Kawai, K., Takayama, H., et al. (2008). Improvement of photoaffinity SPR imaging platform and determination of binding site of p62/SQSTM1 to p38 MAP kinase. *Chem. Asian J.*, *3*, 1607–1612.

24. Katagiri, M., Hayashi, M., Matsuzaki, K., et al. (1995). The neuritogenesis inducer lactacystin arrests cell cycle at both G0/G1 and G2 phases in neuro 2a cells. *J Antibiot.*, *48*, 344–346.

25. Fenteany, G., Standaert, R. F., Lane, W. S., et al. (1995). Inhibition of proteasome activities and subunit-specific amino-terminal threonine modification by lactacystin. *Science*, *268*, 726–731.

26. Adams, J., (2004). The proteasome: a suitable antineoplastic target. *Nat. Rev. Cancer*, *4*, 349–360.

2

TARGET PROFILING OF SMALL MOLECULES

LEONID L. CHEPELEV AND MICHEL DUMONTIER

Department of Biology, Carleton University, Ottawa, Ontario, Canada

2.1 INTRODUCTION

2.1.1 Basics of Small Molecules

Like millions of hard-working citizens in a busy metropolis, the multitudes of our cellular constituents, including small molecules, interact constantly with their environment and each other, participating in countless energy and matter transactions, in the process proliferating what we call *life*. While the attention of much of current biological research may be focused exclusively on biomacromolecules, small molecules (SMs) are also vital to metabolism, biosynthesis, and signaling in regulatory and neuronal networks. Historically, SMs have been called on most widely by the pharmaceutical industry to help combat disease. Although they can be devised to inhibit or promote enzymatic reactions through blockage of the active site, or through allosteric or covalent modifications of the target enzyme or receptor, these mechanisms can be used to inhibit or promote protein–protein or protein–nucleic acid interactions. In addition to this, a portion of pharmaceutically active SMs exert their effect only after subsequent chemical

Protein Targeting with Small Molecules: Chemical Biology Techniques and Applications,
Edited by Hiroyuki Osada

11

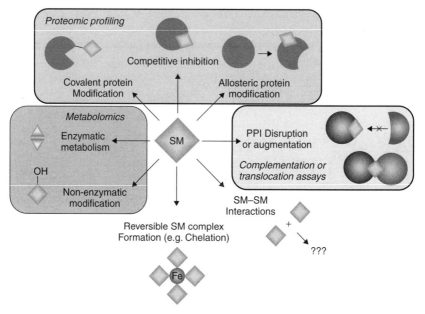

Figure 2-1   Some of the major interaction classes that SM may participate in within a living cell. Major directions in screening for these interaction classes are in italic. Unfortunately, SM–SM interaction detection, be it a direct reaction or the combination of effects of two SMs, resulting in unprecedented behavior, often requires an individualistic approach to assay adequately.

modifications (enzymatic or not) to the form that elicits the effects desired [1]. Because of this vast range of possible interactions (Fig. 2-1), determination of the particular target and the subsequent elucidation of the precise mechanism of action of a specific active small molecule persists as a major problem in drug development. In fact, a number of small-molecule treatments, such as by salicylic acid, were first used without an adequate grasp of the mechanistic background. Indeed, the ability to assess the mechanism of action of a given SM adequately would allow the possibility of improving the compound's specificity and selectivity, eliminating its side effects and improving its efficacy, while uncovering disease mechanisms and alternative targets for treatment. Thankfully, recent advances in quantitative high-throughput genomic, metabolomic, and proteomic analysis methodologies have finally allowed us to begin constructing an entire spectrum of interaction maps for the various cellular constituents as well as to begin to assess the state of a living cell dynamically and quantitatively while bringing us ever closer to an understanding of disease and drug activity mechanisms.

When considering a topic such as target profiling of SMs, we should define the terms used. So for the purposes of this work, a target of a given SM is the cellular constituent upon a reversible or irreversible interaction with which the SM elicits a change in normal cellular behavior or initiates a signaling cascade that elicits such an effect. Most often, the target is a protein, an enzyme to be inhibited selectively (e.g., a parasite's unique enzymes) or a receptor to be incapacitated or activated (e.g., estrogen receptors), although, potentially, some structural proteins can also be targeted. Thus, we will consider primarily SM–protein interaction detection and quantification, although most of the technologies discussed could be applied to screening of binding to other cellular constituents.

The field of quantitative high-throughput screening (HTS) of SM interactions is vast, incorporating diverse assay technologies, both quantitative and qualitative, mechanism- and phenotype-based, approaches that involve a carefully crafted knowledge-based assay and candidate design as well as those that involve the screening of an entire combinatorial chemistry-generated SM library of millions of compounds against thousands of potential targets on a proteomic scale without a priori knowledge, assays that are carried out in silico, in vitro, and in vivo. Indeed, a systematic review of each methodology is a titanic work that could easily span multiple volumes. We have, however, summarize some of the most commonly encountered SM HTS technologies, some of which we cover in more detail in Table 2-1. Here we opt to discuss briefly only a few of the representative technologies of the past that helped guide the field of proteome-scale interaction profiling: ones that we believe to be most promising and deserving of receiving widespread acceptance in the future. We specifically emphasize technologies that identify protein targets of a particular SM rather than the chemical library screening used to identify all SMs that bind to a given protein.

2.1.2 Theoretical Background

Irrespective of the technology used to assay SM interactions and/or metabolic transformations, putting quantum tunneling aside, all biochemical reactions can be characterized by two thermodynamic properties. The first is the standard Gibbs free-energy change for a given reaction (ΔG^0), given by equation (1), which describes the equilibrium position of a reaction, or its feasibility from the thermodynamic standpoint, at standard conditions.

$$\Delta G^0 = \Delta H^0 - T\,\Delta S^0 = -RT \ln K_{eq} \quad \text{where} \quad K_{eq} = \frac{k_f}{k_r} = \frac{\prod_{i=1}^{m} [\text{product}_i]^{n_i}}{\prod_{j=1}^{l} [\text{Reactant}_j]^{n_j}}$$
(1)

Here ΔH^0 is the standard enthalpy change of reaction; ΔS^0 the standard reaction entropy change; R the gas constant; T the temperature; K_{eq} the equilibrium constant for the reaction, given simply by the product of concentrations (activities in reality) of all the products to the power of their stoichiometric coefficients over the same product for reactants; m the number of products; k_f the forward rate constant; k_r the reverse rate constant; n_i the stoichiometric coefficient of species i; and l the number of reactants. A ΔG^0 value below zero indicates a reaction with an equilibrium point where there is an excess of products over reactants, a

TABLE 2-1 Some Common SM Interaction Screening Technologies

Approach	Mode of Action	Main Features	Application Examples
Affinity chromatography	Proteins interacting with a given SM will be retained in the affinity column; weak interactors will be eluted	Mostly for a single target vs. a range of SMs, or vice versa; nonspecific binding a concern	SM affinity ranking
Synthetic genetic array	Comparison of growth of haploid double mutants to that of haploid single mutants treated with SM helps infer drug target	Living cellular milieu; detection of SM targets, and transport and metabolism proteins	Drug target profiling
Drug-induced haploinsufficiency	Strains with reduced fitness upon SM addition indicative of gene deletion being SM target	Living cellular milieu; detection of SM targets, and transport and metabolism proteins	Drug target profiling
Enzyme-linked immunosorbent assay (ELISA)	Various (e.g., interaction of an extraneous SM with fixed protein displaces an SM-reporter complex, resulting in reporter signal drop)	Relatively low cost and great versatility; optimized for routine drug detection	SM detection, interaction profiling [2]
Fluorescent resonant energy transfer (FRET)	Varies; the emission wavelength of isolated fluorescent proteins is different relative to bound proteins (e.g., SM-mediated interaction disruption can be seen directly)	Real-time in vivo SM interaction monitoring; assays only the labeled interaction partners	In vivo interaction monitoring [3]

Method	Description	Advantages/Notes	Application
Isothermal calorimetry (ITC)	Dynamics of temperature change upon SM addition to protein used to derive binding thermodynamics and kinetics	Highly detailed thermodynamic data, at a cost of lower throughput	Elucidation of interaction thermodynamics [4]
Surface plasmon resonance imaging (SPRi)	Local variations in index of refraction upon a binding event trigger a change in resonance of surface plasmons, tracked dynamically	Optimized for array format, providing high quantities of good-quality quantitative data	Proteomic-scale SM interaction profiling [5]; interaction kinetics elucidation
Scintillation proximity assay (SPA)	A fluomicrosphere coated with protein will emit light only when the radiolabeled ligand binds to the protein when coming into proximity	Low cost, especially useful for kinase studies, but involves radioactive isotopes	Kinase substrate profiling; SM interaction profiling [6]
Yeast two-hybrid technology	Varies (e.g., interaction of modified SM to its target activates reporter transcription)	Very low cost, extreme versatility, in vivo assay environment, simple detection	Proteomic interaction mapping [7]

value of zero indicates equality of reactant and product amounts, and a value above zero is indicative of an anticipated lack of reaction progress. In the majority of studies of quantitative interaction characterization, the K_d parameter is used to describe interaction strength. K_d is simply the equilibrium constant for the reaction of dissociation of a binding partner complex, and for the reaction AB $\rightleftharpoons$ A + B, K_d can be found as

$$K_d = \frac{[A][B]}{[AB]} \tag{2}$$

The second factor is the standard activation free energy of the reaction (ΔG^{0*}), which describes reaction kinetics, how fast a reaction will proceed:

$$\Delta G^{0*} = -RT \ln \frac{k_f h}{k_B T} \tag{3}$$

Here k_f is the forward reaction rate constant, which describes directly how fast a reaction will proceed in the direction written, h is Planck's constant, and k_B is the Boltzmann constant. In other words, the smaller the reaction energy barrier, the faster the reaction will proceed. A simplified reaction energy curve is shown in Figure 2-2 for an example of one SM binding to one target, which could be a receptor, for example.

It is these two quantities that are being evaluated directly or indirectly when a given SM interaction is assessed. It is also these two constants that provide a unified and theoretically rigorous framework for comparison of SM interaction

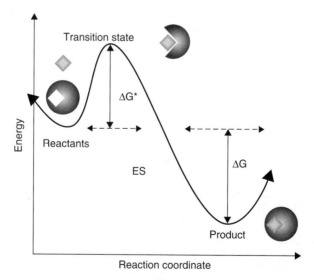

Figure 2-2 Simplified reaction energy diagram for a favorable reaction. As reactants are converted to products, they undergo one or more transformations to transition-state structures, which usually require initial energy input and activation free energy.

strength with different binding partners and across multiple studies. Unfortunately, ΔG^0 and ΔG^{0*} vary with the temperature and the nature of the solvent, making comparisons of studies performed under highly dissimilar conditions somewhat dubious. Although most studies can be carried out under standard conditions, one can easily envision cases where the addition of salt or variation of pH, temperature, or other conditions is unavoidable (e.g., due to protein denaturation).

The first equation provides us with three different ways to assess the ΔG^0 value: the determination of the ratio of products and reactants at equilibrium, the determination of enthalpy and entropy change, possibly with techniques such as isothermal calorimetry, or direct quantification of kinetics of a given reaction in the forward and reverse directions. The latter is the only option to determine the value of ΔG^{0*}. Consider yeast two-hybrid technology as an illustration of an approach that is normally used for qualitative interaction characterization but can actually provide quantitative results when set up appropriately through determination of equilibrium reactant and product levels (see below). On the other hand, the various quantitative interaction characterization technologies, such as surface plasmon resonance imaging (SPRi), aim to obtain the values of SM–protein binding and dissociation rate constants directly simply by monitoring complex formation and dissociation over time and extracting the kinetic constants directly from the resulting time–response curves.

Yet another approach to the determination of ΔG^0 is apparent when considering another definition of the standard Gibbs free-energy change as the difference between the energies of formation of products and reactants:

$$\Delta G^0_{\text{reaction}} = \sum_{i=1}^{m} n_i (\Delta G^0_{\text{formation}})_i - \sum_{j=1}^{l} n_j (\Delta G^0_{\text{formation}})_j \qquad (4)$$

Here i and j are indices of products and reactants, respectively, and n is the stoichiometric coefficient of a given species. In other words, the overall ΔG^0 value can be determined by subtracting standard Gibbs free energies of formation of products from those of reactants. This formulates the basis for the calculation of ΔG^0 for SM interactions in silico. A number of algorithms and software packages have been developed for the accurate calculation of reaction ΔG^0 values, which we consider later in more detail. Additionally, in cases where rigorous ab initio calculations may be prohibitively time consuming, algorithms that employ protein sequence and known SM interaction partner information to approximate SM–target interaction strength have been developed and are considered here.

Whatever the interaction detection technology, it is always important to remember whether a tangible kinetic or thermodynamic constant is provided by an assay or whether we are forced to use a quantitative relative ranking scheme or an outright qualitative interaction assessment. While qualitative interaction detection through yeast two-hybrid approaches is valuable for rapid, low-cost "first approach" screening, it is highly difficult to compare different interaction strengths or to determine the precise role of a given compound in a dynamic kinetic model of the pathway of interest without tangible kinetic data. Without such understanding, we are destined to probe the chemical space blindly in hope of finding the occasional precious active SM while suffering from unanticipated side effects.

2.2 EXPERIMENTAL APPROACHES TO SM INTERACTION PROFILING

2.2.1 Yeast Two-Hybrid System

It is important first to discuss briefly qualitative and semiquantitative interaction screening methodologies, as they have helped pave the way for subsequent quantitative proteome-scale approaches. It was, after all, qualitative techniques that resulted in the first proteome-scale interaction maps [7] and were some of the first to be applied to aid in an understanding of major regulatory and disease pathways as well as drug activity mechanisms [8]. Yeast has been shown to be an excellent model organism for the purpose of SM interaction profiling [9]. Despite the fact that the baker's yeast *Saccharomyces cerevisiae* is a single-celled fungus with little outward resemblance to humans, it combines the versatility and tractability of a prokaryote with the main features of eukaryotes, such as the presence of organelles and eukaryotic post-translational modifications; one-third of human genes involved in disease have yeast orthologs [10]. In addition to this, yeast is one of the best-studied organisms available. Strain collections have been created in which every gene was systematically deleted or fused with, or replaced by, a reporter gene such as GFP [11]. It is perhaps the ease of genetic manipulation of yeast that led to the creation of the versatile yeast two-hybrid technique, which since its introduction in 1989 [12] has also been modified and refined extensively to produce proteome-scale qualitative and semiquantitative information on protein–small molecule interactions [13,14].

The basic two-hybrid approach is quite simple: For two proteins X and Y whose interaction status is to be determined, fusion proteins X-DBD and Y-AD are created, where AD is an activating domain and DBD is a DNA-binding domain specific to an upstream activating sequence of a reporter gene. Successful interaction of X and Y brings DBD in close proximity to AD, resulting in recruitment of RNA polymerase to initiate the transcription of a reporter gene. The reporter gene could be a fluorescent protein (FP) or a chromogenic enzyme, for example, such that observation of its expression does not involve costly detection equipment.

It would seem that this has little to do with SM–protein interaction profiling, and that would be true were it not for the amazing versatility of the approach. To look at two examples of how modified yeast two-hybrid approaches can potentially be used for SM screening, let us consider SM-induced protein–protein interaction inhibition (PPII) and SM–protein binding. To assess the first situation, a modified two-hybrid assay involves a lethal reporter gene and a protein pair that is known to interact (Fig. 2-3). Now, an array of SMs potentially causing PPII

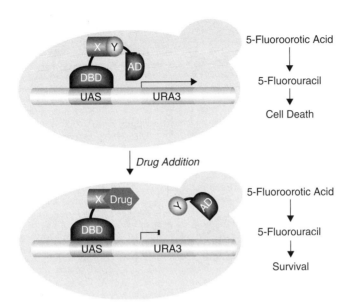

Figure 2-3 SM protein–protein interaction inhibition assay based on the two-hybrid system.

can be screened for activity. In the case of a successful interaction inhibition, the yeast colony with the added SM will survive. This allows for very large-scale SM activity screening.

For the scenario of proteomic-scale target profiling of a single SM, consider a system where the SM of interest (SM_1) is coupled covalently to another SM that tightly binds the Y–AD complex with high specificity (SM_2). In the yeast cell, the SM_2–Y–AD complex will readily form. In the event of SM_1 binding to its protein target X, the DBD–AD complex (actually, DBD–X–SM_1–SM_2–Y–AD) will be formed and transcription from the reporter gene will ensue (Fig. 2-4). Among the first to utilize this approach were Licitra and Liu, who successfully demonstrated its utility [15]. Small molecules FK506 and dexamethasome were covalently linked to form a heterodimer that brought together a rat glucocorticoid receptor fused to a DBD and FK506-binding protein 12 fused to an AD. Cells with the FK506–dexamethasome heterodimer showed reporter activity, while in cells treated with the heterodimer and free FK506, reporter activity was abrogated, demonstrating the utility of the screen for finding SM that can compete with the native receptor ligand. In another more recent study by Henthorn and co-workers, a similar setup was employed to screen a mouse cDNA library to identify targets of the SM methotrexate, which was covalently linked to dexamethasone [16]. As many as 56 protein targets were identified.

Of course, due to the ease of plasmid manipulation, proteins probed for interaction with the SM can be mammalian. If that is the case, it is important to understand that some post-translational modifications of mammalian proteins

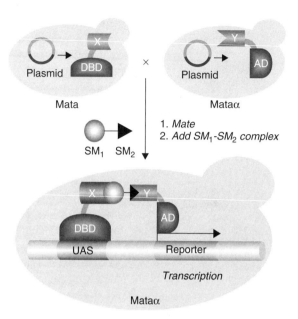

Figure 2-4   SM-protein binding can be assessed with a modified two-hybrid assay.

expressed in yeast will be different from those performed by their native organism, affecting the relevance of interactions detected. Fortunately, two-hybrid approaches are not simply restricted to yeast–mammalian two-hybrid systems that employ transient plasmid transfection, and stem cells are on the rise in recent years—consider the recently reported real-time interaction monitoring in living mice with a two-hybrid system [17].

Unfortunately, both of these techniques are susceptible to false positives and negatives, due to potential SM toxicity and the ability to influence a number of pathways irrelevant to the tested system. However, careful planning and cross-validation using other approaches should help minimize or at least estimate false positive and negative levels. In addition, as with any approach employing protein conjugation or immobilization, the directionality of conjugation of the SM of interest and the interrogated protein library may be important. For instance, protein X conjugated to DBD through a terminus important for SM binding may be falsely unrecognized as an interaction partner of the given SM. For this reason, plasmid libraries exist for N- and C-terminal protein fusion constructs, lowering the chances of such an occurrence.

Although it would seem at first that only qualitative interaction information could be obtained from such screens, in fact, under certain conditions, data obtained could be semiquantitative. For example, one need only determine equilibrium concentrations of interacting species to find the equilibrium constant [see equation (1)]. If the transcription rate for the plasmids used could be determined from FP expression from the plasmid, for instance, and if the transcription rate for the reporter is known, it is possible to deduce SM–protein complex levels along with individual protein levels. With this information in hand, determination of the equilibrium constant from equation (1) becomes trivial. In fact, a similar scheme has already been reported and requires somewhat more handling than a simple qualitative interaction test [13,14]. All in all, the two-hybrid approach combines a number of characteristics that other SM interaction screening technologies would do well to copy: versatility, in vivo assay conditions, extremely low cost, and the potential for quantitative interaction profiling. It is this combination that assures the place of two-hybrid technologies as rapid preliminary screening tools.

2.2.2 Yeast-Based Genetic SM Interaction Screens

As mentioned, yeast is one of the best-studied and best-understood organisms, meaning that large strain collections containing almost every possible nonessential gene-knockout and plasmid collections containing almost every yeast gene are readily available. One of the greatest uses of such collections has been proteome-scale SM target identification through SM-induced growth alteration screens in knockout, haploinsufficient, and gene overexpression strain collections. Such screens are used routinely to help reveal the modes of action of prospective drugs [8]. Unfortunately, only qualitative information is obtained from such screens, and even that information is often subject to interpretation. For instance, a gene identified as important for cell growth alteration in the presence of an added SM is not necessarily an SM target. It is entirely possible that the particular gene product is simply involved in SM transport and metabolism and has little to do with the particular desired effect that the SM has on the cell.

Briefly, in this approach, gene dosage is varied through gene deletion, overexpression, or haploinsufficiency (Fig. 2-5). Libraries of systematic gene dosage perturbations are created. In the case of haploinsufficient strains and gene-deletion strains, the affected gene is replaced by a unique nucleotide sequence, sometimes

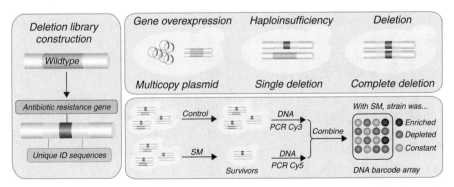

Figure 2-5 Typical SM-induced growth alteration screening procedure. *(See insert for color representation of figure.)*

referred to as a *barcode*. In the case of gene overexpression, the introduced multicopy plasmid carries the identification of the perturbed gene. Now, the entire strain collection is exposed to a SM, either in an array format, which requires a pinning robot and additional handling, or in a pooled format, which makes the different strains compete for survival. In either case, a population of cells is treated with the SM of interest while another is left without treatment. After the end of the treatment, extracted DNA undergoes polymerase chain reaction (PCR) amplification with fluorophore-labeled primers that differ for control and SM-treated samples. Next, the PCR products are combined and competitively hybridized to an oligonucleotide microarray. The dominance of fluorescence of the control sample indicates strain depletion for the particular gene deletion, while SM-treated sample-labeled PCR product fluorescence indicates enrichment for the given strain. The strains whose fitness is changed are identified and their respective mutated genes are inferred to be either drug targets or involved in drug metabolism.

One of the early applications was the screening of a library of 233 heterozygous deletion strains in a pooled format to identify the target of tunicamycin [18]. In another study, 78 growth inhibitors were screened against a collection of 3503 heterozygous deletion strains in a pooled format to identify novel targets and confirm the known pathway involvement of these SMs [19]. Although the data produced in such studies are qualitative, there is nothing more precious than information of SM interactions within the context of a living cell, which could provide SM mode of action information.

2.2.3 Quantitative Label-Free Interaction Profiling Systems

Affinity Chromatography   Affinity chromatography (AC) is one of the first techniques to be adapted for quantitative high-throughput SM interaction detection and screening [20]. Although they do not provide the cellular milieu of the living yeast, AC-based assays remain an important tool in the HTS of SM interactions. The simplicity of these assays is unparalleled (Fig. 2-6). Briefly, an SM is immobilized through a covalent bond to a long linker on a solid substrate in

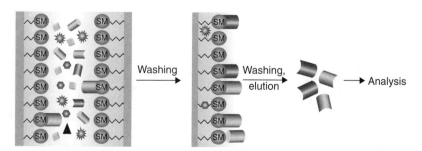

Figure 2-6 Affinity chromatography assay—only the strong interactions will remain after washing.

a chromatographic column, the substrate being column packing beads or column walls. A number of coupling chemistries exist to immobilize a vast majority of SM. A protein solution, which can be a purified protein mixture or a whole-cell extract, is placed into the column. As the solution passes through the column, some constituents interact with the immobilized SM, while others bind nonspecifically to the beads and the linker arm. Subsequent washing of the column is expected to have the beneficial effect of removing constituents involved in nonspecific interactions and the detrimental effect of washing away proteins that interact with the SM transiently or weakly. A final wash, usually with unbound SM, removes the remaining strongly interacting constituents, after which their identity is established, usually with mass spectrometry (MS).

This technology has been miniaturized and coupled to MS identification directly to produce frontal affinity chromatography–mass spectrometry (FAC-MS) [21] (Fig. 2-7). FAC-MS allows for the rapid (thousands of SMs per day) ranking of SM candidates for binding to a particular target protein according to binding affinity. In this setup, the sample travels along a capillary whose walls serve as a substrate or a column containing beads to which the SM or protein are linked. The target is bound to the solid substrate either covalently through a linker (SM) or using a biotin–streptavidin or a similar immobilization system (protein). Unfortunately, as is the case for most techniques where covalent fixing on a solid substrate is used, this approach also often suffers from nonspecific interactions of the sample constituents with the linker as well as from the aforementioned conjugation directionality issues.

As the sample proceeds through the capillary, interacting molecules are retained for longer periods of time than those that do not interact, resulting in a larger elution volume required for the interacting SMs. This allows for

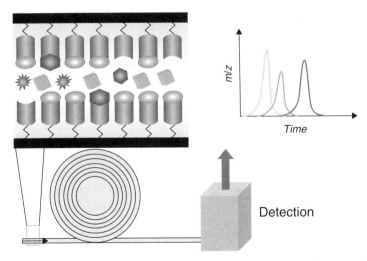

Figure 2-7 FAC-MS: SM affinity is proportional to retention time. *(See insert for color representation of figure.)*

a quantitative ranking of interactions with the various partners through the observation of elution time. Furthermore, while the use of MS as a detector assists in the precise identification of the molecule being tracked, the elution times provide directly information necessary to determine the K_d value of the binding interaction:

$$K_d = \frac{B_t}{V - V_0} - [L] \qquad (5)$$

Here B_t is the number of binding sites on the column, V the elution volume of the analyte for which K_d is being determined, V_0 is the elution volume of a void marker, which does not interact with the immobilized protein, and [L] the ligand concentration in the sample. Needless to say, a FAC-MS assay can be calibrated using SM with known interaction parameters.

The inclusion of weakly interacting reference compounds allows for extension of the FAC-MS detection range to the picomolar level. Although, theoretically, it should be possible to immobilize an SM and identify all protein targets with this approach, given the selection of an appropriate MS protein identification technology, FAC-MS is used almost exclusively to screen and characterize the interactions of one protein with a library of compounds. For example, Ng and co-workers have reported the development of an automated FAC-MS platform that can be used to screen as many as 10,000 SMs for interaction with a protein target within 24 hours [21]. Demonstrating the flexibility of the approach, Sharma and co-workers recently introduced a novel FAC-MS modification which allows for the screening of transient SM–target interactions through the use of a competitively binding indicator compound [22]. Low levels of *Torpedo californica* acetylcholine receptors were used as a target, and in the presence of epibatidine as a competitively binding indicator, the weak interaction to protein target with nicotine was detected and characterized. Developments like these are indicative of the continuing utility of AC approaches, in particular FAC-MS, as an important HTS platform in drug development in the future.

Quartz Crystal Microbalance (QCM)   Although QCM-based systems have appeared on the market relatively recently, it seems that as this technology improves and is adapted for HTS, it would be capable of delivering accurate kinetic information, perhaps even with potentially greater detail than SPRi [23]. Most of us are familiar with the operation of electronic watches that make use of quartz crystal resonance. A QCM works by applying a current to a quartz crystal, causing it to resonate, and subsequently detecting changes in crystal resonance as a result of a binding event by monitoring resonance in real time. The binding occurs on the surface of the QCM electrodes that sandwich a quartz crystal, where a receptor or ligand is immobilized, often with the help of a self-assembled monolayer coating. While monitoring the SM–protein interaction in real time as a change in crystal resonance, a time–response profile similar to that derived with SPRi can be obtained, along with dissociation and association rate constants.

During the binding, not only does the mass on the surface of the microbalance increase, but elasticity and viscosity properties of the ligand–receptor complex also change. Because the resonating crystal emits an acoustic wave that propagates micrometers away from the source, all of these changes can be probed and used for a more detailed interaction characterization than simple binding kinetics (e.g., to probe structural implications of binding). The former, however, are easy to monitor in real time, thanks to the direct correspondence between the change in the frequency of vibration and the weight that is placed on the QCM electrode:

$$\Delta f = \frac{f_0}{A \Pi N} \Delta W \tag{6}$$

Here Δf is the observed frequency change, f_0 the frequency prior to application of a weight ΔW, A the surface area of the cylindrical quartz crystal, Π its density, and N the crystal's frequency constant.

To date, protein–protein, protein–SM [24], and even DNA–SM interactions were characterized successfully using QCM technology [23]. Unfortunately, interaction characterization with QCM is still geared toward detailed and low-throughput interaction analysis, even though the drive toward multiplexing and increasing the throughput of the approach is evident. For instance, the RAP◆id-4 system introduced in 2006 by Akubio Technologies employs the basic principles of QCM along with some variations, and makes possible automated monitoring of SM–protein interactions [25]. The system was demonstrated to rank three cofactors of glucose dehydrogenase correctly according to measured binding response. Other multiplexed QCM arrays incorporating up to 10 sensors have also been developed, but so far have been used only for the detection of analytes in a mixed sample [26]. Whatever the present state of QCM, we believe that sensor miniaturization and consistent improvement in automation may someday make QCM-based multiplexed SM interaction profiling studies a reality.

Electrochemical Interaction Detection Electrochemical interaction detection is a large and rapidly evolving field with a great number of variations in interaction assays [27]. Electrochemical detectors are attractive due to the ease of miniaturization, assay rapidity, low cost of assay hardware, and great sensitivity. It is becoming increasingly easy to produce protein arrays for rapid and quantitative interaction characterization, in part due to the existing advances in electronics, particularly circuit printing, which electrochemical detection technologies can draw upon. In general, an electrochemical detection device involves three electrodes: a reference electrode placed away from the electrochemical reaction and used to maintain a stable potential; an auxiliary electrode that comes in contact with the redox solution and allows for the current flow to the working electrode; and a working electrode in which the biochemical or redox reaction occurs and which acts as a transducer. Connection of these electrodes to an electronic test device allows one to monitor changes in potential, current, conductive properties of the solution medium, and impedance.

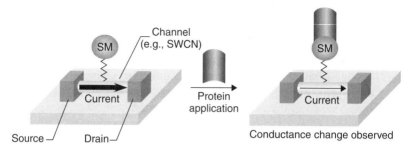

Figure 2-8 One highly simplified mechanism of action of a simple field-effect transistor-based interaction detection device.

Field-effect transistors constitute a very promising and distinct device class which, in basic terms, can measure alterations in conductance of a charge carrier–depleted region (channel) between a source and a drain electrode as a direct or indirect result of biochemical events such as enzymatic reactions or binding events (Fig. 2-8). The channel can be composed of a silicon nanowire or a single-walled carbon nanotube (SWNT), for example. The power of field-effect transistors has been widely demonstrated with aptamer-based protein recognition devices such as the SWNT thrombin detection device [28]. Since it is possible to obtain time–conductance change curves with these systems, quantitative analysis of interactions between SMs and their target proteins is also possible. Furthermore, such devices can be microfabricated or even potentially nanofabricated, resulting in extremely high-density arrays for quantitative multiplexed proteome-scale interaction characterization and protein detection [29]. The channels for such devices can then potentially be derivatized through any of the existing processes, such as photo- or nanolithography [30].

Electrochemical impedance spectroscopy (EIS) is another technology that allows label-free quantitative interaction detection. In this approach, the surface of an electrode is modified through application of a self-assembled monolayer to permit further derivatization that would ultimately lead to protein, DNA, or SM immobilization on the surface. An electrode array can be produced, with each electrode derivatized with a distinct SM or protein. Once the immobilization is complete through the application of an appropriate coupling chemistry, the resulting chip or single device is incubated with the interrogated sample while the impedance of electron transfer to a redox solution is measured (Fig. 2-9). In one setup, in which certain sample constituents bind to the immobilized protein or small-molecule probe, the impedance of the layer immediately adjacent to the working electrode will increase, due to hindrance of electron transfer from the working electrode. In very simple terms, the complex impedance is determined by examining the changes in the current upon application of an ac potential of a given frequency at the working electrode. The parameters measured are the absolute impedance and its phase, which depends on the frequency of applied potential. It is the change in phase that provides the means to detect the changes

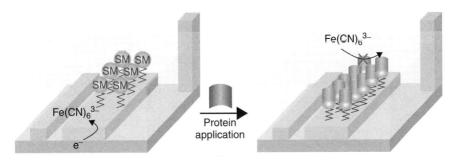

Figure 2-9 Typical EIS device. Auxiliary, working, and reference electrodes are shown in gold, blue, and green, respectively. In a solution (gray transparent cube), a redox compound such as $Fe(CN)_6^{3-}$ will "complete the circuit" between the auxiliary and working electrodes, but upon application of a protein that would bind to a SM immobilized on the working electrode, the access of the redox solute may be severely limited, resulting in impedance changes. *(See insert for color representation of figure.)*

in the layer surrounding the working electrode, and thus to detect and characterize binding events. Impedance phase can be recorded with respect to applied potential frequency, and the resulting spectrum can be monitored for shifts during a binding event.

To date, this approach has been used primarily as the foundation of sensing of target compounds in a solution through the use of electrodes coated with DNA aptamers, allowing for nanomolar to subnanomolar detection. For example, Lee and co-workers have reported the development of an EIS aptamer-based device for thrombin detection whose detector part is no bigger than 25 mm^2 [31]. Since the use of a redox solution poses certain risks to the sample and immobilized molecules, novel methods are being developed that would abolish the need for its use [28].

Additionally, the method was applied to probing protein–protein interactions in an array format by Evans and co-workers [32]. In this study, a selective chemical masking process was used specifically to derivatize 20-μm-wide gold electrodes. The mask was removed by selective application of a potential to the appropriate electrode, after which the appropriate aptamer was applied. Subnanomolar protein detection from cell lysate was attained in a multiplexed format. Although only few electrodes were used in this case, the approach can easily be extended. It is easy to foresee that as this methodology improves, the use of EIS will also allow for highly accurate studies of protein–protein interaction disruption by SMs as well as for dynamic interrogation of SM–protein binding events in a high-throughput format. The possibility of integration of electrochemical detection devices with other instruments (e.g., QCM or SPRi) opens doors to significantly more accurate and data-rich assays [33]. All in all, in terms of HTS, EIS is as yet in an intermediate development stage, but its potential applications in inexpensive and versatile interaction screening will certainly establish it as an important tool for SM target profiling in the very near future.

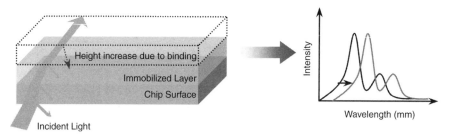

Figure 2-10 Principles of RIfS: A change in the optical thickness (a product of physical thickness and refractive index) of the immobilized layer due to binding will trigger a shift in the interference pattern.

Reflectometric Interference Spectroscopy (RIFS)   Another promising technology developed in recent years is RIfS, which may complement or even replace SPRi as a quantitative HTS method of choice. In very simple terms, RIfS relies on partial reflection of light upon transfer between two materials with different refractive indices combined with thin-film interference effects [34]. In an assay, a portion of incident light will be reflected at each material interface, forming an interference pattern when the path length of light through a particular layer supersedes the coherence wavelength, as is often the case in the immobilized layer (Fig. 2-10). A change in the height or refractive index of the immobilized layer, which occurs upon adsorption of molecules to the layer, will lead to changes in the interference pattern, which can be observed over time with a diode-array detector, for instance. Since RIfS depth of penetration into the solution is about 100-fold greater than that of SPRi [34], this technique is capable of providing more pronounced signals. Further, because RIfS relies on both the refractive index and the path length of light through the immobilized layer, temperature change–induced refraction index changes are not as much of a concern as they are with SPRi.

Although this approach has enjoyed a much lower level of recognition and application for biochemical problems, several interesting studies making use of it have appeared. Of interest is a recent study that utilized RIfS in conjunction with matrix-assisted laser desorption ionization–time of flight (MALDI-TOF) MS not only to study the dynamics of binding of vancomycin-type glycopeptide antibiotics to an immobilized target peptide, but also to identify the bound compounds on the surface of the chip [35]. This study has demonstrated the potential future utility of RIfS as a multiplexed quantitative proteome-scale SM interaction detection technique.

Surface Plasmon Resonance Imaging   Whereas most of the technologies we have discussed so far are only developing, SPRi has been available for quite some time, offering quantitative, label-free, robust, and economical interaction detection in a high-throughput format (Fig. 2-11) [36]. Although RIfS and EIS may provide data types similar to those of this technique, SPRi has been commercialized and subsequently developed for routine practical applications much

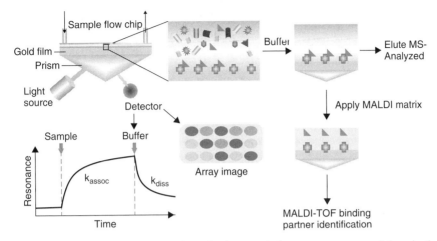

Figure 2-11 Typical SPRi setup, k_{assoc} is the association rate constant and k_{diss} is the rate constant for dissociation. These two parameters can be obtained from fitting the appropriate part of the time–resonance response graph. An entire array can be monitored simultaneously by using appropriate detectors.

earlier than have similar competing technologies. SPRi relies on the observation of surface plasmons, which are simply waves of free-electron oscillation that propagate parallel to an interface between a metal and a dielectric medium. Briefly, the protein of interest (POI) is cross-linked to a thin metal (e.g., gold) film, usually covalently and with the help of a thiol-based self-assembling monolayer. A sample, which can be heterogeneous, is applied to the immobilized molecule of interest through a flow chip. A binding event causes the mass of the complex attached to the film and the local index of refraction to change. This change in refraction index leads to changes in the plasmon resonance, which is recorded in resonance units. The detection of resonance change varies from method to method, but in a fixed-angle SPRi setup, coherent plane-polarized light is reflected off the gold surface. Changes in reflectance upon binding, as observed through a detector, imply changes in plasmon resonance, in turn implying a binding event. Given an adequate detection device, this allows for the simultaneous characterization of multiple interactions on a protein microchip.

As buffer containing an interacting protein flows over the detector surface, complexes are formed at the surface. The resonance signal changes correspondingly, thus providing protein association kinetics [37]. The sample is then purged with buffer containing no protein, providing dissociation kinetics. The ratio of the former over the latter gives the familiar equilibrium constant for the association reaction [equation (1)]. When a mixed sample is used, protein spots that were found to interact may be excised and characterized using MS.

One advantage of SPRi is the fact that no label is necessary, although the targets are immobilized on a support, often through covalent linkage. The interaction of proteins with any other classes of compounds, such as lipids, small molecules, and aptamers or nucleic acid fragments, may be examined. Conveniently, the composition of the flow chamber solution may be altered to characterize the range of conditions in which an interaction occurs. The sensor chip is highly reusable, with some studies using one chip up to 50 times. Additionally, the sample that passes over the sensor surface need not be pure, as MALDI-TOF MS can be used to identify the interacting proteins. One may even envision a scenario in which whole-cell proteomic extract is separated into fractions [e.g., with high-performance liquid chromatography (HPLC)] even before reaching the sensor surface to simplify MALDI analysis by considering HPLC lysate fractions sequentially. SPRi is also highly versatile; in one study, SPRi was used to identify phosphorylated proteins using the biotin–streptavidin system [38], while in another, cells were grown on the detector chip and lysed for interaction analysis [39].

Finally, this technique is highly scalable. Nanohole arrays allow for improved SPRi resolution, making it is theoretically possible to assay the interactions of more than 1 million proteins in 1 mm^2 [40]. Of course, it is imperative for protein printing resolution to correspond to that of analysis resolution to take full advantage of SPRi. Efficient protein printing methods, such as dip-pen nanolithography, have been developed recently [41]. Methods to immobilize proteins with a histidine tag have been available for a long time [42]. Self-assembling protein chips may also be used for high-density protein array production [43]. In this system, cDNA with epitope-tagged sequences of POIs is immobilized on the chip along with a capture agent, such as an aptamer or label capture molecule. As a cell-free expression system synthesizes the POI, it is immediately captured next to the site of its production. In terms of commercially available instruments, a single chip of an area of about 1 cm^2 can contain from 10 to 400 spots (Biacore [44] and GWC Technologies, Inc. [45]) to 1000 to 10,000 spots (Lumera [46]). With these systems it is possible quantitatively to characterize binding in each spot simultaneously with detection limits in the femtogram range and time resolution on the order of 1 second or less.

SPRi is extremely well suited for accurate thermodynamic characterization of binding as well. In a recent study, Yan and co-workers were capable of identifying entropic and enthalpic contributions to free energy of binding of Aβ binding alcohol dehydrogenase to Aβ, considered to be central to Alzheimer's disease [47]. In another study, potential small-molecule ligands were screened for interaction with a human estrogen receptor, providing the authors with kinetic characterization of the binding of these drug candidates [48].

Similar to two-hybrid screens, SPRi can be used to detect conditions under which proteins do not interact. The interaction of human papillomavirus E7 protein and retinoblastoma tumor suppressor protein RB was studied by Jung and co-workers [49]. Fifteen hundred protein spots containing immobilized E7 protein were produced, and an interaction with added RB protein was detected using SPR. The addition of a peptide derived from a motif on E7 along with RB has been observed to clearly disrupt the interaction through lack of an SPR signal. The one peptide discussed in this study can easily be replaced with an array of peptides or SMs, and a number of different proteins can be imprinted or added on this array.

Furthermore, the method was also applied very recently to drug–drug interaction characterization by Kuroda and co-workers [50]. For other applications, Rich and Myszka provide an extensive review of SPR literature that is highly recommended [51]. A large and growing number of other studies that employ SPRi are available, including studies on interactions of membrane proteins, domain–domain interactions, and various protein–ligand interactions. The popularity and the scale of development of SPR and SPRi-based technologies make us believe that it is these technologies that will be the most prominent catalysts of a quantitative revolution in all areas of biology, rapidly providing systems biology and rational drug design fields with much-needed quantitative data on a proteomic scale.

2.3 COMPUTATIONAL APPROACHES

2.3.1 Introduction

Thus far we have considered the derivation of kinetic constants describing interactions from experimental data through either direct observation of dissociation and association kinetics through a time–response curve (Fig. 2-11), relative binding affinity rankings with FAC, or indirect interaction observation with a reporter through yeast two-hybrid techniques. However, many of these technologies suffer from different drawbacks, from the occurrence of false positives and negatives to the incompatibility of assay conditions with successful formation of all potential complexes in a multiplexed assay. The question of whether these experimental complications can be avoided and, if possible, eliminated completely arises naturally. One way to avoid these complications along with the experimental time and material costs is to consider the fourth route to interaction characterization: interaction energy calculations. In fact, virtual interaction screening for large chemical libraries is already helping great numbers of researchers in industry and academia alike to understand and design ligands (e.g., for pharmaceutical applications) [52].

Although many theoretical methodologies, starting with density functional theory and semiempirical geometry optimization and energy calculation algorithms down to protein sequence–based analysis rely inherently on experimental data for parametrization, thus incorporating any experimental errors, other methods, such as ab initio quantum mechanical methods, rely on few, if any, experimental data. The major drawback for the latter algorithms is the dreadful scaling of computational effort required with the number of elements in the system considered. However, with Moore's law–type progression in computational hardware costs and performance, along with developments in nascent technologies such as quantum computing, one can foresee a time when brute-force approaches with all-atom, all-electron, all-interaction complex and transition-state free-energy calculations will become a reality for biomacromolecule-scale problems. For example, a computational power equivalent of the IBM Deep Blue supercomputer built in the early 1990s, used to beat Garry Kasparov at chess, has recently been built by academics for about $1000 using, among other things, four low-cost processors. Whether ab initio methods will provide a level of accuracy sufficiently higher than that of experimentally parametrized ubiquitous low-cost computational chemistry methods such as Austin Method 1 (AM1) is questionable. It is certain, however, that such methods will possess a greater level of general applicability, allowing them to treat accurately compounds outside the parametrization set of modern low-cost methods.

Although first prototypes of quantum computing-based systems are only now being built, sufficiently computationally powerful systems are still out of our reach, forcing researchers to apply simplifying assumptions in their studies. For example, one famous assumption is the treatment of atoms as hard, charged spheres of a given mass connected by springs, used within molecular mechanics force field (MMFF) frameworks, which facilitates calculations on very large systems. Another way to look at interaction prediction and characterization is to consider recurrent protein sequence patterns, which are used by proteins to interact with small molecules. In the following sections we very briefly consider a few interesting applications and trends in theoretical interaction screening technologies. The reader is referred elsewhere for a more in-depth treatment of the subject [53].

2.3.2 Energy-Based SM Interaction Predictions

Most of the methods for SM interaction profiling in use today employ MMFFs, which do not provide the actual Gibbs free energies of formation, but rather the strain energies, which only reflect the instability of the system in question. That is, all the bonds, angles, and dihedral angles are springs, and any distortion of a spring from its equilibrium point will result in the accumulation of potential energy. The presence of electrostatic and van der Waals interactions will further modify this strain energy. In fact, it is the latter two factors that almost entirely determine the energies of interaction between a small molecule and its appropriate binding site. Although the binding site may change its configuration to accommodate an SM distinct from its native ligand, a large number of studies apply a rigid protein assumption, leaving the active site fixed while the optimal configuration of an SM is found within it. In some cases, researchers have gone so far as to suggest abstracting all binding pocket surfaces to form a pocket collection they refer to as the *pocketome* to facilitate the identification of SM protein targets through the calculation of pocket–SM interaction energies [54].

Although many protein x-ray crystal structures are of proteins in a complex with an inhibitor or, in general, an SM interaction partner, making the binding pocket easy to identify, this is not the case for all the three-dimensional structures deposited. For this purpose, several authors successfully identified the binding pockets of a wide range of protein structures from a number of organisms, drawing on protein structure and sequence similarity [55,56]. In one case, of 5616 pockets considered, the algorithm presented identified 96.8% of the binding sites, 85.7% of the sites predicted showing similarity to actual binding sites of 80% or more [54]. Provided that the sequence and geometry of the binding sites can be determined from a crystallographic experiment or from an algorithm such as the one discussed, any of the existing computational chemistry packages, such as AutoDock [57], GROMACS [58], and MOE [59], can be used to estimate binding energies and to evaluate the quality of fit of a particular SM in a particular binding site. (The former two packages are free for academic use, highly accessible, easy to use, and are supported with extensive online documentation and user forums; for a general review, see ref. 60.) This allows, for example, for preliminary SM target profiling for different organisms, which is important in pesticide and antibiotic development. In addition to this, computational interaction profiling is relatively rapid and rather inexpensive relative to the experimental methods mentioned. As computational power and algorithm optimization continue to develop, *in silico* interaction profiling will keep growing in reliability and applicability, providing interaction data on an unprecedented scale.

2.3.3 Sequence-Based SM Interaction Predictions

Another way to predict protein–SM interactions is to assume that similar protein sequences bind similar SMs. This is a simple, yet powerful assumption, providing accurate interaction predictions, demonstrated in a number of studies. The actual implementation of this idea into a computational algorithm is limited only by human imagination and available time, so its discussion is beyond the scope of this chapter. One interesting sequence-based interaction prediction approach reported by Synder and co-workers [61] was used to create the small-molecule interaction database (SMID) (Fig. 2-12). Biologically relevant SMs were determined from three-dimensional x-ray crystal structures containing

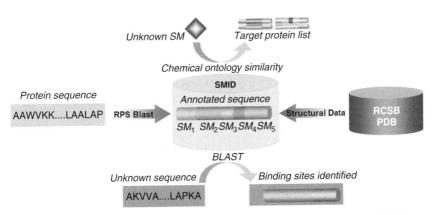

Figure 2-12 Simplified overview of the mechanism of action of SMID-based SM target profiling and protein SM binding site identification.

solvents, cofactors, and extraneous molecules provided by the Research Collaboratory for Structural Bioinformatics Protein Data Bank. Amino acid residues important in binding to the SM were determined from interaction proximity and conservation of binding motif across multiple sequences. All of this information was then used to annotate SM binding sites on structurally determined proteins, generating the SMID.

Embodied as the SMID-BLAST tool, a protein of interest is compared against the SM-binding proteins in the SMID database so as to identify portions of the protein anticipated to interact with a particular small molecule. Interestingly, this approach was then extended to allow for the profiling of interactions of a given small molecule within SMID [62]. An ontological-based semantic similarity measure was chosen to identify close structural or drug analogs to the queried molecule, although SMILES string or InChI-based searches would also have been possible. Thus, accurate SM target identification is possible *in silico* using very little preliminary sequence information. Although protein–SM interaction energy calculations from first principles may someday become commonplace and highly reliable in the short to medium term, there is nothing as elegant and as computationally inexpensive as sequence-based interaction prediction.

2.4 CONCLUSIONS

Indeed, in the past decade, there has been a remarkably rapid evolution of qualitative interaction profiling based on yeast two-hybrid methods to proteome-scale quantitative dynamic characterization of SM–protein binding events based on, for example, SPRi, augmented with in silico interaction prediction. We cannot remain unmoved both by the present state of the field, where only a few years ago quantitative information on a proteomic scale seemed like a dream, and by the exciting paths of scientific discovery unlocked by this information. Much like the revolutionary sequencing of the human and other organisms' genomes has provided vast quantities of information regarding the essence of life and underlying mechanisms, the present-day quantitative revolution will leave a mark on all fields of biochemical enquiry. Although experimental and theoretical SM target profiling methodologies are presently very well developed, we are looking forward to the next stages of interaction characterization: the determination of precise binding mechanisms in a high-throughput format.

Additionally, since most of the techniques discussed here can monitor a large range of reaction dynamics besides SM–protein target interaction, including enzymatic reaction dynamics, complete dynamic cell-scale process description will become available. Among other things, these breakthroughs will enable us to develop quantitative, mechanistic models of the cell, which in turn would lead to the discovery of side effect–free drugs to combat entrenched diseases, more effective pesticides and antibiotics, and more productive and novel bioindustrial processes. Indeed, the SM target profiling field is ripe for a bountiful harvest of knowledge.

REFERENCES

1. Rautio, J., Kumpulainen, H., Heimbach, T., et al. (2008). Prodrugs: design and clinical applications. *Nat. Rev. Drug Discov.*, *7*, 255–270.

2. Ling, M. M., Ricks, C., Lea, P. (2007). Multiplexing molecular diagnostics and immunoassays using emerging microarray technologies. *Expert Rev. Mol. Diagn.*, *7*, 87–98.

3. Lohse, M. J., Hein, P., Hoffmann, C., Nikolaev, V. O., Vilardaga, J. P., Bünemann, M. (2008). Kinetics of G-protein-coupled receptor signals in intact cells. *Br. J. Pharmacol.*, *153*, S125–S132.

4. Freyer, M. W., Lewis, E. A. (2008). Isothermal titration calorimetry: experimental design, data analysis, and probing macromolecule/ligand binding and kinetic interactions. *Methods Cell Biol.*, *84*, 79–113.

5. Nelson, R. W., Nedelkov, D., Tubbs, K. A. (2000). Biosensor chip mass spectrometry: a chip-based proteomics approach. *Electrophoresis*, *21*, 1155–1163.

6. Evans, D. B., Rank, K. B., Sharma, S. K. (2002). A scintillation proximity assay for studying inhibitors of human tau protein kinase II/cdk5 using a 96-well format. *J. Biochem. Biophys. Methods*, *50*, 151–161.

7. Stanyon, C. A., Liu, G., Mangiola, B. A., et al. (2004). A *Drosophila* protein-interaction map centered on cell-cycle regulators. *Genome Biol.*, *5*, R96.

8. Parsons, A. B., Lopez, A., Givoni, I. E., et al. (2006). Exploring the mode-of-action of bioactive compounds by chemical-genetic profiling in yeast. *Cell*, *126*, 611–625.

9. Becker, F., Murthi, K., Smith, C., et al. (2004). A three-hybrid approach to scanning the proteome for targets of small molecule kinase inhibitors. *Chem. Biol.*, *11*, 211–223.

10. Botstein, D., Chervitz, S. A., Cherry, J. M. (1997). Yeast as a model organism. *Science*, *277*, 1259–1260.

11. Giaever, G., Flaherty, P., Kumm, J., et al. (2004). Chemogenomic profiling: identifying the functional interactions of small molecules in yeast. *Proc. Natl. Acad. Sci. USA*, *101*, 793–798.

12. Fields, S., Song, O. (1989). A novel genetic system to detect protein–protein interactions. *Nature*, *340*, 245–246.

13. Titz, B., Thomas, S., Rajagopala, S. V., Chiba, T., Ito, T., Uetz, P. (2006). Transcriptional activators in yeast. *Nucleic Acids Res.*, *34*, 955–967.

14. Ehlert, A., Weltmeier, F., Wang, X., et al. (2006). Two-hybrid protein–protein interaction analysis in *Arabidopsis* protoplasts: establishment of a heterodimerization map of group C and group S bZIP transcription factors. *Plant J.*, *46*, 890–900.

15. Licitra, E. J., Liu, J. O. (1996). A three-hybrid system for detecting small ligand–protein receptor interactions. *Proc. Natl. Acad. Sci. USA*, *93*, 12817–12821.

16. Henthorn, D. C., Jaxa-Chamiec, A. A., Meldrum, E. (2002). A GAL4-based yeast three-hybrid system for the identification of small molecule–target protein interactions. *Biochem. Pharmacol.*, *63*, 1619–1628.

17. Ray, P., Pimenta, H., Paulmurugan, R., et al. (2002). *Proc. Natl. Acad. Sci. USA*, *99*, 3105–3110.

18. Giaever, G., Shoemaker, D. D., Jones, T. W., et al. (1999). Genomic profiling of drug sensitivities via induced haploinsufficiency. *Nat. Genet.*, *21*, 278–283.

19. Lum, P. Y., Armour, C. D., Stepaniants, S. B., et al. (2004). Discovering modes of action for therapeutic compounds using a genome-wide screen of yeast heterozygotes. *Cell*, *116*, 121–137.

20. Katayama, H., Oda, Y. (2007). Chemical proteomics for drug discovery based on compound-immobilized affinity chromatography. *J. Chromatogr. B*, *855*, 21–27.

21. Ng, W., Dai, J. R., Slon-Usakiewicz, J. J., Redden, P. R., Pasternak, A., Reid, N. (2007). Automated multiple ligand screening by frontal affinity chromatography–mass spectrometry (FAC-MS). *J. Biomol. Screen.*, *12*, 167–174.

22. Sharma, J., Besanger, T. R., Brennan, J. D. (2008). Assaying small-molecule-receptor interactions by continuous flow competitive displacement chromatography/mass spectrometry. *Anal. Chem.*, *80*, 3213–3220.

23. Cooper, M. A., Singleton, V. T. (2007). A survey of the 2001 to 2005 quartz crystal microbalance biosensor literature: applications of acoustic physics to the analysis of biomolecular interactions. *J. Mol. Recognit.*, *20*, 154–184.

24. Nishino, H., Nihira, T., Mori, T., Okahata, Y. (2004). Direct monitoring of enzymatic glucan hydrolysis on a 27-MHz quartz-crystal microbalance. *J. Am. Chem. Soc.*, *126*, 2264–2265.

25. Godber, B., Frogley, M., Rehak, M., et al. (2007). Profiling of molecular interactions in real time using acoustic detection. *Biosens. Bioelectron.*, *22*, 2382–2386.

26. Luo, Y., Chen, M., Wen, Q., et al. (2006). Rapid and simultaneous quantification of four urinary proteins by piezoelectric quartz crystal microbalance immunosensor array. *Clin. Chem.*, *52*, 2273–2280.

27. Grieshaber, D., MacKenzie, R., Vörös, J., Reimhult, E. (2008). Electrochemical biosensors: sensor principles and architectures. *Sensors*, *8*, 1400–1458.

28. Radi, A. E., Acero Sánchez, J. L., Baldrich, E., O'Sullivan, C. K. (2006). Reagentless, reusable, ultrasensitive electrochemical molecular beacon aptasensor. *J. Am. Chem. Soc.*, *128*, 117–124.

29. Chen, C., Zhang, Y. (2006). Carbon nanotube multi-channeled field-effect transistors. *J. Nanosci. Nanotechnol.*, *6*, 3789–3793.

30. Lee, J. Y., Shah, S. S., Zimmer, C. C., Liu, G. Y., Revzin, A. (2008). Use of photolithography to encode cell adhesive domains into protein microarrays. *Langmuir*, *24*, 2232–2239.

31. Lee, J. A., Hwang, S., Kwak, J., Park, S. I., Lee, S. S., Lee, K.-C. (2007). An electrochemical impedance biosensor with aptamer-modified pyrolyzed carbon electrode for label-free protein detection. *Sens. Actuat. B*, *1*, 372–379.

32. Evans, D., Johnson, S., Laurenson, S., Davies, A. G., Ko Ferrigno, P., Wälti, C. (2008). Electrical protein detection in cell lysates using high-density peptide-aptamer microarrays. *J. Biol.*, *7*, 3.

33. Jia, X., Xie, Q., Zhang, Y., Yao, S. (2007). Simultaneous quartz crystal microbalance–electrochemical impedance spectroscopy study on the adsorption of anti-human immunoglobulin G and its immunoreaction at nanomaterial-modified Au electrode surfaces. *Anal. Sci.*, *23*, 689–696.

34. Proll, G., Steinle, L., Pröll, F., et al. (2007). Potential of label-free detection in high-content-screening applications. *J. Chromatogr. A*, *1161*, 2–8.

35. Mehlmann, M., Garvin, A. M., Steinwand, M., Gauglitz, G. (2005). Reflectometric interference spectroscopy combined with MALDI-TOF mass spectrometry to determine quantitative and qualitative binding of mixtures of vancomycin derivatives. *Anal. Bioanal. Chem.*, *382*, 1942–1948.

36. Yu, X., Xu, D., Cheng, Q. (2006). Label-free detection methods for protein microarrays. *Proteomics*, *6*, 5493–5503.

37. Rich, R. L., Cannon, M. J., Jenkins, J., et al. (2008). Extracting kinetic rate constants from surface plasmon resonance array systems. *Anal. Biochem.*, *373*, 112–120.

38. Inamori, K., Kyo, M., Nishiya, Y., et al. (2005). Detection and quantification of on-chip phosphorylated peptides by surface plasmon resonance imaging techniques using a phosphate capture molecule. *Anal. Chem.*, *77*, 3979–3985.

39. Kim, M., Park, K., Jeong, E. J., Shin, Y. B., Chung, B. H. (2006). Surface plasmon resonance imaging analysis of protein–protein interactions using on-chip-expressed capture protein. *Anal. Biochem.*, *351*, 298–304.

40. De Leebeeck, A., Kumar, L. K., de Lange, V., Sinton, D., Gordon, R., Brolo, A. G. (2007). On-chip surface-based detection with nanohole arrays. *Anal. Chem.*, *79*, 4094–4100.

41. Liu, X., Yue, J., Zhang, Z. (2008). Generation of F(0)F(1)-ATPase nanoarray by dip-pen nanolithography and its application as biosensors. *Arch. Biochem. Biophys.*, (in press).

42. Gershon, P. D., Khilko, S. (1995). Stable chelating linkage for reversible immobilization of oligohistidine tagged proteins in the BIAcore surface plasmon resonance detector. *J. Immunol. Methods*, *183*, 65–76.

43. Ramachandran, N., Hainsworth, E., Bhullar, B., et al. (2004). Self-assembling protein microarrays. *Science*, *305*, 86–90.

44. Biacore. http://www.biacore.com/lifesciences/index.html.

45. GWC Technologies, Inc. http://www.gwctechnologies.com.

46. Lumera. http://www.lumera.com.

47. Yan, Y., Liu, Y., Sorci, M., et al. (2007). Surface plasmon resonance and nuclear magnetic resonance studies of ABAD–Abeta interaction. *Biochemistry*, *46*, 1724–1731.

48. Rich, R. L., Hoth, L. R., Geoghegan, K. F., et al. (2002). Kinetic analysis of estrogen receptor/ligand interactions. *Proc. Natl. Acad. Sci. USA*, *99*, 8562–8567.

49. Jung, S. O., Ro, H. S., Kho, B. H., Shin, Y. B., Kim, M. G., Chung, B. H. (2005). Surface plasmon resonance imaging-based protein arrays for high-throughput screening of protein–protein interaction inhibitors. *Proteomics*, *5*, 4427–4431.

50. Kuroda, Y., Saito, M., Sakai, H., Yamaoka, T. (2008). Rapid characterization of drug–drug interaction in plasma protein binding using a surface plasmon resonance biosensor. *Drug Metab. Pharmacokinet.*, *23*, 120–127.

51. Rich, R. L., Myszka, D. G. (2007). Survey of the year 2006 commercial optical biosensor literature. *J. Mol. Recognit.*, *20*, 300–366.

52. Charifson, P. S., Walters, W. P. (2002). Filtering databases and chemical libraries. *J. Comput.-Aided Mol. Des.*, *16*, 311–323.

53. Kukol, A. (ed.) (2008). *Molecular Modeling of Proteins*. Humana Press, Totowa, NJ.

54. An, J., Totrov, M., Abagyan, R. (2005). Pocketome via comprehensive identification and classification of ligand binding envelopes. *Mol. Cell. Proteom.*, *4*, 752–761.

55. Henschel, A., Winter, C., Kim, W. K., Schroeder, M. (2007). Using structural motif descriptors for sequence-based binding site prediction. *BMC Bioinformatics*, *8* (Suppl. 4), S5.

56. Laurie, A. T., Jackson, R. M. (2005). Q-SiteFinder: an energy-based method for the prediction of protein–ligand binding sites. *Bioinformatics*, *21*, 1908–1916.

57. AutoDock. http://autodock.scripps.edu/.

58. GROMACS. http://www.gromacs.org/.

59. MOE. http://www.chemcomp.com/.

60. Perola, E., Walters, W. P., Charifson, P. S. (2004). A detailed comparison of current docking and scoring methods on systems of pharmaceutical relevance. *Proteins*, *56*, 235–249.

61. Snyder, K. A., Feldman, H. J., Dumontier, M., Salama, J. J., Hogue, C. W. (2006). Domain-based small molecule binding site annotation. *BMC Bioinformatics*, *7*, 152.

62. Feldman, H. J., Dumontier, M., Ling, S., Haider, N., Hogue, C. W. (2005). CO: a chemical ontology for identification of functional groups and semantic comparison of small molecules. *FEBS Lett.*, *579*, 4685–4691.

3

NOVEL APPLICATIONS OF AFFINITY BEADS

YASUAKI KABE, MAMORU HATAKEYAMA, SATOSHI SAKAMOTO, KOSUKE NISHIO, AND HIROSHI HANDA

Graduate School of Bioscience and Biotechnology, Tokyo Institute of Technology, Yokohama, Kanagawa, Japan

3.1 INTRODUCTION: BIOLOGICAL BACKGROUND

A large number of drugs are currently used therapeutically without knowledge of their target molecules or precise pharmacological mechanisms of action. Occasionally, this results not only in complicated side effects and adverse reactions, but also in limited efficacy or treatment failure in certain patients. Identifying the target molecule and its function can aid in predictions about a drug's efficacy and the likelihood of adverse reactions or individual sensitivity, even before clinical trials begin. In addition, the optimization of seed compounds depends on increasing the binding affinity of the target protein with drug.

Affinity purification is a well-known technique for identification of ligand-binding proteins [1], but its widespread use is limited by the inefficiency and instability of conventional matrices. Moreover, nonspecific binding of proteins to the solid support complicates identification of the specific target protein. Often, low purification efficiency requires that source material should

Protein Targeting with Small Molecules: Chemical Biology Techniques and Applications,
Edited by Hiroyuki Osada
Copyright © 2009 John Wiley & Sons, Inc.

be partially purified before utilizing affinity chromatography, and the unstable character of conventional matrices narrows the spectrum of suitable ligands. The recent development of combinatorial chemistry has produced many drugs and drug candidates of interest. Once a candidate compound is selected, further rational drug design depends on identifying the drug receptor or target. Thus, a simple receptor identification system is required for advancing highly efficient drug development. Despite the pressing need to identify drug targets, conventional matrices for affinity purification have changed little in more than three decades. To facilitate drug target identification, we have developed latex beads adaptable to small-scale purification of drug receptors [2].

Nanobeads are useful as a matrix material because they offer a large surface area for binding, are highly mobile, and are easily resuspended and recovered. If the nanobead material is a polymer, its surface structure can easily be modified for application optimization, which is an advantage over nanobeads composed of silica or titanium particles. For these reasons, polymer nanobeads have been used widely for bioseparation and bioassays. In the field of chemical biology, for example, polymer nanobeads on which chemical compounds are immobilized are used for drug target screening.

3.2 DEVELOPMENT OF MAGNETIC BEADS

3.2.1 Affinity Latex Beads

Using seed polymerization methods, we have developed a novel affinity matrix composed of latex beads (SG beads). The SG particle consists of a polystyrene (poly-St)/polyglycidyl methacrylate (poly-GMA) core and a surface covered completely with poly-GMA. SG beads exhibit a number of advantageous properties as an affinity carrier: submicrometer size, monodispersion in water, very low nonspecific protein adsorption, and reactive epoxide groups for ligand binding (Fig. 3-1). These properties offer a solution to the disadvantages of agarose beads. SG beads have been used successfully to purify a variety of biomolecules directly from crude extracts [3].

However, SG beads are very difficult to separate from the dispersion solution. Centrifugation is generally used to precipitate polymer beads; however, it is undesirable because the molecule of interest may be denatured by the process. To address the problem, polymer beads that include a magnetic substance have been developed. Ugelsted et al. developed a submicrometer-sized polymer bead composed of nonpolar and monodispersed magnetic particles [4]. This improved affinity matrix was used successfully in affinity separation of cells, and the technology of polymer beads separated by their magnetic property was immediately useful in a wide range of applications.

Magnetic particles have been used for affinity purification of DNA-binding enzymes and receptors. In addition, made-to-order magnetic polymer nanobeads can be prepared for specific applications by introducing desired functional groups to the bead surface. Technologies utilizing magnetic separation are important and

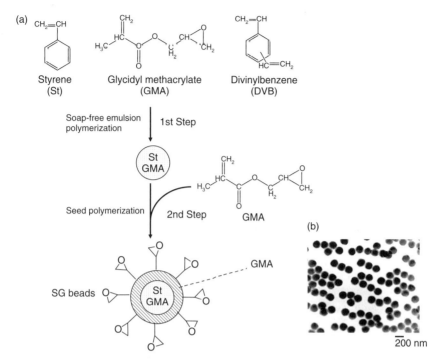

Figure 3-1 (a) Preparation of SG beads; (b) TEM view of SG beads.

essential tools for medical research: for example, magnetic separation of nucleic acids, cell separation and identification using antibodies, and separation and detection of viruses and bacteria. Of these applications, separation of immune cells is used frequently, because separating cells using magnetic nanobead technology yields cells with a greater probability of survival, so that smaller samples are required.

Magnetic separation technology is well known; however, its use is often limited because of the low binding activity of ligands and the relatively small surface area of the beads. Therefore, we designed and produced nanobeads that have a greater surface area in order to improve their effectiveness of ligand binding.

3.2.2 Preparation of Size-Controlled (30–100 nm) Magnetite Nanoparticles for Biomedical Applications

Magnetite (Fe_3O_4) nanoparticles (MNPs) have been used as magnetic carriers in a variety of biomedical applications. Recently, we developed magnetic nanobeads of submicrometer size composed of a core–shell structure of MNPs (40 nm in size)/polymer for use as magnetic carriers in bioscreening. To attain high throughput and high accuracy in bioscreening, core MNPs must be relatively large and uniform in size, controlled precisely within several dozen nanometers. MNPs synthesized by the conventional wet process, or coprecipitation method, are small, less than 8 nm, and range widely in size. Sugimoto and Matijevic synthesized MNPs ranging in size from 30 nm to 1.1 μm using sophisticated chemical reactions conducted at 90°C [5]. However, the size of these MNPs was widely distributed with a coefficient of variation exceeding 15%. While subsequent methodological progress has enabled the synthesis of MNPs within a narrow size distribution, the particle size has remained smaller than 20 nm. Furthermore, since organic solvents and/or surfactants are used in their production, residual organic solvent and/or surfactant are found on the MNP surface, which makes encapsulation in a hydrophilic polymer shell difficult. Here we present the preparation of residual-free MNPs with tightly controlled sizes, between 30 and 100 nm, produced in aqueous solution at relatively low temperatures (4 to 37°C).

MNPs were synthesized using a surfactant-free oxidation process in an alkaline aqueous solution of deaerated 21 mM NaOH (pH 12 to 13) and 8.80 mmol NaNO$_3$ oxidant. Deaerated 0.1 M ferrous chloride (25 mL) was added to the alkaline solution, and the mixture was held at 4, 15, 25, or 37°C for 24 hours. The resulting MNPs were separated from solution with a magnet and washed several times with pure water. We investigated the morphology and size, crystal phase, and magnetization of these MNPs using transmission electron microscopy, x-ray diffractometer, and VSM, respectively.

The MNPs exhibited x-ray diffraction lines assigned only to the spinel structure (Fig. 3-2). The nanoparticles showed physical properties of magnetite with a lattice constant value very close to that reported for bulk Fe$_3$O$_4$, which was confirmed by a saturation magnetization value very close to that reported for bulk Fe$_3$O$_4$ (92 emu/g) (Table 3-1). All MNPs possessed residual magnetization because their sizes were larger than the superparamagnetic limit of Fe$_3$O$_4$ (26 nm).

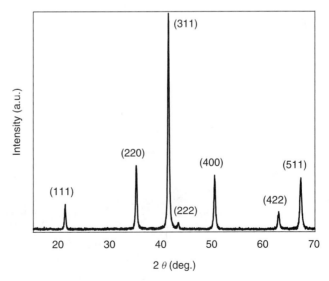

Figure 3-2   XRD diagram for MNPs synthesized at 4°C.

TABLE 3-1 Size [(d (average) ±Δd (standard deviation)], Size Distribution (Δd/d), Saturation Magnetization (M_s), and Residual Magnetization (M_r) for Magnetite Particles at Various Temperatures

$T(°C)$	Size (nm)	$\Delta d/d$	M_s (emu/g)	M_r (emu/g)
4	102.1 ± 5.6	5.5	91.9	28.5
15	46.1 ± 5.3	11.5	90.2	27.4
25	40.4 ± 5.7	14.1	87.8	31.8
37	31.7 ± 4.9	15.5	81.6	13.0

Characterization data (Table 3-1 and Fig. 3-3) showed that increasing synthesis temperature from 4°C to 37°C caused a decrease in the MNP size (from 102 ± 5.6 nm to 31.7 nm) and changed the shape from octahedral to nearly spherical. In addition, the MNP size distribution was observed to increase as the temperature increased; that is, $\Delta d/d$ increased from 5.5% to 15.5%. This indicates that higher reaction temperatures accelerate both nucleation and growth of magnetite crystals in aqueous solution.

This simple chemical method using a surfactant-free, alkaline, aqueous solution produces discrete size-controlled MNPs ($d = 30$ to 100 nm, $\Delta d/d < 15\%$). These MNPs are suitable as a magnetic core for subsequent coating with functional molecules to accomplish quick and effective response in magnets [6].

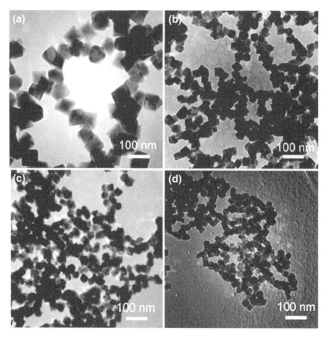

Figure 3-3 TEM images for MNPs synthesized at (a) 4°C, (b) 15°C, (c) 25°C, and (d) 37°C.

3.2.3 Magnetic Nanocarriers for Affinity Purification

Bioseparation utilizing magnetic carriers is a facile and convenient method for a large number of biological applications, including simultaneous analysis of multiple bioanalytes. Various types of magnetic carriers have been developed and reported, but almost all are relatively large (micrometer size) and are designed for cell separation, collection of DNA, and concentration of tagged fusion proteins. For the efficient capture of drug targets, which are often expressed in very low abundance, carriers of nanometer size and high dispersion properties are desirable. It is also important that carriers retain their dispersibility and polymer coating in organic solvent, since most drugs require dissolution in organic solvents because they are water insoluble. Furthermore, carrier surfaces must exhibit low nonspecific binding of proteins, because nonspecific background signal can interfere with the identification of low-abundance targets. Previously we showed that poly-GMA-covered latex beads suppress nonspecific binding of proteins and are useful for isolation and identification of novel drug targets. Here we report the development of a novel, magnetic, nanometer-sized affinity carrier coated with poly-GMA, which demonstrates high performance for drug target isolation.

Magnetite nanoparticles of approximately 10 nm have generally been used as the magnetic core for carriers. Their superparamagnetic property excludes magnetic aggregation and facilitates regulation of their dispersibility. However, magnetically collectable carriers of nanometer size are difficult to produce because of their small magnetization value. Therefore, we designed magnetic carrier beads composed of a core with several dozen particles of magnetite to achieve both a nanometer size and magnetic property. Our initial effort was focused on preparing uniformly sized magnetite particles (MP_1) greater than 10 nm in size, because larger MP_1 have potentially higher magnetization and respond more rapidly to a permanent magnet then do the magnetite nanoparticles commonly used. We prepared size-controlled MPs of several dozen nanometers using a wet process according to our previous study [6]. MPs were separated from solution with a magnet and washed several times with pure water. MPs of approximately 40 nm were obtained as single-crystalline MPs, and VSM (Riken Denshi custom model) analyses revealed that they exhibited a high magnetization saturation value ($M_s = 88.6$ emu/g) (Fig. 3-4).

To encapsulate magnetite nanoparticles, mini-emulsion polymerization was used. However, self-aggregation of large magnetites induced large clusters as a function of their strong magnetic dipole interaction, which prevented the formation of stable mini-emulsions. Therefore, we selected a method of admicellar polymerization reported previously by Glatzhofer et al. [7] for encapsulation. MPs were covered with a double layer of surfactant: The first (primary) layer was adsorbed onto the MPs' surface and changed the surface property from hydrophilic to hydrophobic; the second (secondary) layer covered the first layer and facilitated dispersion in a water-based medium. After examination of many surfactant candidates, ionic surfactant was selected as the hydrophobic primary layer. The ionic surfactant is soluble in alkali solution and was adsorbed uniformly to the MPs. The surfactant sodium salt (0.1 M) was mixed with MPs

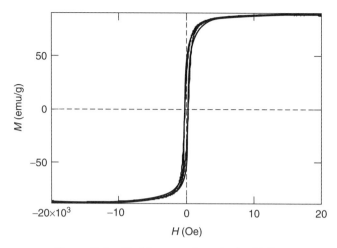

Figure 3-4 $H-M$ curve of synthesized MPs.

(180 mg) dispersed in distilled water. After addition of 0.1 M HCl, MPs were flocculated by overlaying with the surfactant.

The second surfactant layer is important to facilitate dispersion of MPs coated with first-layer surfactant. Since ionic surfactants such as sodium dodecyl sulfate and sodium dodecyl benzene sulfonate did not achieve MP dispersion, Emulgen 1150S-60 (Kao Corp.), a nonionic surfactant with polyoxyethlene, was selected, because it minimized ionic effects and increased steric repulsion. Emulgen 1150S-60 in aqueous solution was added to the MPs coated with primary-layer surfactant. Resulting MPs with double-layer surfactant (MP-DL) were obtained after ultrasonic treatment. The size distribution of MP-DL was 91.3 ± 18.3 nm, determined by dynamic laser scattering (DLS) analysis (FPAR-1000, Otsuka Electronics Co., Ltd.). The MPs' crystal size, surfactant thicknesses, and hydrodynamic size indicate that each micelle is composed of several MPs. High dispersibility was sustained for at least several dozen minutes.

Admicellar polymerization was carried out according to the following procedure. Divinylbenzene (DVB) was used as a bridged reagent, because a bridged reagent renders a polymer structure rigid, suppresses swelling in organic solvent, and inhibits decapsulation of magnetic cores. Monomer mixtures of styrene, GMA, and DVB, were added into continuous aqueous phase MP-DL at room temperature. After the mixture was stirred at 200 rpm for 20 minutes at 70°C, water-soluble radical initiator 2,2'-azobis(2-methylpropionamidine) dihydrochloride was added to the suspension. After 16 hours of stirring at 70°C, synthesized magnetic beads were collected and washed with distilled water. After redispersion into distilled water, bead surfaces were covered additionally with poly-GMA by seed polymerization (Fig. 3-5a). The amount and ratio of styrene, GMA, and DVB are important for preparing monodispersed magnetic beads, since these parameters define the polymer thickness, which in turn regulates the magnetic dipole interaction of the bead cores.

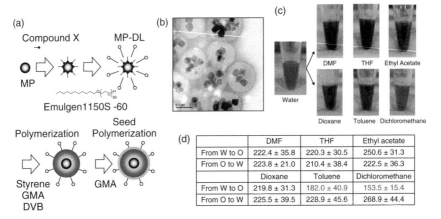

Figure 3-5 (a) Synthetic scheme of FG beads; (b) FE-TEM image of the isolated FG beads; (c) photo image of FGNE beads dispersed in DMF, THF, ethyl acetate, 1,4-dioxane, toluene, and dichloromethane (these dispersions contain 0.4 mg of FGNE beads); (d) dynamic light-scattering (DLS) analyses of FGNE beads in DMF, THF, ethyl acetate, 1,4-dioxane, toluene, and dichloromethane and water-resuspended beads from each organic solvent. *(See insert for color representation of figure.)*

Figure 3-5b shows an FE-TEM (Tecnai F20; FEI Company, Hillsboro, Oregon) image of the magnetic beads (FG beads), and several MPs encapsulated by the homogeneous polymer shell. The magnetic beads are small and show a discrete size distribution (184.5 ± 9.0 nm). Monodispersibility was confirmed by DLS measurement of their hydrodynamic size in distilled water (203.6 ± 28.7 nm), and analyses of thermogravimetry (TG) showed that the MP content was approximately 30%.

For affinity purification of drug targets, linkers are immobilized on carrier beads to prevent steric hindrance between affinity partners. After amination of epoxy groups on the surface of FG beads, ethylene glycol diglycidyl ether (EGDE) was immobilized as the linker (FGNE beads) [8]. High dispersibility was confirmed in a variety of organic solvents: N,N-dimethylformamide (DMF), tetrahydrofuran (THF), ethyl acetate, 1,4-dioxane, toluene, and dichloromethane. Some shrinkage and swelling were observed (Fig. 3-5c and d), but the bead core–shell structure and dispersibility were conserved after returning beads to aqueous conditions (data not shown), confirming the excellent amphiphilic property of FG beads in various types of organic solvent.

3.3 CHALLENGES TO SMALL-MOLECULE BINDING PROTEIN IDENTIFICATION

3.3.1 Affinity Purification of DNA-Binding Proteins Using SG Beads

Prior to our work in methods of identification of small-molecule binding proteins, we developed a purification system for DNA-binding proteins that uses

DNA-fixed SG beads. SG beads are composed of a styrene core and poly(glycidyl methacrylate) surface to which DNA oligomers are immobilized by means of covalent bonds formed between epoxy and amino groups [9]. In our studies, double-stranded DNA oligomers with protruding ends were bound to beads five times more efficiently than were those with blunt ends, suggesting that amino groups of the protruding ends react with epoxy groups. As shown in Figure 3-6a, a DNA fragment was prepared that has five nucleotides protruding at the 5'-ends and consensus sequences for DNA-binding transcription factor ATF/CREB family proteins. For efficient purification of DNA-binding factors, the DNA oligomers added should contain tandem repeats of the DNA-binding sequence. Oligomer ends were phosphorylated using T4 polynucleotide kinase, and ligated using T4 DNA ligase. Ligated DNA ranging in length from 5-mers to 15-mers proved suitable for efficient purification. The DNA oligomers prepared were easily coupled to SG beads. SG beads were incubated with approximately 200 ng/mL of DNA at 50°C for a period ranging from 5 hours to overnight. Usually, 20 to 30% of input DNA was immobilized on the beads. The DNA-coupled beads were stable for several months of storage at 4°C.

One advantage of using DNA immobilized to SG beads is the possibility of one-step purification of DNA-binding proteins from crude extracts. Several points must be considered when purifying proteins directly from crude extracts: nuclease activity in cellular extract, binding conditions such as reaction temperature and buffer conditions, and the effect of adding nonspecific DNA such as single-stranded DNA or poly(dI-dC). Such information greatly aids

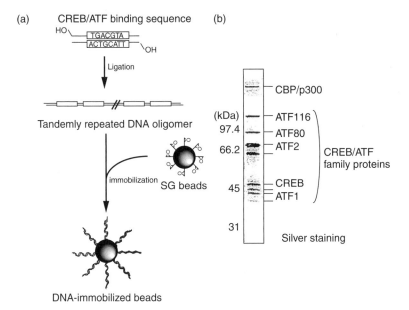

Figure 3-6 (a) Construction of DNA-immobilized beads; (b) purification of DNA-binding proteins using SG beads.

efficient purification of DNA-binding proteins. We examined direct purification of ATF/CREB family transcription factors from crude nuclear extracts. Beads bearing approximately 5-μg DNA oligomers containing ATF/CREB-binding sequences were placed in a 1.5-mL tube and incubated for 30 minutes with 1 ml of HeLa cell nuclear extract. Bound proteins were washed and eluted with 20 mL of buffer containing 1.0 M KCl. About 70% of bound proteins were recovered in one step elution. Eluates were subjected to SDS-PAGE, and several distinct bands were observed (Fig. 3-6b). These bands were analyzed by MS, and resulting amino acid sequences indicated the presence of proteins known to bind directly to the ATF/CREB consensus sequence, ATF1, ATF2, ATF80, ATF116, and CREB. In addition, a band of approximately 300 kDa was identified as CBP/p300 protein, which is a transcriptional coactivator that interacts with ATF/CREB family proteins. This indicates that oligo-DNA-immobilized SG beads are useful for the isolation not only of DNA-binding proteins targeted by specific oligo-DNAs but also for co-purification of associated proteins such as those comprising large DNA–protein macromolecules of transcription complexes. The binding capabilities of SG beads will be helpful in developing a comprehensive understanding of transcriptional machinery. In addition, purified proteins could be analyzed for sequence-specific DNA-binding activity by gel retardation assay, because affinity-bound proteins retain their native confirmation when recovered under nondenaturing conditions.

3.3.2   Affinity Purification of Small-Molecule Binding Proteins

SG beads proved useful for purification of DNA-binding proteins and their associated proteins; however, this system was unsuccessful in affinity purification of small-molecule binding. We reasoned that small-molecule target proteins were prevented from binding chemicals moieties on the SG bead surface due to steric hindrance. To solve this problem, the divalent epoxide spacer EGDE was introduced (Fig. 3-7b). Briefly, amino groups were reacted with the epoxy groups on the SG bead surface by mixing with 3 M NH_4OH for 24 hours at 70°C (SGN beads). Excess amounts of EGDE were coupled with SGN beads at 30°C for 24 hours (SGNEGDE beads).

To test the efficacy of SGNEGDE beads in targeting drug receptors, immunosuppressive agent FK506 (Fig. 3-7a) was used as a model ligand in a comparison of purification by the bead system and the conventional agarose resin system. After incubation with Jurkat cell cytoplasmic extracts, beads were washed and bound proteins were analyzed by SDS-PAGE (Fig. 3-7c). SGNEGDE beads yielded a major 12-kDa protein exhibiting high specificity directly from crude cell lysate. In contrast, FK506-fixed agarose beads showed significantly reduced specificity and minimal detection of the protein of interest. Immunoblot analysis indicated that the protein purified by the SGNEGDE beads was previously identified as FK506-binding protein (FKBP12). FKBP12 was also present in the agarose-purified fraction, but recovery was lower. These results demonstrate that SG beads allow higher binding efficiency of a target protein and less nonspecific binding. The FKBP12 yield using SG beads was between 8 and 10 times

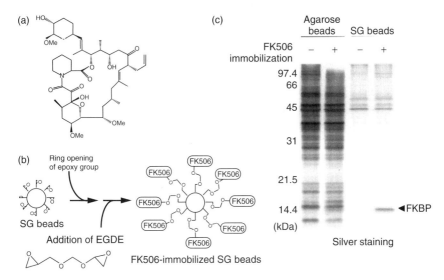

Figure 3-7 (a) Structure of FK506; (b) construction of FK506-immobilized beads; (c) purification of FKBP12 using agarose beads or SG beads.

higher than that achieved with either previous affinity methods or conventional purifications [2].

Next, we determined to identify the novel target protein of quinone derivative E3330, a compound developed originally as an anti-inflammatory drug which specifically suppresses NF-κB-mediated transcriptional activation of Jurkat cells [10]. To identify the molecular target of E3330, an amino derivative of E3330 (NH$_2$-E3330) was immobilized on SG beads (Fig. 3-8a). Jurkat cell nuclear extracts were incubated with NH$_2$-E3330-immobilized SG beads, and bound proteins were analyzed by SDS-PAGE. Three protein bands of 55, 38, and 27 kDa showed specific binding (Fig. 3-8b). Amino acid sequencing with Q-TOF MS analysis identified the 38-kDa protein as redox-related factor Ref-1. We confirmed that E3330 binding to Ref-1 was strong with drug far-western analysis and surface plasmon resonance (SPR) analysis. From biochemical analyses, Ref-1 was shown to enhance DNA-binding activity of NF-κB through direct interaction with the p50 subunit of NF-κB and subsequent regulation of the NF-κB redox state. E3330 specifically inhibited Ref-1-mediated NF-κB activation. Thus, we demonstrated that modified SG beads can be used to identify the small-molecular protein target of a drug compound. Such information is an important contribution to understanding the mechanism of drug action and the network of related biological reactions.

3.4 BIOLOGICAL AND MEDICAL APPLICATIONS

For analysis of chemical biology, the most important question to answer is how target molecules for chemical compounds can be identified with convenience and

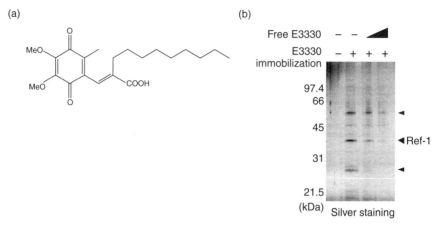

Figure 3-8 (a) Structure of E3330; (b) purification of the E3330-binding proteins using SG beads.

reliability. We have developed an application for high-performance affinity latex bead, which enables rapid and efficient purification of target proteins for a wide range of chemical compounds from complex sources such as protein libraries and cellular extracts. Using this purification technology, we have successfully purified various drug target proteins, such as immunosuppressors, antibiotics, and anticancer and anti-inflammation agents. In this chapter we present our analysis of chemical biology using affinity nanobeads.

3.4.1 Identification of the Mitochondrial Transporter for Porphyrins

Some metalloporphyrin derivatives are fluorescence- or phosphorescence-emitting molecules. Palladium *meso*-tetra(4-carboxyphenyl)porphyrin (PdTCPP) and palladium *meso*-tetra(4-aminophenyl)porphyrin (PdTAPP) (Fig. 3-9a) emit a long-lived phosphorescence in response to an excitation wavelength of approximately 400 nm at room temperature, as its photoexcited triplet-state decays. Oxygen quenches phosphorescence by decreasing the lifetime of phosphor. Based on their phosphorescent properties, porphyrin derivatives have been used as optical oxygen sensors in aerodynamics applications and in testing of normal tissues, cancer cells, and blood. However, the physiological properties of PdTCPP are still unclear. Heme (Fe-protoporphyrin IX), an endogenous porphyrin derivative, is essential to aerobic organisms and participates in a variety of physiological processes, such as oxygen transport, respiration, and signal transduction. Biosynthesis of heme requires mitochondrial accumulation of its biosynthetic precursor porphyrin from the cytosol. The mechanism of porphyrin accumulation in the mitochondrial inner membrane remains unclear.

We analyzed the mechanism for mitochondrial translocation of porphyrin derivatives [11] and found that PdTCPP and PdTAPP accumulate in the mitochondria of several cell lines. To investigate the mechanism of mitochondrial

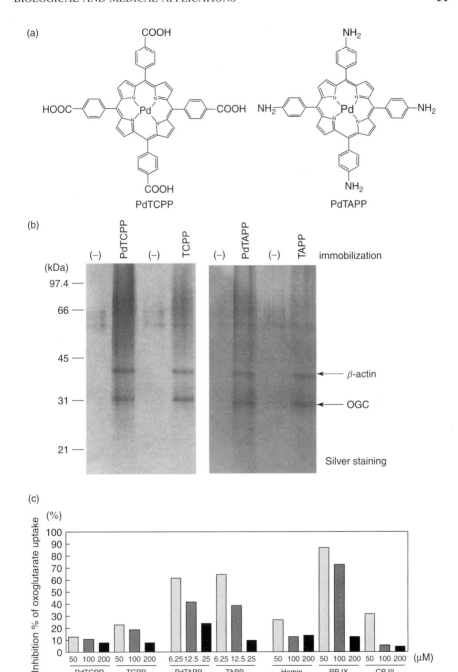

Figure 3-9 (a) Structures of PdTCPP and PdTAPP; (b) purification of porphyrin-binding proteins using SG beads; (c) inhibition of 2-oxoglutarate mitochondrial uptake by porphyrin derivatives (PP IX, protoporphyrin IX; CP III, coproporphyrin III); (d) model of mitochondrial accumulation of porphyrins. (*Continued*)

(d)

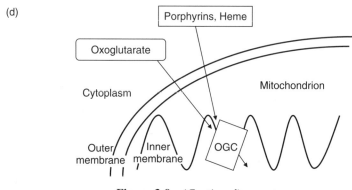

Figure 3-9   (*Continued*)

porphyrin derivative accumulation, we used SG beads to search for the mitochondrial target protein of PdTCPP. To prepare PdTCPP fixed beads, amino-modified SG beads were incubated with PdTCPP in one equivalent of N-hydroxysuccimide (NHS) to propagate a reaction using the four reactive carboxyl groups of PdTCPP. PdTCPP-fixed SG beads were used to purify PdTCPP-binding proteins from HeLa cell mitochondrial extracts. Two proteins, sized approximately 45 and 32 kDa on a silver-stained SDS–polyacrylamide gel, were bound specifically to PdTCPP-fixed beads (Fig. 3-9b). Interestingly, these bands were also observed using carboxysuccinate-modified SG beads fixed with PdTAPP or with the non-Pd-chelated derivatives TCPP and TAPP (Fig. 3-9b). These results suggest that tetraphenyl porphyrin derivatives may recognize the same mitochondrial target proteins. These proteins were identified using PdTCPP-fixed beads in SDS-PAGE with CBB staining and proteolytic digestion followed by amino acid identification with Q-TOF MS. The 45-kDa protein was identified as β-actin, and the 32-kDa protein as mitochondrial transporter 2-oxoglutarate carrier (OGC), which catalyzes the transport of 2-oxoglutarate for malate in an electroneutral exchange.

To assess the effect of PdTCPP on the transporter activity of OGC, we analyzed the incorporation activity of the OGC substrate 2-oxoglutarate using mitochondria fractions isolated from rat liver (Fig. 3-9c). PdTCPP inhibited 2-oxoglutarate incorporation in a dose-dependent manner. A Lineweaver–Burk plot of 2–oxoglutarate incorporation showed competitive inhibition by PdTCPP; the inhibitory constant (K_i) of PdTCPP was 15 µM. We analyzed the inhibitory effect of the tetraphenyl porphyrin derivatives PdTAPP, TAPP, and TCPP on 2-oxoglutarate. PdTAPP inhibited 91% of 2-oxoglutarate incorporation. Interestingly, the endogenous porphyrin derivatives hemin and heme precursors protoporphyrin IX and coproporphyrin III also inhibited 2-oxoglutarate mito-chondrial uptake. Inhibition by hemin was also competitive with a K_i value of 56 µM. These results suggest that porphyrins are assembled in the mitochondria through interactions with OGC and competitively inhibit 2-oxoglutarate uptake (Fig. 3-9d). Because heme is synthesized in a multistep process that is completed in the mitochondria, mitochondrial uptake of heme and related porphyrin derivatives via OGC could play a role in hcmc biosynthesis or the salvage of heme in the mitochondria. Additionally, it is known that porphyrias are inherited disorders of heme biosynthesis in which abnormalities in enzymes of the heme biosynthetic pathway cause generalized clinical defects. Seven enzymes implicated in causing porphyria have been identified; our data suggest that OGC could be a candidate for causing heme defects leading to disease.

3.4.2 Identification of a Novel Target Protein of Methotrexate

Methotrexate (MTX) is a folate antagonist used to treat malignancies and rheumatoid arthritis. It is known to target dihydrofolate reductase (DHFR), but evidence suggests that it may have other molecular targets as well. To identify novel MTX-binding proteins, we prepared two types of SG beads to which MTX was immobilized through different functional groups, either using the amino group of an MTX amino derivative or using the two carboxyl groups of MTX itself (Fig. 3-10) [12]. An MTX amino derivative bearing an amino group instead of the γ-carboxyl group was synthesized and fixed to SGNE-COOH beads, the SG beads carrying carboxyl groups, through the amino group by amidation (Fig. 3-10a). After incubation with cytoplasmic extracts, beads were washed,

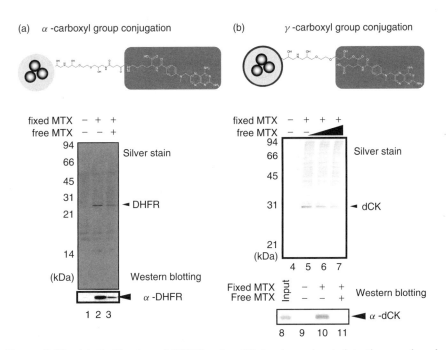

Figure 3-10 (a) Purification of DHFR using SG beads conjugated to the α-carboxyl group of MTX; (b) dCK purification dusing SG beads conjugated to the γ-carboxyl group of MTX.

and bound proteins were analyzed by SDS-PAGE and silver staining. A 23-kDa protein was purified, and immunoblot analysis revealed that the protein was DHFR. Authentic MTX was also fixed to SGNE-OH beads, the SG beads carrying hydroxyl groups, through the two carboxyl groups by esterification (Fig. 3-10b). To our surprise, the same procedure resulted in purification of an abundant 30-kDa protein and a small amount of DHFR detectable only by immunoblotting. The interaction between the 30-kDa protein and MTX was shown to be specific, as the protein did not bind to control beads, the binding interaction was inhibited by preincubation of the extracts with excess amounts of free MTX, and the protein was eluted from the beads with free MTX. Because a hydroxyl group generally reacts with equal efficiency with either acyl group of a carboxylic anhydride, both the α- and γ-carboxyl groups probably participated in fixation, which may account for the low yield of DHFR. The 30-kDa band was identified as dCK by Q-TOF MS analysis. dCK identity was further confirmed by immunoblot analysis using anti-dCK antibody (Fig. 3-10b).

dCK is an enzyme in the salvage pathway of nucleic acid biosynthesis that catalyzes phosphorylation of nucleosides such as dCyd, dAdo, and dGuo. The antitumor nucleoside araC is also phosphorylated by dCK, which allows it to exert its antiproliferative effect primarily against leukemia cells. It is known that the combination of araC and MTX is synergistically effective in the clinical treatment of Burkitt's lymphoma, and it has been shown that the presence of MTX enhances araC-induced cytotoxicity and elevates the amounts of intracellular araC triphosphate. Thus, it is possible that MTX activates endogenous dCK activity. To examine this hypothesis, we analyzed the effect of MTX on the activity of purified recombinant dCK. Remarkably, MTX enhanced dCK activity at high inhibitory concentrations of nucleoside substrates in a dose-dependent manner (Fig. 3-11a). For comparison, we investigated the effects of folate, leucovorin, and tetrahydrofolate, which are structurally similar to MTX, on dCK activity. These molecules had no apparent effect on dCK activity, which suggests that the regulatory effect on dCK is specific to MTX.

We next analyzed whether MTX actually regulates endogenous dCK activity in Burkitt's lymphoma HS-Sultan cells. AraC-induced cytotoxicity was significantly enhanced by MTX in HS-Sultan cells. The results of our kinetic assays predicted that araC phosphates would increase in cells when high concentrations of araC were co-administered with MTX in response to the MTX-derived up-regulation of dCK enzyme activity. To examine this hypothesis, araC phosphate incorporated into genomic DNA was quantified in the presence of MTX and various concentrations of araC. After HS-Sultan cells were incubated in media containing various con-centrations of MTX and [^{3}H]araC, genomic DNA was prepared from cell extracts, and the radioactivity of genome-incorporated araC phosphate in each fraction was measured with a liquid scintillation counter. MTX dose-dependently elevated the amount of incorporated araC phosphate (Fig. 3-11b). This suggests that MTX actually regulates dCK enzyme activity through direct interaction in vivo (Fig. 3-11c).

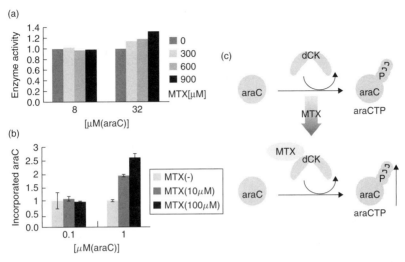

Figure 3-11 (a) Activation of dCK-mediated phosphrylation of araC by MTX; (b) enhancement of araC incorporation into the chromosome by MTX; (c) model of the dCK-mediated cytotoxicity by MTX.

MTX target proteins were isolated differentially depending on the method of ligand fixation. For example, the use of MTX-fixed SGNE-OH beads resulted in efficient purification of dCK but not of DHFR. Although the reason is not clear at present, steric hindrance caused by MTX fixation is probably responsible for the low yield of DHFR. In addition, there may be competition between dCK and DHFR for binding to MTX. Further analysis of MTX's interactions with dCK and DHFR may facilitate the development of a novel MTX derivative that selectively binds and modulates dCK without affecting DHFR. Such an MTX derivative would be clinically useful, because when administered in combination with araC, the MTX derivative would have negligible DHFR-mediated cytotoxic effects in most cells and still enhance araC-induced cytotoxicity through dCK, especially in lymphoma/leukemia cells expressing high levels of dCK. Moreover, the ability of a MTX derivative to specifically enhance dCK activity would also be particularly useful for the treatment of araC-resistant patients in which dCK expression has been reduced.

3.5 CONCLUSIONS

We have developed an application for high-performance affinity particles, SG beads and FG beads, and demonstrated their usefulness for purification of various proteins, including transcription factors and drug receptors. Identification of novel target proteins for chemical compounds gives valuable information about biochemical reactions and also provides new insight for developing therapeutic agents by analysis of structure–activity relationships. Recent advances in human genomics have predicted 5000 to 10,000 proteins as potential drug targets; combinatorial chemistry allows rapid and comprehensive development of chemical libraries for drug screening; and various chemogenomic methods for determining drug target interactions are being explored. Together these approaches have the potential to powerfully advance drug discovery and development. To this end, we have developed a drug screening system that can seek out hit compounds interacting with target proteins, or functional domains of target proteins, from chemical libraries or mycobacterial extracts [13]. We expect that our affinity nanobead technology will provide high-quality chemogenomic information.

REFERENCES

1. Cuatrecasas, P., Wilchek, M., Anfinsen, C. B. (1968). Selective enzyme purification by affinity chromatography. *Proc. Natl. Acad. Sci. USA*, *61*, 636–643.

2. Shimizu, N., Sugimoto, K., Tang, J., et al. (2000). High-performance affinity beads for identifying drug receptors. *Nat. Biotechnol.*, *18*, 877–881.

3. Inomata, Y., Kawaguchi, H., Hiramoto, M., Wada, T., Handa, H. (1992). Direct purification of multiple ATF/E4TF3 polypeptides from HeLa cell crude nuclear extracts using DNA-carrying affinity latex particles. *Anal. Biochem.*, *206*, 109–114.

4. Ugelstad, J., El-Aasser, M. S., Vanderhoff, J. W. (1973). Initiation of polymerization in monomer droplets. *J. Polym. Sci. Polym. Lett. Ed.*, *11*, 503–513.

5. Sugimoto, T., Matijevic, E. (1980). Formation of uniform spherical magnetite particles by crystallization from ferrous hydroxide gels. *J. Colloid Interface Sci.*, *74*, 227–243.

6. Nishio, K., Ikeda, M., Gokon, N., et al. (2007). Preparation of size-controlled (30–100nm) magnetite nanoparticles for biomedical applications. *J. Magn. Magn. Mater.*, *310*, 2408–2410.

7. Glatzhofer, D. T., Cho, G., Lai, C. L., et al. (1993). Polymerization and copolymerization of sodium 10-undecane-1-yl sulfate in micelles and in admicelles on the surface of alumina. *Langmuir*, *9*, 22949–22954.

8. Hatakeyama, M., Nishio, K., Nakamura, M., et al. (2007). Polymer particles as the carrier for affinity purification. *Koubunshi Ronbunshu*, *1*, 9–20.

9. Kawaguchi, H., Asai, A., Ohtsuka, Y., Watanabe, H., Wada, T., Handa, H. (1989). Purification of DNA-binding transcription factors by their selective adsorption on the affinity latex particles. *Nucleic Acids Res.*, *17*, 6229–6240.

10. Hiramoto, M., Shimizu, N., Sugimoto, K., et al. (1998). Nuclear targeted suppression of NF-kappa B activity by the novel quinone derivative E3330. *J. Immunol.*, *160*, 810–819.

11. Kabe, Y., Ohmori, M., Shinouchi, K., et al. (2006). Porphyrin accumulation in mitochondria is mediated by 2-oxoglutarate carrier. *J. Biol. Chem.*, *281*, 31729–31735.

12. Uga, H., Kuramori, C., Ohta, A., et al. (2006). A new mechanism of methotrexate action revealed by target screening with affinity beads. *Mol. Pharmacol.*, *70*, 1832–1839.

13. Ohtsu, Y., Ohba, R., Imamura, Y., et al. (2005). Selective ligand purification using high-performance affinity beads. *Anal. Biochem.*, *338*, 245–252.

4

RECENT DEVELOPMENTS AND ADVANCES IN CHEMICAL ARRAYS

Naoki Kanoh

Graduate School of Pharmaceutical Sciences, Tohoku University, Sendai, Miyagi, Japan

Hiroyuki Osada

Chemical Biology Department, RIKEN Advanced Science Institute, Wako, Saitama, Japan

4.1 INTRODUCTION

One of the major tasks in the postgenomic era is to understand the cellular and molecular functions of thousands of predicted gene products. These gene products are thought to have important and diverse roles in biological systems. For example, a considerable number of these proteins are expected to be involved in diseases, and the disease-specific genes and their protein products are especially important for understanding the molecular mechanism of diseases and developing clinical medicies and public health policies.

Historically, classical genetics has been used primarily to study the gene function by removing or mutating the gene product directly from the organisms and cells. But recently an approach called *chemical genetics* [1–4] has gained recognition as an alternative way to study biological process. Instead of gene mutation and gene knockout, chemical genetic study uses small molecules,

Protein Targeting with Small Molecules: Chemical Biology Techniques and Applications,
Edited by Hiroyuki Osada
Copyright © 2009 John Wiley & Sons, Inc.

sometimes referred to as *bioprobes* [5], to modulate gene products. There are several advantages of chemical genetics over classical genetics: That is, small molecules (1) can cause rapid and reversible effects on cells or animal models, (2) can modulate products of multicopy genes simultaneously, (3) can be used to study an essential gene product at any developmental stage, (4) can be used in a dose–dependent manner, thus producing dose–response data, (5) can modulate the specific domain of a gene product, and so on. Note that most of these goals cannot be accomplished by classical genetics.

However, chemical genetics also have a critical disadvantage over genetics at present: It lacks generality. In other words, we do not have a sufficient set of small molecules to modulate every gene product. Therefore, to make the chemical genetic approach more generalizable to the full range of possible gene products, it will be necessary to develop screening methods to identify small-molecule modulators for any protein of interest [6]. These methods are also important for the drug development process because they offer ways to obtain specific modulators for disease-related proteins.

To refine the chemical genetics process, a new technology called *chemical arrays* have been developed as a high-throughput method for the identification of protein-binding small molecules (Fig. 4-1) [7–13]. Chemical arrays, also referred to as small-molecule microarrays, chemical chips, ligand chips, and so on, are defined as solid surfaces on which a spatially addressable and high-density array of small molecules is introduced. When a chemical array is treated with a protein of interest, which is often labeled with a reporter group such as a fluorescent dye, the signals observed on the array indicate the presence of ligands specific to the protein. The ligands identified are thought to be good candidates of small-molecule modulators of the protein's function. To date, high-precision

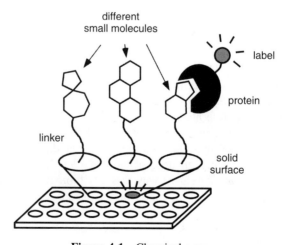

Figure 4-1   Chemical arrays.

instruments for printing thousands of microspots have been made available commercially, and consequently, thousands of binding assays can be performed in a simultaneous fashion.

The key elements of the chemical array technology can be classified as follows: (1) design and synthesis of a small-molecule library, (2) attachment of the library molecules to chemically modified surfaces, (3) detection of small molecule–protein interactions, and (4) data analysis. Items (1) and (2) can be integrated when each small molecule is prepared on a solid surface (i.e., during in situ synthesis) and used directly in the binding assay. Since this technology was introduced by MacBeath et al. in 1999 [14], a large portion of the advances in this field have been devoted to developing efficient ways to introduce small-molecule libraries as microarrays. Indeed, small-molecule immobilization is the specific issue to be addressed in chemical array technology, whereas library design and data analysis have been developed for wide-ranging applications in the chip-based and solution-phase high-throughput screens [15]. Methods for the detection of small molecule–protein interaction are thought to be similar to those developed for DNA and protein microarrays. However, technological breakthroughs in this area are still needed, since methods of detecting low-affinity and/or fragile binding are still in their infancy.

Accordingly, we focus here on recent advances in the array fabrication strategies: introduction of small molecules to the solid surfaces as well as surface modifications that allow the surfaces to incorporate small molecules. We included peptides and carbohydrates in the category of small molecules due to their technical similarities when introduced on the substrates, although the resulting peptide and carbohydrate microarrays have their own peculiarities. We then focus on signal detection methods. In most chemical array platforms, signals observed originate from small molecule–protein interactions on the solid surface, whereas some research groups in this area have recently started to use the chemical array platform for other purposes, such as cell-based assay and enzyme assay. In these assay systems, many different signals can be obtained, and we therefore highlight these advances. Finally, some of this significant progress has culminated in the identification of small-molecule ligands that have been applied successfully in chemical genetic studies, and these discoveries are discussed in the final section.

It should be noted that chemical array platforms have also been utilized in discovering and analyzing new organic reactions [16,17]. Although these studies include outstanding ideas and successful applications, they are not mentioned in this chapter because these platforms are intended to address different chemical issues.

4.2   ARRAY FABRICATION STRATEGIES

To date, several tactics and strategies have been described for the fabrication of chemical arrays. These strategies can be classified in the following manner based on how small molecules are introduced into the array: (1) site-specific covalent immobilization, (2) site-nonspecific covalent immobilization, (3) site-specific noncovalent immobilization, (4) site-nonspecific noncovalent immobilization, and (5) in situ synthesis.

4.2.1 Site-Specific Covalent Immobilization

In the late 1990s, the group led by Schreiber introduced the concept of diversity-oriented synthesis (DOS) [18–20] into the field of organic chemistry. In combination with the split-pool synthesis [21,22] and tagging strategies [23,24], DOS allowed chemists to prepare large collections of structurally diverse small molecules. The library compounds are usually prepared on polymeric beads, and thus each compound contains a common functional group that mediates covalent attachment to the solid surface.

In the first publication describing chemical arrays, MacBeath et al. utilized thiol as the common functional group on small molecules that readily attach to the maleimide-functionalized surface via a Michael addition [14] (Fig. 4-2a). Each compound immobilized through a thioether linkage was found to be recognized successfully by its binding proteins. The same type of strategy utilizing a Michael reaction was used for the fabrication of carbohydrate microarrays [25,26]. These platforms were employed to analyze the binding profiles of oligosaccharides and their binding proteins [27].

Alcohol-containing small molecules are readily synthesized using particular polymeric resins having silicon-based linking elements [28]. Hergenrother et al. found that a standard glass slide of the type used for microscopic observation can be activated for the reaction with alcohols [29] (Fig. 4-2b). They activated glass slides with thionyl chloride and catalytic N,N-dimethylformamide to give chlorinated slides. Compounds having a primary alcohol attached effectively to the slide surface, whereas the reaction with secondary or phenolic alcohols was less effective. In combination with a primary alcohol-containing library synthesized from a DOS pathway, this chemical array platform was used successfully in protein–ligand screening, which culminated in the discovery of several protein modulators. These topics are discussed later.

Another type of attachment chemistry can be used for immobilization of alcoholic compounds on solid surfaces. Isocyanate is known to have high reactivity toward alcohols, and this functional group has also been used to fabricate chemical arrays [15]. The DOS pathway can, of course, lead not only to alcohol-containing compounds but also to compounds having other functional groups. Indeed, Barnes-Seeman et al. reviewed the literature and discovered that the number of synthetic routes yielding libraries of carboxylic acids or phenols was greater than that of those yielding primary alcohols [30]. Because alcohol-reactive slides are not suitable for the efficient capture of phenolic compounds, they developed a different strategy that utilizes a diazobenzylidene group for covalent capture of phenols and functional groups having comparable acidity (Fig. 4-2c).

Various methods for immobilizing amine-containing small molecules have been reported. These molecules can be introduced on aldehyde-functionalized

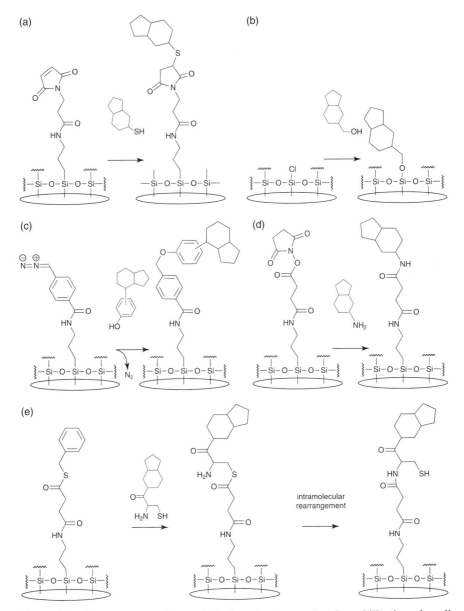

Figure 4-2 Some representative methods for selective covalent immobilization of small molecules.

slide surfaces via a Schiff base formation [31]. Primary amines are generally present in the N-terminus of peptides and proteins, thereby making this strategy useful for fabrication of peptide and protein macroarrays. Compounds having alkoxyamine also react with an aldehyde on the solid surface via an oxime linkage. Salisbury et al. have developed peptide microarrays for the determination of protease substrate specificity by utilizing this chemistry for immobilizing a library of peptide–coumarin conjugates [32]. Alkokyamine-containing compounds have also been immobilized on surface functionalized with glyoxyl group [33]. On the other hand, to immobilize compounds having a glyoxyl group, semicarbazide-functionalized surfaces have been used [34,35].

Activated ester functionalities such as *N*-hydroxysuccinimide (NHS) ester are useful chemical entities to immobilize amine-containing small molecules (Fig. 4-2d). Chang et al. constructed microarrays of an amine-terminated tri-azine library by using NHS ester–introduced glass slides [36]. This strategy has been applied in aminoglycoside microarrays [37].

Gregorius et al. have introduced peptides that have only one good nucle-ophilic group (the N-terminal α-amino group) on microtiter plates coated with 2,2,2-trifluoroethanesulfonyl (tresyl)-activated dextran polymer through nucle-ophilic substitution [38,39]. To eliminate the possibility of binding through the lysine side chain (ε-amino group) and cysteine side chain (thiol group), the pep-tides were synthesized and purified with protection groups in the lysine and cysteine side chains and immobilized covalently through the N-terminal α-amino groups on the microtiter plates. The protecting groups were cleaved off in situ after immobilization.

N-terminal cysteine residues on peptides are also used for immobilization. Falsey et al. utilized the reaction of cysteine with a glyoxylyl group [33]. Upon printing of cysteine-containing small molecules on glyoxylyl group–functionalized glass slides, the molecules were immobilized on the surface via a thiazoline ring.

Lesaicherre et al. have developed an approach to immobilize cysteine-containing small molecules by using the native chemical ligation method (devised originally by Dawson et al. [40]) to allow the preparation of proteins with native backbone structures [41]. They used thioester to functionalize the surfaces of glass slides (Fig. 4-2e). Upon printing, the thiol group in the cysteine-containing small molecules reacts with thioester via ester exchange to give the initial thioester intermediate. This intermediate spontaneously under-goes intramolecular reaction because of the favorable geometric arrangement involving a five-membered ring of the α-amino group of cysteine to yield the ligation product. The presence of other reactive groups, including internal cysteines, is tolerated in this method.

In the examples described above, the small molecules to be immobilized usu-ally possess (or are designed to have) a specific functional group such as OH, NH_2, COOH, or SH. However, the ideal small-molecule ligand of interest should possess a number of different functional groups because multisite interactions are usually needed to make the protein–ligand interaction specific. Therefore, Lee and Shin proposed that the major requirement of techniques used to fabricate chemical arrays was that immobilization of the diversely functionalized com-pounds on the modified surfaces must be highly selective for a specific functional group in the presence of others [42]. To this end, they have explored such reac-tions and found that in the presence of amine and/or thiol, a hydrazine group on the small molecule could react selectively with an epoxide-functionalized slide surface (Fig. 4-3a).

Polar and/or nucleophilic functional groups on a small molecule are thought to be important not merely for interaction with proteins but also for the biological activity of the compounds. Fazio et al. [43] and Köhn et al. [44] have independently sought a new coupling strategy that does not use these functional groups and determined that azide is an almost ideal functional group because it (1) tolerates a diverse array of other functionalities, (2) reacts efficiently with certain functional groups under mild reaction conditions in an atmosphere containing oxygen and water, and (3) is compatible with the efficient strategies of solid-phase combinatorial chemistry. Fazio et al. have explored 1,3-dipolar cycloaddition between azide-containing oligosaccharides and alkyne-containing hydrocarbon chains that are noncovalently absorbed on microtiter well surfaces [43] (Fig. 4-3b). Reaction of azides and alkynes in the presence of catalytic CuI proceeded regioselectively in 8 hours to give a 1,3-dipolar cycloaddition product (i.e., 1,4-disubstituted 1,2,3-triazole) in good yield, whereas thermal cycloaddition gave a mixture of 1,4- and 1,5-disubstituted 1,2,3-triazole. The Cu(I)-catalyzed reaction was applied successfully to fabricate carbohydrate arrays in microtiter plates [45–47].

As an alternative strategy, Köhn et al. utilized a Staudinger ligation between azide-containing compounds and solid surfaces covered with appropriately substituted phospholanes in which a chemically stable amide bond is formed [44] (Fig. 4-3c). Moreover, to achieve maximum surface coverage by small molecules, slide surfaces were modified with a multigeneration layer of polyamidoamine (PAMAM) dendrimers [48] before the introduction of small molecules. Soellner et al. have also utilized the Staudinger ligation strategy to fabricate peptide and protein microarrays [49].

Houseman et al. have developed a Diels–Alder reaction strategy for fabricating microarrays of carbohydrates and peptides on gold surfaces [50,51] (Fig. 4-3d). They employed self-assembled monolayers of mixed alkanethiolates, one of which terminated in a hydroquinone group. The hydroquinone groups were oxidized chemically or electrochemically to benzoquinone groups, which then reacted with cyclopentadiene-conjugated ligands by way of a Diels–Alder reaction to immobilize the ligand covalently. In combination with photo-cleavable nitroveratryloxycarbonyl (NVOC)-protected hydroquinone, a method for photochemical patterning of ligands was also developed [52].

4.2.2 Site-Nonspecific Covalent Immobilization

In general, the site-specific immobilization methods described above are especially effective when the protein recognition site on ligand molecules is known or can be predicted. Small molecules immobilized on solid surfaces through a specific functional group uniformly present a certain area of their molecular surfaces, and thus the interaction with proteins that recognize this surface can be analyzed in a highly sensitive and quantitative manner. However, except in the case of certain protein-reactive small molecules [53], it is difficult to predict the

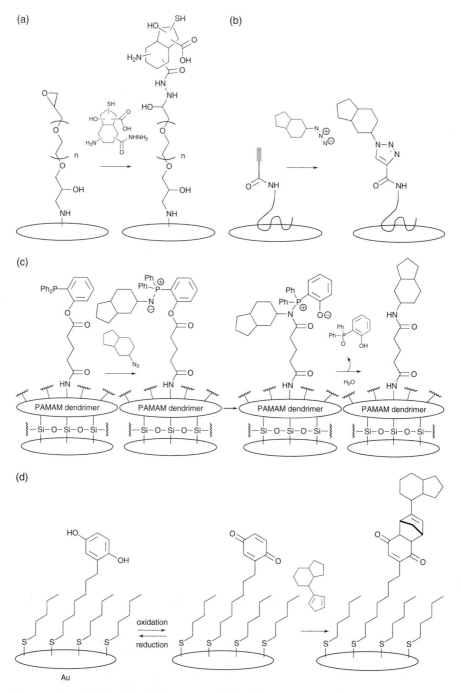

Figure 4-3 Additional representative methods for selective covalent immobilization of small molecules.

binding domain of small molecules that interact with unknown proteins. In addition, there are bifunctional small molecules that bind to multiple proteins through different interaction sites [54,55]. If such an interaction site were to be used for immobilization, the resulting small molecule would be totally useless. It is also possible that the presence of a linker connecting the small molecules to the solid surfaces reduces the number of binding modes available to each compound [14].

Another drawback to using the selective coupling approach is that small molecules have to possess a certain functional group to be attached to a solid surface. The combinatorial chemical libraries from solid-phase organic syntheses inherently have such functional groups, but libraries consisting of natural products contain a variety of structures and functional groups that are not always compatible with the selective immobilization process.

Based on these considerations, two approaches using nonselective (or highly functional group compatible) covalent immobilization have been developed to date. Kanoh et al. reported that reactive carbene species generated from photoreactive aryltrifluoromethyldiazirine could be used to capture a variety of small molecules in a functional group–independent manner [56] (Fig. 4-4a). The photoreactive groups introduced to the solid surface are promoted upon ultraviolet irradiation to highly reactive carbene species, which in turn bind irreversibly with the proximal small molecules present on the surfaces. This method eliminates the need for incorporating specific functional groups among library members and is thought to be suitable for complex natural products, which are otherwise difficult to use in chemical arrays. Further applications of this methodology have also been documented [57–59]. Angeloni et al. utilized the same strategy to immobilize glycoconjugates and lectins [60].

A platform developed by Bradner et al. utilizes an isocyanate-mediated covalent capture strategy [61] (Fig. 4-4b). An isocyanate-functionalized surface has been utilized by the same group to capture hydroxyl groups [15], but in their

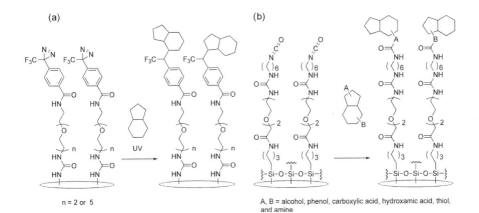

Figure 4-4 Representative methods for highly functional group–compatible covalent immobilization of small molecules.

more recent study they demonstrated that the surface could be used to capture not only alcohols but also a number of nucleophilic functional groups, such as phenols, carboxylic acids, hydroxamic acids, thiols, and amines. Importantly, the reaction was found to be tolerant to water. This water tolerance is an important consideration for chemical array preparation because compound stock solutions in an organic solvent such as dimethylformamide or dimethyl sulfoxide (DMSO) appear to take on water over time as they are moved in and out of freezer storage.

4.2.3 Site-Specific Noncovalent Immobilization

Noncovalent interactions are usually thought to be weaker than covalent interactions. However, some noncovalent interactions are quite useful for fabricating chemical arrays. Among them, Lesaicherre et al. have taken advantage of the biotin–avidin interaction, one of the strongest noncovalent interactions ($K_D = 10^{-15}$ M) known to date, to immobilize N-terminally biotinylated peptides onto a glass slide functionalized with avidin [41]. The resulting peptide microarrays were developed as a platform for high-throughput identification of protein kinase substrates [62].

Another type of useful noncovalent interaction is the self-assembly of oligonucleotides. When small molecules are tagged with oligonucleotides or their congeners, the molecules can be immobilized on a solid surface that was functionalized with the complementary oligonucleotides. Each nucleotide tag introduced on small molecules can be used both as a coding sequence unique for each tethered small molecule (e.g., to provide a synthetic history of the small molecule) and as a decoding sequence responsible for identifying the small molecule by its location upon hybridization to an oligonucleotide microarray (decoding array).

Winssinger et al. have developed an intriguing method for the preparation of chemical arrays using positionally encoded libraries [63,64] (Fig. 4-5). Instead of oligonucleotides, they used peptide nucleic acids (PNAs) as encoding tags because of their desirable hybridization properties, the flexibility of their synthesis, and their chemical robustness. The PNA-tagged small-molecule ligands were synthesized with a fluorophore to enable not only fluorescent detection without protein labeling, but also a solution-phase protein-binding assay without small-molecule immobilization (Fig. 4-5a). After incubation with the protein sample, the resulting mixture was passed through a size-exclusion column to separate PNA-tagged ligands bound to a protein from the unbound PNA-tagged ligands. Hybridization of the protein-bound ligands to a decoding array identifies the ligand structure (Fig. 4-5c).

Another example utilizing the self-assembling nature of nucleic acids is the encoded self-assembling chemical (ESAC) library developed by Melkko et al. [65,66]. When two different pharmacophores are introduced at the end of two complementary nucleotides, the resulting pair of ligands after self-assembly can be regarded as a bidentate ligand (Fig. 4-6). Similarly, a large library of the bidentate ligands can be formed from hybridizing two sublibraries of oligonucleotide–pharmacophore conjugates. Mixing the ESAC library with a

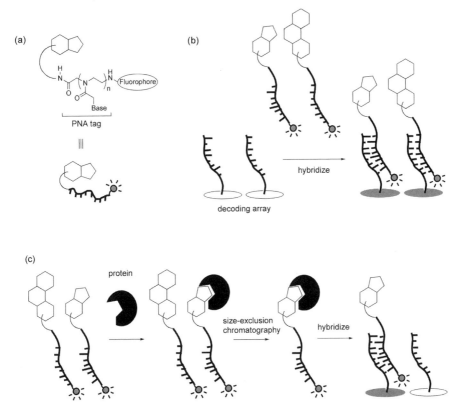

Figure 4-5 PNA-encoded self-assembling chemical arrays. (a) Each small molecule was labeled with a PNA tag as well as a fluorophore. (b) The tagged molecules were hybridized with their complementary sequences on a decoding array. (c) Incubation of the tagged molecule library with a protein sample generated a protein–small molecule conjugate, which could be separated by using size-exclusion chromatography. Structural information on the bound small molecule was decoded by the decoding array.

protein immobilized on affinity beads afforded a bidentate ligand that specifically binds to the protein. Amplification of the oligonucleotide region by using polymerase chain reaction (PCR) as well as its fluorescent labeling gave a labeled oligonucleotide. After decoding the pharmacophore information contained within the oligonucleotide by means of a decoding array, the oligonucleotide moiety of the ESAC compound selected may be replaced by chemical linkers that covalently connect the two pharmacophores, thereby producing a chemically tethered bidentate ligand. Three different sublibraries, of course, can be assembled to form a combinatorial triple helix library and can be used for the same purpose.

Noncovalent fluorous-based interactions have also been used to construct chemical arrays. The original concept of fluorous-based chemical arrays was demonstrated by Ko et al. with unprotected carbohydrate ligands [67]. Fluorous tags are originally developed to simplify purification steps during the synthesis of small molecules [68], but they demonstrated the strength of the

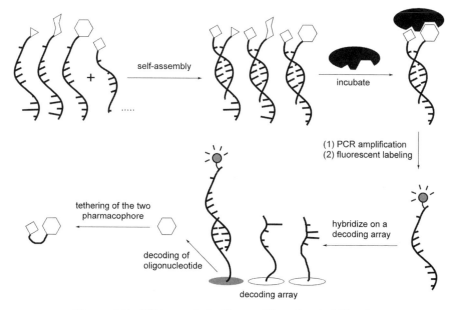

Figure 4-6   DNA-encoded self-assembling chemical libraries.

fluorous interactions in the direct formation of carbohydrate microarrays for binding assay. Quite recently, Vegas et al. [69] and Nicholson et al. [70] independently utilized the interactions to immobilize fluorous-tagged small molecules such as biotin [70] and candidates for histone deacetycase inhibitors [69] on fluorous-derivatized glass slides.

4.2.4   Site-Nonspecific Noncovalent Immobilization

With the exception of molecules inherently having a tether moiety [71], structural modification (including introduction of a tether) is usually needed to immobilize small molecules on the solid surface. Strictly speaking, these modified (or immobilized) molecules differ from the parent molecule in terms of their mobility and affinity toward every possible binding protein. The example described in Section 4.2.2 originated from a concept in which if small molecules were immobilized in a truly functional group-independent manner to produce a variety of conjugates, some but not all of the small molecules immobilized are expected to retain binding affinity toward all possible binding proteins. Nevertheless, the best way to remove the influence of the tethering effect is, of course, removing the tether. However, the small molecules must retain their original positions on the solid surface even after probing and extensive washing.

To this end, Wang et al. utilized physical absorption for introducing polysaccharides on solid surfaces [72]. By using nitrocellulose-coated glass slides, they demonstrated that even after extensive washing, a dextran preparation of 20 kDa or larger was stably immobilized on the slide without chemical conjugation, and the immobilized dextran was recognized successfully by their antibodies.

Bryan et al. have also constructed microarrays of carbohydrates using physical absorption [73]. They demonstrated that a library of glycosyl lipids containing saturated alkyl chains at the anomeric position was mostly retained on a polystyrene surface and could be used in biological assays. However, the hydrophobicity of the alkyl chain interfered significantly with the synthesis and purification of the glycosyl lipids. They searched for an alternative method to overcome this problem and developed the 1,3-dipolar cycloaddition strategy described in Section 4.2.1 [43].

Instead of physical absorption, Gosalia et al. have developed a technique in which small molecules were microarrayed as individual nanoliter droplets of a DMSO/glycerol solution [74,75]. They have utilized this platform for a variety of enzymatic reactions [75]. The compounds are always in solution without any chemical linking to the surface, making the microarray screen compatible with that in the microtiter plates. Another unique strategy utilizes biodegradable polymers to "embed" small molecules. Bailey et al. developed a microarray platform in which small molecules were impregnated in 200-μm-diameter disks composed of biodegradable poly-(D),(L)-lactide/glycolide (PLGA) copolymer [76]. This platform was used for testing the effects of small molecules on mammalian cells. Both platforms represent new directions in chemical array technologies and pave the way forward in high-throughput chemical genomic exploration.

4.2.5 In Situ Synthesis

Instead of printing or immobilizing libraries of presynthesized small molecules on the solid surface, many strategies in which a library of small molecules is constructed in situ (in particular places on a chemical array) have been developed. In fact, the earliest example of an array of organic compounds is a peptide array that was synthesized on polyethylene rods [77]. Geysen et al. manufactured a polyethylene holder into which acrylic acid–functionalized polyethylene rods were assembled with the format and spacing of a microtiter plate. By using conventional methods of solid-phase peptide chemistry, 208 hexapeptides were prepared on the polyethylene rods in a "one-rod one-compound" fashion. These peptides were used, as they stood, for an enzyme-linked immunoabsorbent assay (ELISA). Also, Lam et al. have created a large peptide library consisting of millions of beads in a "one-bead one-peptide" fashion by utilizing split and pool synthesis, and used the resulting library of peptide-immobilized beads for the screening of acceptor-binding ligands [22].

Thereafter, Frank developed a novel technical concept for the cellulose paper sheet–supported simultaneous parallel synthesis utilizing conventional Fmoc/tBu chemistry [78]. As he carried out different coupling reactions simultaneously on distinct areas of the same sheet of paper instead of using separate pieces of paper for individual peptides, this approach is named SPOT synthesis. Recent advances in this field have expanded the SPOT technology beyond peptidic systems into the realm of complex small-molecule synthesis [79]. These platforms are thought to be very suitable for chemical arrays.

In 1991, these in situ solid-phase chemical techniques were combined with photo-deprotection chemistry and photolithography to achieve light-directed, spatially addressable parallel peptide synthesis, yielding a highly diverse set of small molecules on the solid surface [80]. The technology developed by Fodor and co-workers was then applied to in situ nucleotide synthesis, and was later used widely in the DNA microarray field [81]. Ironically, widespread use of the original peptide arrays generated by in situ solid-phase peptide synthesis and the photolabile protection-photolithography technology has been hampered by practical limitations such as the expensive and time-consuming photomask preparation, difficulties in the preparation of amino acids containing a photolabile protecting group, and the inefficiency of the photochemistry used. The pursuit of technical developments to solve these limitations is ongoing [82].

4.3 SIGNAL DETECTION METHODS

4.3.1 Signal Detection in the Protein-Binding Assay

The use of fluorescent dye for labeling nucleotides is a popular method of DNA microarray analysis. Indeed, fluorescence is one of the most sensitive signals used in the life sciences, and technologies for detecting and analyzing the fluorescently labeled nucleotides bound on a DNA microarray have been developed intensively in the last decade. Since the technologies of chemical arrays were derived from those of the DNA microarray, it was a matter of course for the pioneers in this area to use fluorescent dye for labeling proteins [14,83]. Usually, in the protein-binding assay, a chemical array is thus treated with the purified or unpurified [61] fluorescently labeled protein of interest, and after washing and drying the slide, the fluorescent signals remaining on the array are detected using a fluorescent slide scanner. If the protein of interest is not amenable to labeling because of its instability toward the labeling condition and inactivation during the labeling, indirect and multistep labeling can be used; that is, a chemical array is probed first with nonlabeled protein and binding is detected using a fluorescently labeled antibody against the protein. Also, as is commonly done in in situ hybridization and Western blotting analysis, the binding signal can be amplified on the array by using tyramide signal amplification (TSA) reagents [58,84].

Although this approach has worked well in several successful examples, a fundamental limitation is associated with this protocol. To detect the interactions between a fluorescently labeled protein and an immobilized small molecule with a fluorescent scanner, the array has to be washed and dried prior to scanning. In contrast to the relatively uniform DNA–DNA and DNA–RNA interactions, which are tolerant to washing and drying, small molecule–protein interactions vary in associative strength [85], and a considerable portion of them may be so weak and fragile that the interaction will be lost in the washing and drying processes [9].

An important advance toward solving these limitations is the use of surface plasmon resonance (SPR). As SPR measures the change in refractive index occurring near the surface of a metal when a protein binds to a surface-immobilized compound, it does not require a label on the proteins. Moreover, the change of SPR signal can be monitored in real time. Instruments for two-dimensional SPR imaging measurement have recently been developed [59,86–88]. Although there is some room for improvement in the sensitivity and density limit, this surface-based technology allows real-time high-throughput detection of weak interactions between proteins and ligand molecules, which have been difficult to observe using other platforms. As an alternative method in which no chemical modification of the protein is needed, Kurosu et al. utilized white light interferometry [89].

As described in Section 4.2.3, there are conceptually different strategies. In the case of PNA-encoded self-assembling chemical arrays, the binding signals (i.e., protein-bound ligands) were selected by size-exclusion column chromatography, and the signals (the fluorescent label introduced on the ligand) were detected and decoded based on the positional information on the decoding array (Fig. 4-5). Also, in the case of ESAC selection, binding events were selected by the pull-down assay, and the information encoded in the DNA tags was amplified and decoded (Fig. 4-6). Note that these strategies cleverly circumvent the drawbacks associated with the usual chemical array ligand screen.

4.3.2 Signal Detection in Other Types of Assays

Chemical arrays have been used primarily to identify ligands for several proteins, but several new screening strategies, which include protein activity assays and phenotype assays, have recently emerged. Gosalia and Diamond developed a unique enzymatic assay system by using a platform in which small molecules in DMSO/glycerol are arrayed as nanodroplets [74]. They adopted an aerosol deposition technology that converts protein and reagent solutions into a fine mist which can cover the top surface of the entire microarray simultaneously. In the case of proteases, enzyme and fluorogenic substrate are added sequentially into each nanodroplet by aerosol deposition technology. The enzymatic reaction is monitored using a fluorescent scanner. When the fluorescent intensity is significantly altered in a nanodroplet as compared with others, the compound in the nanodroplet may be a modulator of the enzymatic activity. This screening system can be applied to a variety of enzyme assays by using appropriate fluorogenic- or FRET-based substrates [75].

Bailey et al. have developed a format for cell-based high-content screening by using their chemical array platform, in which small molecules were impregnated in biodegradable PLGA copolymer [76]. Mammalian cells were seeded on the top of the small molecule–impregnated microarray. During the incubation, impregnated compounds were slowly diffused out to affect proximal cells. After fixing and permeabilization, the cells were stained by an antibody toward a target protein or an organelle marker, both of which were labeled fluorescently.

Analysis of the fluorescent image provides simultaneous information on multiple parameters for a given target, or even for multiple target proteins in a biologically relevant setting.

4.4   LIGAND AND INHIBITOR DISCOVERY USING CHEMICAL ARRAYS AND ITS APPLICATION TO CHEMICAL GENETIC STUDY

A growing number of successful results from chemical array screening have been appearing in the literature. A landmark paper appeared in 2002 [90]. Kuruvilla et al. screened a ligand for a particular yeast protein, Ure2p, which represses the transcription factors Gln3p and Nil1p, from an unbiased, 3780-member 1,3-dioxane small-molecule DOS library that was arrayed on a functionalized glass slide. By probing the chemical array with fluorescently labeled Ure2p, they identified eight compounds that bind to Ure2p. To determine the cellular activity, the compounds were then screened in a *PUT1–lacZ* reporter system, because *PUT1* expression is repressed by *URE2*. Among the eight compounds, one compound, uretupamine A (**1**), activated the *PUT1–lacZ* reporter in a dose-dependent manner (Fig. 4-7). They synthesized a focused library that consisted of eight uretupamine derivatives and found that a derivative named uretupamine B (**2**) was more potent in terms of Ure2p binding (18.1 μM for **1**; 7.5 μM for **2**), *PUT1* expression (**2** is approximately twice as active as **1**), and solubility in aqueous medium. Whole-genome transcription profiling in the wild type and *ure2* Δ mutant showed that both **1** and **2** up-regulated only a subset of genes known to be under the control of Ure2p. By using this selective modulation of Ure2p function by uretupamines, which cannot be accomplished by physiological stimulus or genetic deletion, they were able to demonstrate a functional connection between Ure2p, Nil1p, and the glucose signaling pathway. This study clearly demonstrated the utility of chemical arrays in protein–ligand screening, as well as the possibility of using a chemical genetic approach in protein function analysis.

In the second example, Koehler et al. screened a ligand for a protein Hap3p, a subunit of the yeast Hap2/3/4/5p transcription factor complex involved in aerobic respiration and the nutrient-response signaling network [91]. They prepared chemical arrays containing 12,396 compounds derived from three different DOS pathways by using the alcohol-capture strategy [29]. When the microarrays were screened with the purified Hap3p-glutathione *S*-transferase (GST) fusion protein, and binding was detected using a Cy5-labeled antibody against GST, they identified two reproducible positives from a dihydropyrancarboxamide library. One compound appeared as a positive when the library was screened with GST, revealing that the compound binds to the GST portion of Hap3p-GST. Another hit compound named haptamide A (**3**) was found to bind to Hap3p with a K_D value of 5.03 μM by SPR studies. Moreover, the compound inhibited the expression of *GDH1*, which requires the Hap2/3/4/5p complex, with an IC_{50} value of 42.1 μM in a *GDH–lacZ* reporter gene assay. The expression levels approached those of

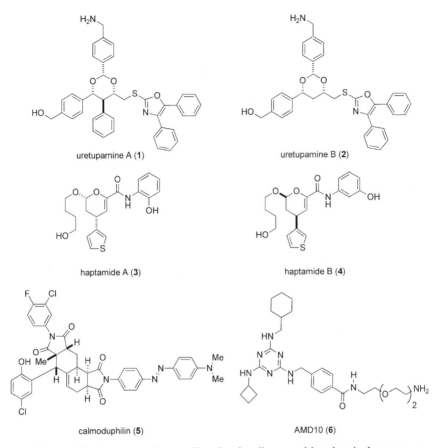

uretupamine A (1)

uretupamine B (2)

haptamide A (3)

haptamide B (4)

calmoduphilin (5)

AMD10 (6)

Figure 4-7 Representative small molecules discovered by chemical arrays.

the untreated sample when the haptamide A–treated cells were washed in fresh medium. These results suggested that haptamide A is a reversible inhibitor of Hap3p-mediated transcription. Subsequent SAR studies resulted in the discovery of a more potent inhibitor, haptamide B (**4**), which has a K_D value of 0.33 μM and an IC_{50} value of 23.8 μM. Whole-genome transcription profiling analysis provided evidence that haptamide B (**4**) selectively inhibits Hap2/3/4/5p-mediated transcriptions in cells.

Barnes-Seeman et al. utilized their chemical array platform containing a 6336 DOS-derived phenol-containing fused bicycle/tetracycle library for screening of calmodulin ligands. The binding experiment using Cy5-labeled calmodulin gave 16 positive hits. The subsequent secondary screen using SPR confirmed that 13 out of 16 compounds showed qualitative binding to surface-immobilized calmodulin. They resynthesized several compounds, including calmoduphilin (**5**), and determined that calmoduphilin has a K_D value of 0.121 ± 0.03 μM. Wong et al.

also identified a calmodulin binder, which has a K_D value of 10 to 20 μM, using an chemical array containing a diverse, 18,000-member 1,3-dioxane library [15].

Uttanchamdani et al. screened human IgG binders from a 2688-member tagged combinatorial triazine library [36]. Known high-molecular-weight IgG ligands such as staphylococcal protein A and streptococcal protein G have been used for purification of the immunoglobulins, but they have problems, including their potential pyrogenic effects as well as their low biological and chemical stability. By using Cy3-labeled human IgG, they identified three strong IgG binders, including AMD10 (**6**), and one moderately effective binder (Fig. 4-7). SPR analysis revealed that compound **6** has a K_D value of 4.35 μM toward human IgG.

Melkko et al. have generated bidentate ligands to human serum albumin (HSA) and bovine carbonic anhydrase II (CA) using ESAC selection and affinity maturation [65] (Fig. 4-8). They found that aminocoumarin (**8**) potentiated the binding affinity of dansylamide (**7**) toward HSA ($K_D = 146$ μM) and that tethering these ligands by means of a 1,6-diaminohexane linker gave an optimized bidentate ligand (**9**) with a K_D value of ≤ 4 μM. Using the same strategy, they created a bidentate CA ligand (**12**) which has a K_D value of 12 nM, from an original sulfonamide lead compound (**10**, $K_D = 430$ nM) and a binding potentiator (**11**). The bidentate ligand (**12**) was also shown to inhibit CA more efficiently (IC$_{50} \leq 25$ nM) than the parent compound (**10**) (IC$_{50} = 1$ μM).

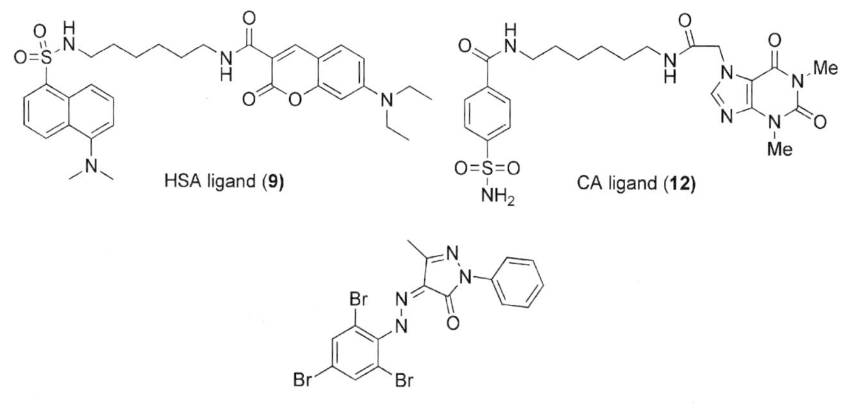

Figure 4-8 Representative small molecules discovered by chemical arrays (continued). R = oligonucleotide tag.

Chemical arrays have also been used in the screening of enzyme inhibitors. In an application of the chemical array platform in which small molecules are arrayed as individual glycerol/DMSO nanodroplets, Gozalia and Diamond identified an inhibitor of caspases [74]. That is, when a microarray of 352 commercially available compounds was sprayed sequentially with human caspase **4** and then its fluorogenic substrate, spots containing the heterocycle **13** on the microarray displayed low substrate conversion. This compound was then tested using a fluorescence plate reader for inhibitory activity in a standard buffer (no glycerol and < 0.4 vol% of DMSO present). In this assay, compound **13** displayed an IC_{50} value of about 0.5 mM.

4.5 CONCLUSIONS

Since 1999, enormous progress has been made in the development and application of chemical arrays, enabling a variety of high-throughput screens, including protein-binding assays, enzyme assays, and phenotypic assays. Thousands to tens of thousands of small molecules can currently be introduced as a high-density array on a solid support in an addressable fashion. However, despite several successful results, chemical arrays are still thought to be in their infancy, and most of the technologies described in this chapter are at the level of proof of concept. To make the chemical array technologies truly useful for chemical genetic studies as well as for the development of clinical medicines, many important problems, including array sensitivity and surface design, must be solved. For example, use of fully automated mass spectrometry will be a candidate for solving the sensitivity problem, and some preliminary results have already been reported in the literature [92,93]. At the same time, efforts toward standardization and commercialization of chemical array–specific technology will be needed for chemical arrays to become a workable platform for chemical genetic study.

REFERENCES

1. Schreiber, S. L. (1998). Chemical genetics resulting from a passion for synthetic organic chemistry. *Bioorg. Med. Chem.*, *6*, 1127–1152.
2. Schreiber, S. L. (2003). The small-molecule approach to biology. *Chem. Eng. News*, *81*, 51–61.
3. Spring, D. R. (2005). Chemical genetics to chemical genomics: small molecules offer big insights. *Chem. Soc. Rev.*, *34*, 472–482.
4. Walsh, D. P., Chang, Y.-T. (2006). Chemical genetics. *Chem. Rev.*, *106*, 2476–2530.
5. Osada, H. (ed.) (2000). *Bioprobes: Biochemical Tools for Investigating Cell Function*. Springer-Verlag, Tokyo, 2000.
6. Stockwell, B. R. (2004). Exploring biology with small organic molecules. *Nature*, *432*, 846–854.
7. Lam, K. S., Renil, M. (2002). From combinatorial chemistry to chemical microarray. *Curr. Opin. Chem. Biol.*, *6*, 353–358.

8. Walsh, D. P., Chang, Y. T. (2004). Recent advances in small moelcule microarrays: applications and technology. *Comb. Chem. High Throughput Screen.*, *7*, 557–564.

9. He, X. G., Gerona-Navarro, G., Jaffrey, S. R. (2005). Ligand discovery using small molecule microarrays. *J. Pharmacol. Exp. Ther.*, *313*, 1–7.

10. Uttamchandani, M., Walsh, D. P., Yao, S. Q., Chang, Y. T. (2005). Small molecule microarrays: recent advances and applications. *Curr. Opini. Chem. Biol.*, *9*, 4–13.

11. Ma, H., Horiuchi, K. Y. (2006). Chemical microarray: a new tool for drug screening and discovery. *Drug Discov. Today*, *11*, 661–668.

12. Uttamchandani, M., Wang, J., Yao, S. Q. (2006). Protein and small molecule microarrays: powerful tools for high-throughput proteomics. *Mol. Biosyst.*, *2*, 58–68.

13. Wang, J., Uttamchandani, M., Sun, H. Y., Yao, S. Q. (2006). Small molecule microarrays: applications using specially tagged chemical libraries. *Qsar Comb. Sci.*, *25*, 1009–1019.

14. MacBeath, G., Koehler, A. N., Schreiber, S. L. (1999). Printing small molecules as microarrays and detecting protein–ligand interactions en masse. *J. Am. Chem. Soc.*, *121*, 7967–7968.

15. Wong, J. C., Sternson, S. M., Louca, J. B., Hong, R., Schreiber, S. L. (2004). Modular synthesis and preliminary biological evaluation of stereochemically diverse 1,3-dioxanes. *Chem. Biol.*, *11*, 1279–1291.

16. Korbel, G. A., Lalic, G., Shair, M. D. (2001). Reaction microarrays: a method for rapidly determining the enantiometroc excess of thousands of samples. *J. Am. Chem. Soc.*, *123*, 361–362.

17. Kanan, M. W., Rozenman, M. M., Sakurai, K., Snyder, T. M., Liu, D. R. (2004). Reaction discovery enabled by DNA-templated synthesis and in vitro selection. *Nature*, *431*, 545–549.

18. Tan, D. S., Foley, M. A., Stockwell, B. R., Shair, M. D., Schreiber, S. L. (1999). Synthesis and million evaluation of a library of polycyclic small molecules for use in chemical genetic assays. *J. Am. Chem. Soc.*, *121*, 9073–9087.

19. Schreiber, S. L. (2000). Target-oriented and diversity-oriented organic synthesis in drug discovery. *Science*, *287*, 1964–1969.

20. Burke, M. D., Berger, E. M., Schreiber, S. L. (2003). Generating diverse skeletons of small molecules combinatorially. *Science*, *302*, 613–618.

21. Houghten, R. A., Pinilla, C., Blondelle, S. E., Appel, J. R., Dooley, C. T., Cuervo, J. H. (1991). Generation and use of synthetic peptide combinatorial libraries for basic research and drug discovery. *Nature*, *354*, 84–86.

22. Lam, K. S., Salmon, S. E., Hersh, E. M., Hruby, V. J., Kazmierski, W. M., Knapp, R. J. (1991). A new type of synthetic peptide library for identifying ligand-binding activity. *Nature*, *354*, 82–84.

23. Ohlmeyer, M. H. J., Swanson, R. N., Dillard, L. W., et al. (1993). Complex synthetic chemical libraries indexed with molecular tags. *Proc. Nat. Acad. Sci. USA*, *90*, 10922–10926.

24. Nestler, H. P., Bartlett, P. A., Still, W. C. (1994). A general method for molecular tagging of encoded combinatorial chemistry libraries. *J. Org. Chem.*, *59*, 4723–4724.

25. Park, S., Shin, I. (2002). Fabrlication of carbohydrate chips for studying protein–carbohydrate interactions. *Angew. Chem. Int. Ed.*, *41*, 3180–3182.

26. Park, S., Lee, M.-R., Pyo, S.-J., Shin, I. (2004). Carbohydrate chips for studying high-throughput carbohydrate–protein interactions. *J. Am. Chem. Soc.*, *126*, 4812–4819.

27. Adams, E. W., Ratner, D. M., Bokesch, H. R., McMahon, J. B., O'Keefe, B. R., Seeberger, P. H. (2004). Oligosaccharide and glycoprotein microarrays as tools in HIV glycobiology: glycan-dependent gp120/protein interactions. *Chem. Biol.*, *11*, 875–881.

28. Lee, D., Sello, J. K., Schreiber, S. L. (2000). Pairwise use of complexity-generating reactions in diversity-oriented organic synthesis. *Org. Lett.*, *2*, 709–712.

29. Hergenrother, P. J., Depew, K. M., Schreiber, S. L. (2000). Small-molecule microarrays: covalent attachment and screening of alcohol-containing small molecules on glass slides. *J. Am. Chem. Soc.*, *122*, 7849–7850.

30. Barnes-Seeman, D., Park, S. B., Koehler, A. N., Schreiber, S. L. (2003). Expanding the functional group compatibility of small-molecule microarrays: discovery of novel calmodulin ligands. *Angew. Chem. Int. Ed.*, *42*, 2376–2379.

31. Newman, J. R. S., Keating, A. E. (2003). Comprehensive identification of human bZIP interactions with coiled-coil arrays. *Science*, *300*, 2097–2101.

32. Salisbury, C. M., Maly, D. J., Ellman, J. A. (2002). Peptide microarrays for the determinaiton of protease substrate specificity. *J. Am. Chem. Soc.*, *124*, 14868–14870.

33. Falsey, J. R., Renil, M., Park, S., Li, S., Lam, K. S. (2001). Peptide and small molecule microarray for high throughput cell adhesion and functional assays. *Bioconjug. Chem.*, *12*, 346–353.

34. Olivier, C., Hot, D., Huot, L., et al. (2003). α-Oxo semicarbazone peptide or oligodeoxynucleotide microarrays. *Bioconjug. Chem.*, *14*, 430–439.

35. Duburcq, X., Olivier, C., Malingue, F., et al. (2004). Peptide–protein microarrays for the simultaneous detection of pathogen infections. *Bioconjug. Chem.*, *15*, 307–316.

36. Uttamchandani, M., Walsh, D. P., Khersonsky, S. M., Huang, X., Yao, S. Q., Chang, Y.-T. (2004). Microarrays of tagged combinatorial triazine libraries in the discovery of small-molecule ligands of human IgG. *J. Comb. Chem.*, *6*, 862–868.

37. Disney, M. D., Magnet, S., Blanchard, J. S., Seeberger, P. H. (2004). Aminoglycoside microarrays to study antibiotic resistance. *Angew. Chem. Int. Ed.*, *43*, 1591–1594.

38. Gregorius, K., Mouritsen, S., Elsner, H. I. (1995). Hydrocoating: a new method for coupling biomolecules to solid phases. *J. Immunol. Methods*, *181*, 65–73.

39. Gregorius, K., Theisen, M. (2001). In situ deprotection: a method for covalent immobilization of peptides with well-defined orientation for use in solid phase immunoassays such as enzyme-linked immunosorbent assay. *Anal. Biochem.*, *299*, 84–91.

40. Dawson, P. E., Muir, T. W., Clark-Lewis, I., Kent, S. B. H. (1994). Synthesis of proteins by native chemical ligation. *Science*, *266*, 776–779.

41. Lesaicherre, M. L., Uttamchandani, M., Chen, G. Y. J., Yao, S. Q. (2002). Developing site-specific immobilization strategies of peptides in a microarray. *Bioorg. Med. Chem. Lett.*, *12*, 2079–2083.

42. Lee, M.-R., Shin, I. (2005). Fabrication of chemical microarrays by efficient immobilization of hydrazide-linked substances on epoxide-coated glass surfaces. *Angew. Chem. Int. Ed.*, *44*, 2881–2884.

43. Fazio, F., Bryan, M. C., Blixt, O., Paulson, J. C., Wong, C.-H. (2002). Synthesis of sugar arrays in microtiter plate. *J. Am. Chem. Soc.*, *124*, 14397–14402.

44. Köhn, M., Wacker, R., Peters, C., et al. (2003). Staudinger ligation: a new immobilization strategy for the preparation of small-molecule arrays. *Angew. Chem. Int. Ed.*, *42*, 5830–5834.

45. Bryan, M. C., Fazio, F., Lee, H.-K., et al. (2004). Covalent display of oligosaccharide arrays in microtiter plates. *J. Am. Chem. Soc.*, *126*, 8640–8641.

46. Bryan, M. C., Lee, L. V., Wong, C.-H. (2004). High-throughput identification of fucosyltransferase inhibitors using carbohydrate microarrays. *Bioorg. Med. Chem. Lett.*, *14*, 3185–3188.

47. Calarese, D. A., Lee, H.-K., Huang, C.-Y., et al. (2005). Dissection of the carbohydrate specificity of the broadly neutralizing anti-HIV-1 antibody 2G12. *Proc. Natl. Acad. Sci. USA*, *102*, 13372–13377.

48. Benters, R., Niemeyer, C. M., Drutschmann, D., Blohm, D., Wöhrle, D. (2002). DNA microarrays with PANAM dendritic linker systems. *Nucleic Acids Res.*, *30*, e10.

49. Soellner, M. B., Dickson, K. A., Nilsson, B. L., Raines, R. T. (2003). Site-specific protein immobilization by Staudinger ligation. *J. Am. Chem. Soc.*, *125*, 11790–11791.

50. Houseman, B. T., Huh, J. H., Kron, S. J., Mrksich, M. (2002). Peptide chips for the quantitative evaluation of protein kinase activity. *Nat. Biotechnol.*, *20*, 270–274.

51. Houseman, B. T., Mrksich, M. (2002). Carbohydrate arrays for the evaluation of protein binding and enzymatic modification. *Chem. Biol.*, *9*, 443–454.

52. Dillmore, W. S., Yousaf, M. N., Mrksich, M. (2004). A photochemical method for patterning the immobilization of ligands and cells to self-assembled monolayers. *Langmuir*, *20*, 7223–7231.

53. Drahl, C., Cravatt, B. F., Sorensen, E. J. (2005). Protein-reactive natural products. *Angew. Chem. Int. Ed.*, *44*, 5788–5809.

54. Choi, J., Chen, J., Schreiber, S. L., Clardy, J. (1996). Structure of the FKBP12–rapamycin complex interacting with the binding domain of human FRAP. *Science*, *273*, 239–242.

55. Jin, L., Harrison, S. C. (2002). Crystal structure of human calcineurin complexed with cyclosporin A and human cyclophilin. *Proc. Nat. Acad. Sci. USA*, *99*, 13522–13526.

56. Kanoh, N., Kumashiro, S., Simizu, S., et al. (2003). Immobilization of natural products on glass slides by using photoaffinity reaction and the detection of protein–small-molecule interactions. *Angew. Chem. Int. Ed.*, *42*, 5584–5587.

57. Kanoh, N., Honda, K., Simizu, S., Muroi, M., Osada, H. (2005). Photo-cross-linked small-molecule affinity matrix for facilitating forward and reverse chemical genetics. *Angew. Chem. Int. Ed.*, *44*, 3559–3562.

58. Kanoh, N., Asami, A., Kawatani, M., et al. (2006). Photo-cross-linked small-molecule microarrays as chemical genomic tools for dissecting protein–ligand interactions. *Chem. Asian J.*, *1*, 789–797.

59. Kanoh, N., Kyo, M., Inamori, K., et al. (2006). SPR imaging of photo-cross-linked small-molecule arrays on gold. *Anal. Chem.*, *78*, 2226–2230.

60. Angeloni, S., Ridet, J. L., Kusy, N., et al. (2005). Glycoprofiling with micro-arrays of glycoconjugates and lectins. *Glycobiology*, *15*, 31–41.

61. Bradner, J. E., McPherson, O. M., Mazitschek, R., et al. (2006). A robust small-molecule microarray platform for screening cell lysates. *Chem. Biol.*, *13*, 493–504.

62. Lesaicherre, M. L., Uttamchandani, M., Chen, G. Y. J., Yao, S. Q. (2002). Antibody-based fluorescence detection of kinase activity on a peptide array. *Bioorg. Med. Chem. Lett.*, *12*, 2085–2088.

63. Winssinger, N., Harris, J. L., Backes, B. J., Schultz, P. G. (2001). From split-pool libraries to spatially addressable microarrays and its application to functional proteomic profiling. *Angew. Chem. Int. Ed.*, *40*, 3152–3154.

64. Harris, J. L., Winssinger, N. (2005). PNA encoding (PNA = peptide nucleic acid): from solution-based libraries to organized microarrays. *Chem. Eur. J.*, *11*, 6792–6801.

65. Melkko, S., Scheuermann, J., Dumelin, C. E., Neri, D. (2004). Encoded self-assembling chemical libraries. *Nat. Biotechnol.*, *22*, 568–574.

66. Scheuermann, J., Dumelin, C. E., Melkko, S., Neri, D. (2006). DNA-encoded chemical libraries. *J. Biotechnol.*, *126*, 568–581.

67. Ko, K.-S., Jaipuri, F. A., Pohl, N. A. (2005). Fluorous-based carbohydrate microarrays. *J. Am. Chem. Soc.*, *127*, 13162–13163.

68. Zhang, W. (2003). Fluorous technologies for solution-phase high-throughput organic synthesis. *Tetrahedron*, *59*, 4475–4489.

69. Vegas, A. J., Bradner, J. E., Tang, W., et al. (2007). Fluorous-based small-molecule microarrays for the discovery of histone deacetylase inhibitors. *Angew. Chem. Int. Ed.*, *46*, 7960–7964.

70. Nicholson, R. L., Ladlow, M. L., Spring, D. R. (2007). Fluorous tagged small molecule microarrays. *Chem. Commun.*, 3906–3908.

71. Khersonsky, S. M., Jung, D.-W., Kang, T.-W., et al. (2003). Facilitated forward chemical genetics using a tagged triazine library and zebrafish embryo screening. *J. Am. Chem. Soc.*, *125*, 11804–11805.

72. Wang, D., Liu, S., Trummer, B. J., Deng, C., Wang, A. (2002). Carbohydrate microarrays for the recognition of cross reactive molecular markers of microbes and host cells. *Nat. Biotechnol.*, *20*, 275–281.

73. Bryan, M. C., Plettenburg, O., Sears, P., Rabuka, D., Wacowich-Sgarbi, S., Wong, C.-H. (2002). Saccharide display on microtiter plates. *Chem. Biol.*, *9*, 713–720.

74. Gosalia, D. N., Diamond, S. L. (2003). Printing chemical libraries on microarrays for fluid phase nanoliter reactions. *Proc. Nat. Acad. Sci. USA*, *100*, 8721–8726.

75. Ma, H., Horiuchi, K. Y., Wang, Y., Kucharewicz, S. A., Diamond, S. L. (2005). Nanoliter homogeneous ultra-high throughput screening microarray for lead discoveries and IC_{50} profiling. *Assay Drug Dev. Technol.*, *3*, 177–187.

76. Bailey, S. N., Sabatini, D. M., Stockwell, B. R. (2004). Microarrays of small molecules embedded in biodegradable polymers for use in mammalian cell-based screen. *Proc. Nat. Acad. Sci. USA*, *101*, 16144–16149.

77. Geysen, H. M., Meloen, R. H., Barteling, S. J. (1984). Use of peptide synthesis to probe vital antigens for epitopes to a resolution of a single amino acid. *Proc. Nat. Acad. Sci. USA*, *81*, 3998–4002.

78. Frank, R. (1992). SPOT-synthesis: an easy technique for the positionally addressable, parallel chemical synthesis on a membrane support. *Tetrahedron*, *48*, 9217–9232.

79. Blackwell, H. E. (2006). Hitting the SPOT: small-molecule macroarrays advance combinatorial synthesis. *Curr. Opin. Chem. Biol.*, *10*, 203–212.

80. Fodor, S. P. A., Read, J. L., Pirrung, M. C., Stryer, L., Lu, A. T., Solas, D. (1991). Light-directed, spatially addressable parallel chemical synthesis. *Science, 251*, 767–773.

81. Lipshutz, R. J., Fodor, S. P. A., Gingeras, T. R., Lockhart, D. J. (1999). High density synthetic oligonucleotide arrays. *Nat. Genet., 21*, 20–24.

82. Li, S., Bowerman, D., Marthandan, N., et al. (2004). Photolithigraphic synthesis of peptoids. *J. Am. Chem. Soc., 126*, 4088–4089.

83. MacBeath, G., Schreiber, S. L. (2000). Printing proteins as microarrays for high-throughput function determination. *Science, 289*, 1760–1763.

84. Bobrow, M. N., Harris, T. D., Shaughnessy, K. J., Litt, G. J. (1989). Catalyzed reporter deposition, a novel method of signal amplification. *J. Immnol. Methods, 125*, 279–285.

85. Houk, K. N., Leach, A. G., Kim, S. P., Zhang, X. (2003). Binding affinities of host–guest, protein–ligand, and protein–transition-state complex. *Angew. Chem. Int. Ed., 42*, 4872–4897.

86. McDonnell, J. M. (2001). Surface plasmon resonance: towards an understanding of the mechanisms of biological molecular recognition. *Curr. Opin. Chem. Biol., 5*, 572–577.

87. Vetter, D. (2002). Chemical microarrays, fragment diversity, label-free imaging by plasmon resonance: a chemical genomics approach. *J. Cell. Biochem., Suppl., 39*, 79–84.

88. Metz, G., Ottlesen, H., Vetter, D. (2003). In *Protein–Ligand Interactions: From Molecular Recognition to Drug Design*, G. Schneider, ed. Wiley-VCH, Weinheim, germany, pp. 213–236.

89. Kurosu, M., Mowers, W. A. (2006). Small molecule microarrays: development of novel linkers and an efficient detection method for bound proteins. *Bioorg. Med. Chem. Lett., 16*, 3392–3395.

90. Kuruvilla, F. G., Shamji, A. F., Sternson, S. M., Hergenrother, P. J., Schreiber, S. L. (2002). Dissecting glucose signaling with diversity-oriented synthesis and small-molecule microarrays. *Nature, 416*, 653–657.

91. Koehler, A. N., Shamji, A. F., Schreiber, S. L. (2003). Discovery of an inhibitor of a transcription factor using small molecule microarrays and diversity-oriented synthesis. *J. Am. Chem. Soc., 125*, 8420–8421.

92. Su, J., Mrksich, M. (2002). Using mass spectrometry to characterize self-assembled monolayers presenting peptides, proteins, and carbohydrates. *Angew. Chem. Int. Ed., 41*, 4715–4718.

93. Min, D.-H., Tang, W.-J., Mrksich, M. (2004). Chemical screening by mass spectrometry to identify inhibitors of anthrax lethal factor. *Nat. Biotechnol., 22*, 717–723.

5

USE OF THE PHAGE DISPLAY TECHNIQUE TO IDENTIFY THE TARGET PROTEIN

YOON SUN CHO AND HO JEONG KWON

Chemical Genomics Laboratory, Department of Biotechnology, College of Life Science and Biotechnology, Yonsei University, Seoul, Republic of Korea

5.1 INTRODUCTION

The biological functions of the proteins that were newly discovered from the human genome project remain largely uncovered. Small molecules that control protein functions leading to phenotype changes in particular biological systems could be valuable tools not only to decipher the biological functions of these new proteins but to develop the basis of medical applications. In this respect, small molecule–target protein information is of great help. In addition, when a biologically active small molecule is achieved through a phenotype-based screen, the information on target protein toward the small molecule is also crucial for understanding the mode of action of the small molecule. Accordingly, systematic approaches to identifying target proteins of small molecules are of great interest in related fields. This information will eventually link the gap between a small molecule–target protein and a small molecule–biological phenotype [8,13].

Protein Targeting with Small Molecules: Chemical Biology Techniques and Applications,
Edited by Hiroyuki Osada
Copyright © 2009 John Wiley & Sons, Inc.

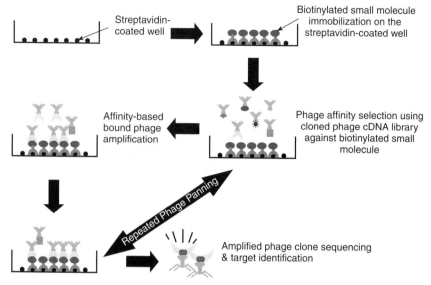

Figure 5-1   Overall phage display biopanning scheme. Identification of a target protein of a small molecule. *(See insert for color representation of figure.)*

Identification of the target protein of a small molecule has been developed in various ways. Among them, affinity- and yeast genetics–based target identification methods have been mainly utilized for this aim. Particularly, owing to the generation of direct physical binding information between a small molecule and a target protein, a number of new technologies based on affinity-based target identification have been developed: (1) the affinity chromatography- and mass analysis–associated quantitative proteome method [9], (2) protein and compound microarray method [20], (3) phage display method, and so on [2,12,15,16]. A number of groups, including ours, have developed a phage display method to identify the target proteins of small molecules with biological relevancy. Several target proteins of small molecules have been identified using this strategy, as summarized in Table 5-1. Among them, Ca^{2+}/calmodulin (Ca^{2+}/CaM) was identified as a target protein of hydrazinobenzoicurcumin (HBC), one of the derivatives of curcumin, using this method [15]. The phage display biopanning method for the identification of target proteins of small molecules is the focus of this chapter (Fig. 5-1).

5.2   PHAGE DISPLAY TECHNIQUE

Phage display is a method of displaying (expressing) protein or peptide on the surface of the phage using single-stranded DNA bacteriophage-coated protein on the fiber of a cloned peptide or antigen gene that uses M13 or fd *Escherichia coli* as the host, or displaying the fused protein of T7 and λ-phage capsid proteins

TABLE 5-1 Representative Examples of Small Molecules and Their Target Proteins Identified by the Phage Display Method

Small Molecule	Target Protein	Reference
FK506	FKBP	Sche et al. [12]
AP1497 (synthetic analog of FK506)	FKBP1a, FKBP1b, FKBP2	McKenzie et al. [7]
Biotin-YZZZ (biotinylated phosphotyrosine diversity peptide)	Clones with phosphotyrosine-binding (SH2) domains	Videlock et al. [20]
HBC (synthetic curcumin derivative)	Ca^{2+}/CaM	Shim et al. [15]
Doxorubicin	hNopp140	Jin et al. [4]
Palau'amine	hNopp140	Piggott and Karuso [10]
β-Lactamase	RRGHYY (TEM-1 β-lactamase inhibitor)	Huang et al. [3]
Stenotrophomonas maltophilia GN12873 L-1 metallo-β-lactamase	Cys–Val–His–Ser–Pro–Asn–Arg–Glu–Cys (peptide inhibitor of class B β-lactamase L-1 and FEZ-1)	Sanschagrin and Levesque [11]
Pg-ompA1	Minimal epitope requirements of the MAb (96)IALDQTLGIP(105) (87)WPRVGQLFIALDQTL-GIPTFSVCRME(116)	Chang et al. [1]
Colon carcinoma cells (HT29, HCT116)	CPIEDRPMC (RPMrel)	Kelly and Jones [5]
Endothelial protein C receptor (EPCR)	Leukolysin (MT-MMP6) and cerastocytin	White et al. [21]

[2,16]. These phage particles could be utilized not only as protein or peptide that fused with the phage but provide the genetic information of expressed protein or peptide at the gene level, due to the gene information in the phage DNA. In other words, phage display is a new technology-expressing peptide and antigen library and is a screening method that allows the identification of specific proteins with high affinity to small molecules or peptides at the genome level. This method first screens the displayed protein as the rod. Then the DNA that codes the protein is extracted, decoding the nucleotide sequence and thus making it possible to identify the molecule's nucleotide sequence.

Phage display is able to express on most protein coats, but g3p and g8p have generally been used. When g3p, which has five molecules per phage, is used as the expression base, it is possible to express one to five molecules of polypeptide with molecular weight up to 50,000 per phage. This relatively small number of molecules allows the separation of polypeptides with strong affinity toward specific interacting molecules (in this case, small molecules). But when g8p, a phage with about 3000 molecules per phage, is used as an expression base, it is possible to express about five to eight amino acid polypeptides with 3000 molecules, thus making it possible to separate only polypeptides with weak affinity. In addition, a new display system using T7 phage has been developed recently by Novagen (Madison, Wisconsin). T7 phage is made up of a 415-molecule coat protein (g10p) covering the capsid. In the case of a small peptide on a g10p N- or C-terminal, it expresses 415 molecules and 5 to 15 molecules in the case of big polypeptides of 1200 amino acid residues. Different from fiber phages, T7 phage is able to express on the C-terminal, due to the reorganized frame of the coat protein from induction of foreign genes. Moreover, this T7 phage is more effective to make up the cDNA library, as there is no stop codon on the coat protein. In addition, different from other phages, T7 infection is possible even in the presence of protein denaturant (1% sodium dodecyl sulfate, 4 M urea, 2 M guanidine-HCl). Therefore, this unique feature is applied as a denaturant during biopanning elution and makes it possible to differentiate polypeptides that are unique or have higher affinity than other fiber phages. Here, biopanning is one of the crucial steps that enriches and discriminates the target protein of the phage from the polypeptide library.

The basic idea of biopanning is to bind the phage polypeptide library with an immobilized small molecule onto the solid surface and remove the noninteracting phage through washing. Then the bound phage is collected and infected into *E. coli*. This step is repeated several times to enrich the phage that expresses a protein with specific binding to the small molecule. To that extent, biopanning makes possible the differentiation of cloned high-affinity polypeptide by increasing the interaction between each phage.

5.3   IDENTIFICATION OF THE TARGET PROTEIN OF A SMALL MOLECULE USING PHAGE DISPLAY

As described above, phage display biopanning is a prominent method of discovering new target proteins for small molecules. As phage display starts from the transcriptome of a cDNA library as an expression system and the tagged proteome is identified toward a small molecule, it has also been viewed as reverse chemical proteomics [10]. In this section, the key steps of the phage display method and its application to the target identification of a small molecule, HBC, are demonstrated.

5.3.1   Small-Molecule Immobilization

To efficiently monitor the interaction between a small molecule and a protein expressed on the phage surface, the small molecule is to be immobilized on the solid surface. Generally, biotin–avidin interaction has been used for this purpose. Biotin is covalently linked to the specific functional group of a small molecule that does not affect its biological effect on cells. Small molecules can be linked with a longer linker or rod to increase the solubility and binding affinity of a small molecule toward a target protein. After generation of the biotinylated small molecule, it is immobilized on the streptavidin-coated well. The biotin linked to the small molecule binds to the streptavidin on the well and forms a strong binding affinity, immobilizing the small molecule at a specific orientation.

5.3.2 Binding of a Phage-Expressed Proteome Library to a Small Molecule

After small-molecule immobilization, a number of phage particles ($1 \times 10^{10-11}$ phages) encoding human cDNA libraries are added to the well. With incubation at room temperature, the human protein expressed on the phage surface from a cDNA library of phages interacts with the small molecule on the basis of its affinity. Then the well is washed with buffer several times to wash out weak-affinity protein-expressing phages. The remaining binding phages provide stronger-affinity protein-expressing phages that are subjected to biopanning for enrichment of the strongest-affinity protein-expressing phages to the small molecule.

5.3.3 Elution of Strong-Affinity Protein-Expressing Phage to a Small Molecule

To obtain strongly interacting protein-expressing phage to a small molecule, the bound phage is eluted with a cold small molecule (a small molecule not linked with biotin) with 10- to 100-fold concentration. By introducing competition between a biotinylated small molecule and a cold small molecule to protein-expressing phage, the interacting protein-expressing phage to the small molecule can be eluted selectively.

5.3.4 Panning the Eluted Phage

With the eluted phage, the number of phages is determined by panning the phage. First, the host strain *E. coli* strain BLT5615 is freshly cultured for the panning. The eluted phage and the control wash are diluted 10-fold, 100-fold, or more, and each dilution is infected into fresh BLT5615, mixed with top agar, and panned on LB agar plated supplemented with ampicillin. Then the number of phages bound to the small molecule is calculated by counting the number of phage particles appearing on the plate. Second, a small amount of eluted phage is infected into fresh BLT5615 to amplify the phage. This amplified phage is then used in the next round of biopanning to obtain concentrated strong-affinity binding phage. Biopanning rounds are carried out at least four or five times where the concentrated pattern appears between the wash and elute plates in each round (Fig. 5-2).

5.3.5 Target Protein Identification and Validation

After biopanning, the target protein of a small molecule is identified and validated. An appropriate number of phage is isolated from the last round of biopanning, and

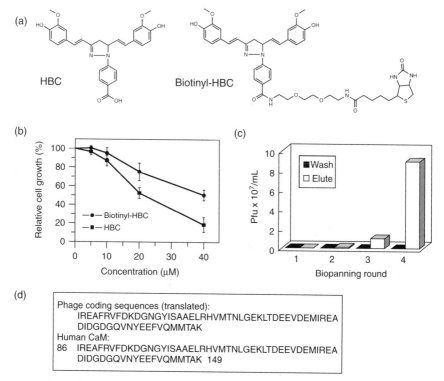

Figure 5-2   Identification of CaM as a novel cell-cycle inhibitor HBC target protein through the phage display technique. (a) Structures of HBC and biotinyl-HBC. (b) Effects of HBC and biotinyl-HBC on the proliferation of HCT15 cells. The cells were treated with each compound for 72 hours and an MTT assay was carried out to evaluate the biological activity of the compounds. (c) Analysis of HBC-binding phage particles eluted after each round of biopanning. A wash indicates nonspecific phages bound to biotinyl-HBC immobilized on streptavidin-coated well. (d) Sequence homology between human CaM and the coding protein of HBC-binding phage. The phage sequences were 100% identical to the C-terminal (86–149) of human CaM. (From ref. 15.)

each is infected into freshly cultured BLT5615 and amplified. Around 40 to 60 phages are selected arbitrary and each phage cDNA size is determined by polymerase chain reaction (PCR). Concentrated cDNAs are selected and sequenced to identify the target protein using BLAST sequence homology search. The target protein identified will constitute most of the sequenced cDNAs.

5.3.6   Target Protein Validation

When the target protein of a small molecule is determined, it goes under validation with a number of biophysical, biochemical, and biological experiments, such as forward-affinity binding assay, reverse-affinity binding assay, small molecule–target protein docking model, surface plasmon resonance (SPR) analysis, and cell-based assays. Forward-affinity binding assay consists of phage-binding assay and phage competition assay with the phage that expresses the concentrated target protein. Reverse-affinity binding assay is also used to further validate the binding affinity of the phage and the small molecule [17]. Afterward, when the target protein is selected from the several candidates, SPR, protein docking modeling, and cell-based assays using siRNA and antibodies could be conducted to validate the biological relevancy of the binding of a small molecule to the target protein identified by the phage display method.

5.3.7 Target Protein Identification of HBC, an Inhibitor of Cell-Cycle Progression

Recently, our group identified the Ca^{2+}/calmodulin (Ca^{2+}/CaM) as a biologically relevant target protein of HBC (hydrazinobenzoicurcumin), a new curcumin analog, based on the phage display technique described here. HBC inhibited the proliferation of colon cancer cell lines without showing any inhibitory activity toward aminopeptidase N (APN), a target protein of curcumin, the mother compound of HBC. To fully understand the underlying mechanism of how HBC exhibits its cellular activity, phage display biopanning was conducted with HBC for the identification of target protein of the compound. From the fourth round of biopanning, 17 phage plaques were isolated and sequenced. Among them, 12 matched exactly with the C-terminal end of human CaM, which constitutes almost 70% of the sequenced phages. Phage-binding assay and phage competition assay resulted in a specific binding of HBC-and CaM-expressing phage. Moreover, SPR analysis between HBC and purified Ca^{2+}/CaM showed dose-dependent binding, validating that Ca^{2+}/CaM is the direct target protein of HBC. In addition, a docking model of HBC and the C-terminal of Ca^{2+}/CaM showed the best possible interacting position of the two molecules (Fig. 5-3). As a result, the mechanism of HBC-inhibiting cell proliferation is clearly explained: Ca^{2+}/CaM activity of tumor cell proliferation is inhibited by its antagonist HBC, resulting in the inhibition of cell-cycle progression.

5.4 CONCLUSIONS

The use of small molecules to explore and control the phenotype of a given biological system is the core concept of chemical biology and drug discovery. Of the various methods to identify the target of small molecules, phage display biopanning might be one of the easiest and most effective techniques. A number of advantages of this technique have been implicated as follows. First, the method can be applied to the identification of protein targets with low expression level. Not only human but also other organisms and synthetic peptide libraries can be generated, and these diverse libraries are easily amplified for screening of targets of small molecules. Second, control and maintenance are simple, followed by the accuracy of the binding affinity between the target protein on the phage and the

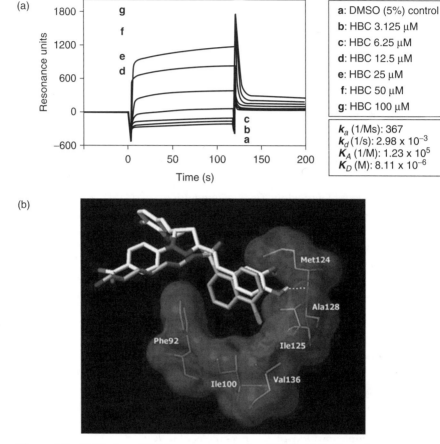

Figure 5-3   Validation of CaM as a target protein of HBC. (a) Surface plasmon resonance analysis of interaction between HBC and Ca^{2+}/CaM. Purified Ca^{2+}/CaM was immobilized on a CM5 sensor chip and various concentrations of HBC were loaded into the sensor cell. Binding sensor grams were obtained from the BIAcore evaluation software. Kinetic parameters of k_a, k_d, K_A, and K_D are shown. (b) Docking model of HBC in a complex with the C-terminal Ca^{2+}/CaM domain. The docking mode of HBC (gray carbon) and W7 (orange carbon) obtained from FlexX. The Connolly molecular surface of the active site is shown in purple with amino acid residues occupying the active site. Hydrogen atoms are not shown for clarity. The yellow dotted line indicates the hydrogen-bonding interaction ($d = 1.244$Å). (From ref. 15.) *(See insert for color representation of figure.)*

small molecule. Third, rapid and convenient identification of selected proteins can be achieved by easy and rapid decoding of the protein by DNA sequence analysis. Finally, identification and validation using forward- and reverse-affinity binding assays are reliable and can be reproduced with accuracy. However, there are also some drawbacks to phage display biopanning. In the process of making cDNA libraries on the phage coat protein, random primers are used to expand the number of cDNAs. This can result in severed proteins without considering their frames from the nucleotide sequence, a frameshift, which results in a different protein expression on the phage coat protein. Additionally, read-through of the stop codon on the cDNA results in abnormal protein expression. These factors interfere with the identification of the target protein, making it possible to catch real proteins instead of synthetic peptides. It is also noteworthy that post-translational modifications of proteins are not fully achieved in a phage display system. Therefore, the binding protein information from this approach is better to be validated using other biochemical and biological validation experiments described above. Even with these drawbacks, phage display biopanning is an effective and rapid means of discovering target proteins or peptides, due to its technical advantages in selecting the high-affinity binding targets of the small molecule.

Acknowledgments

This study was supported by grants from the National R&D Program for Cancer Control, Ministry of Health & Welfare, and from the Brain Korea 21 Project, Republic of Korea.

REFERENCES

1. Chang, W. J., Kishikawa-Kiyama, M., Shibata, Y., Lee S. Y., Abiko, Y. (2004). Inhibition of *Porphyromonas gingivalis* hemagglutinating activity by synthetic peptides derived from phage display selection using MAb against the recombinant outer membrane protein. *Hybrid Hybridom.*, *23*, 352–356.

2. Hoess, R. H. (2001). Protein design and phage display. *Chem. Rev.*, *101*, 3205–3218.

3. Huang, W., Beharry, Z., Zhang, Z., Palzkill, T. (2003). A broad-spectrum peptide inhibitor of β-lactamase identified using phage display and peptide arrays. *Protein Eng.*, *16*, 853–860.

4. Jin, Y., Yu, J., Yu, Y. G. (2002). Identification of hNopp140 as a binding partner for doxorubicin with a phage display cloning method. *Chem. Biol.*, *9*, 157–162.

5. Kelly, K. A., Jones, D. A. (2003). Isolation of a colon tumor specific binding peptide using phage display selection. *Neoplasia*, *5*, 437–444.

6. Keresztessy, Z., Csosz, E., Harsfalvi, J., et al. (2006). Phage display selection of efficient glutamine-donor substrate peptides for transglutaminase 2. *Protein Sci.*, *15*, 2466–2480.

7. McKenzie, K. M., Videlock, E. J., Splittgerber, U., Austin, D. J. (2004). Simultaneous identification of multiple protein targets by using complementary-DNA phage display and a natural-product-mimetic probe. *Angew. Chem. Int. Ed. Engl.*, *43*, 4052–4055.

8. Mitchison, T. J. (1994). Towards a pharmacological genetics. *Chem. Biol.*, *1*, 3–6.

9. Oda, Y., Owa, T., Sato, T., et al. (2003). Quantitative chemical proteomics for identifying candidate drug targets. *Anal. Chem.*, *75*, 2159–2165.

10. Piggott, A. M., Karuso, P. (2005). Quality, not quantity: the role of marine natural products in drug discovery and reverse chemical proteomics. *Mar. Drugs*, *3*, 36–63.

11. Sanschagrin, F., Levesque, R. C. (2005). A specific peptide inhibitor of the class B metallo-beta-lactamase L-1 from *Stenotrophomonas maltophilia* identified using phage display. *J. Antimicrob. Chemother.*, *55*, 252–255.

12. Sche, P. P., McKenzie, K. M., White, J. D., Austin, D. J. (2001). Display cloning: functional identification of natural product receptors using cDNA-phage display. *Chem. Biol.*, *8*, 399–400.

13. Schreiber, S. L. (1998). Chemical genetics resulting from a passion for synthetic organic chemistry. *Bioorg. Med. Chem.*, *6*, 1127–1152.

14. Shadidi, M., Sioud, M. (2003). Identification of novel carrier peptides for the specific delivery of therapeutics into cancer cells. *FASEB J.*, *17*, 256–258.

15. Shim, J. S., Lee, J., Park, H. J., Park, S. J., Kwon, H. J. (2004). A new curcumin derivative, HBC, interferes with the cell cycle progression of colon cancer cells via antagonization of the Ca^{2+}/calmodulin function. *Chem. Biol.*, *11*, 1455–1463.

16. Smith, G. P., Petrenko, V. A. (1997). Phage display. *Chem. Rev.*, *97*, 391–410.

17. Takakusagi, Y., Kobayashi, S., Sugawara, F. (2005). Camptothecin binds to a synthetic peptide identified by a T7 phage display screen. *Bioorg. Med. Chem. Lett.*, *15*, 4850–4853.

18. Takakusagi, Y., Ohta, K., Kuramochi, K., et al. (2005). Synthesis of a biotinylated camptothecin derivative and determination of the binding sequence by T7 phage display technology. *Bioorg. Med. Chem. Lett.*, *15*, 4846–4849.

19. Takagi, T., Arisawa, T., Yamamoto, K., Hirata, I., Nakano, H., Sawada, M. (2007). Identification of ligands binding specifically to inflammatory intestinal mucosa using phage display. *Clin. Exp. Pharmacol. Physiol.*, *4*, 286–289.

20. Videlock, E. J., Chung, V. K., Mohan, M. A., Strok, T. M., Austin, D. J. (2004). Two-dimensional diversity: screening human cDNA phage display libraries with a random diversity probe for the display cloning of phosphotyrosine binding domains. *J. Am. Chem. Soc.*, *126*, 3730–3731.

21. White, S. J., Simmonds, R. E., Lane, D. A., Baker, A. H. (2005). Efficient isolation of peptide ligands for the endothelial cell protein C receptor (EPCR) using candidate receptor phage display biopanning. *Peptides*, *26*, 1264–1269.

22. Xu, Q., Lam, P. S. (2003). Protein and chemical microarrays:powerful tools for proteomics. *J. Biomed. Biotechnol.*, *2003*, 257–266.

23. Zhang, Y., Chen, J., Zhang, Y., et al. (2007). Panning and identification of a colon tumor binding peptide from a phage display peptide library. *J. Biomol. Screen.*, *12*, 429–435.

6

DEVELOPMENT OF FLUORESCENT PROBES FOR SMALL MOLECULES

ADRIAN P. NEAL AND CARSTEN SCHULTZ

European Molecular Biology Laboratory, Heidelberg, Germany

6.1 INTRODUCTION

Fluorescent sensors that detect small molecules have a wide range of applications in industry, environmental monitoring, and biomedical approaches. They are able to detect a diverse range of analytes, such as toxic compounds or changes of signaling molecules inside living cells, to name two prominent examples. In this chapter we focus on fluorescence-based tools for detecting small-molecule analytes. There are two fundamentally different concepts for designing such sensors. The majority of sensing molecules recognize the analyte by engulfing it in a way that permits a sufficiently large change in the sensor to induce a change in fluorescence. Hence, these types of sensors (type I sensors) consist of a detector unit attached to a reporter unit, usually a fluorophore (Fig. 6-1a) or a pair of fluorophores that communicate to each other via fluorescence or Foerster resonance energy transfer (FRET; Fig. 6-1b). The second group of sensors (type II sensors) bind the analyte in a way that leads to a chemical reaction, which in turn changes the fluorescent properties of the reporter unit (Fig. 6-1c and d). Provided that the chemical reaction is irreversible, the latter type of sensor leads

Protein Targeting with Small Molecules: Chemical Biology Techniques and Applications,
Edited by Hiroyuki Osada

91

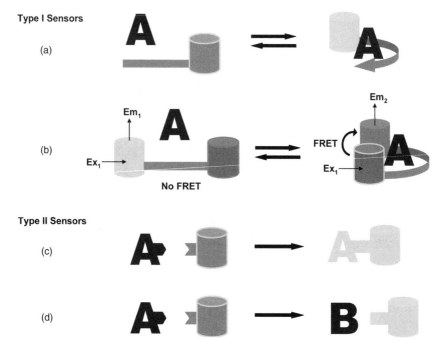

Figure 6-1   Type I and II sensors for analyte A. Type I sensors bind the analyte reversibly, leading to a change in fluorescence (a) or a change in FRET between a pair of fluorophores (b). Type II sensors react chemically either to incorporate the analyte into a fluorescent product (c) or to produce fluorescent product B (d). These reactions are usually irreversible and lead to an accumulation of fluorescent products.

to an accumulation of the signal and hence an enormous sensitivity even when very little analyte is abundant at any given point in time.

Type I sensors need to be sufficiently large to wrap around the analyte. This can easily be achieved for ions with relatively small molecules, such as multidentate chelators and crown ethers [1]. The literature is heavy with examples of fluorescent sensors for metal and nonmetal ions, with increasing emphasis on measuring ions in living biological tissues [2]. The designs of metal ion probes share many key concepts with those of small-molecule sensors, and these are discussed here; however, the metal ion probes themselves have been reviewed extensively very recently [3] and are therefore not included in this chapter. The smallest molecules of interest here are NO and CO, and several sensors based on heme complexes have been introduced for NO [4,5]. Any larger molecule usually requires a significantly larger detector unit. Most of the latter are designed after proteins or protein domains or proteins are used directly for analyte recognition.

As mentioned above, type II sensors have the advantage of accumulating the signal. In return, however, these sensors do not respond easily to dynamic changes in analyte concentration over time. The number of examples is to date fairly limited but useful sensors for copper(I) ions [6], phosphor organics [7], and H_2O_2 [8–10] have been introduced recently. Overall, the number of sensors for small molecules is surprisingly small. There are probably no more than 100 fluorescent sensors of this type described. This is a marginal number considering the large number of small molecules existing in higher organisms and the vast number of other naturally occurring and synthetic organics. The reason for this is that the synthesis of an artificial reporter unit is extraordinarily cumbersome. Only with the help of molecular biology for preparing proteins and protein fragments as reporters and, subsequently, the introduction of genetically encoded fluorophores (i.e., fluorescent proteins from jellyfish and corals) has the preparation of sensors become sufficiently easy. Hence, the preparation of fluorescent sensors for small molecules is now a wide open field of applying a combination of chemistry and biology with a good squirt of inspiration and imagination.

6.2 PROBE DESIGN

6.2.1 Ligand-Binding Sensors

The design of fluorescence-based detection systems for small molecules is influenced immediately by the fact that the vast majority of these target analytes are nonfluorescent. Therefore, sensors are required that probe the analyte and incorporate fluorophores which respond sensitively to the probe–analyte interaction. As stated above, in type I sensors these fluorophores are coupled to the sensor unit of the reporter. A vast array of fluorescent molecules is known. Here we give a brief overview of these fluorophores by family—small molecules, proteins, and the new fluorescent nanomaterials—and comment on how they may be incorporated into sensors for small molecules. The description of the sensor units is usually unique to the analyte of interest and is detailed in Section 6.3.

Small-Molecule Fluorophores Low-molecular-weight fluorophores (< 600 Da) are emissive across the visible spectrum, and together with their sensitivity to pH, metal ions, and solvent, their use offers a range of multicolored responses which when harnessed as part of a larger sensing unit may be adapted to serve as reporters of small molecules. An excellent review on small-molecule fluorophores for biological applications has just been published [11]. Common fluorophore motifs are coumarins [12], xanthenes [13,14], cyanines [15], boradiazaindacenes [16], and lanthanide metal ion complexes [6,17]. The best known of these, the xanthenes, include fluoresceins and their red-shifted nitrogenous cousins, the rhodamines. Syntheses of simple coumarin and xanthene dyes are relatively straightforward condensations [14]. Fine tuning the fluorescence output of small-molecule dyes has been the focus of considerable effort and has required some innovative chemistry. These efforts have established rules by which emission of a fluorophore may be shifted or even switched on/off and hence have yielded some interesting novel molecules [18]. For example, coumarins have been transformed from toxic natural products (and now banned food additives) into sophisticated fluorescent photoreleasable protecting groups [19] and transition metal ion sensors [20]. More recently, by furnishing the 3-position with an electron-withdrawing group (azide), fluorogenic coumarin reagents for "click" chemistry were produced [21,22].

The choice of fluorophores of this type are often limited by their commercial availability since bespoke fluorophore design further burgeons the task of constructing probes by adding difficult synthetic steps. Labels are introduced by means of reaction, usually (but not exclusively) with amines to form amides or carbamates. Amine-reactive fluorophore reagents for labeling applications are thus predominantly commercially available as carboxylic acid derivatives or, for ease of synthesis (but at significantly elevated prices), may be provided as the N-hydroxysuccinimidyl ester. Alternatively, thiols are alkylated with fluorescent iodoacetamides or maleimides. For organic syntheses of small-molecule probes these reactions are ideal; however, for labeling proteins, on their own they offer little, if any, site specificity.

Fluorescent Proteins   Fluorescent proteins are derived from naturally occurring proteins found in jellyfish (especially *Aequorea victoria*) and other species. The best known example, green fluorescent protein (GFP), consists of a beta barrel surrounding (and protecting from the quenching environment) a fluorescent imidazolone ($\lambda_{ex} = 488$ nm, $\lambda_{em} = 509$ nm) formed from the post-translational cyclization–oxidation of sequential serine–tyrosine–glycine residues [23–25]. Mutation of these and other amino acids as well as circular permutations have provided access to spectral variants of GFP, which are now used routinely as genetically encoded fluorescent labels of proteins [23,26–30]. Site-specific labeling and superior photostability are some of the advantages over small-molecule fluorophores, although their large size (around 27 kDa) needs to be considered so that they do not impair normal protein function. The biggest incentive to use fluorescent proteins, however, comes from the fact that no bench chemistry is needed to prepare fluorophore and sensor.

Quantum Dots   Quantum dots (QDs) are semiconducting nanosized crystals (2 to 50 nm in diameter) of group II–VI elements. Typical examples of such materials are zinc sulfide–coated cadmium selenide (ZnS–CdSe) particles [31]. Emission wavelength is dependent on particle size, larger QDs producing red-shifted fluorescence. Production may be tuned to produce QDs in a range of sizes, and hence they are commercially available with emissions across the visible spectrum. QDs have been functionalized to confer water solubility and such that they may serve as labels for biomolecules [32]. With high quantum yields, good photostability, and dimensions similar to those of fluorescent proteins, QDs have emerged as alternative fluorophores for use in microscopy [33]. Applications for measuring analytes in living cells have recently been introduced [34].

Single-Walled Carbon Nanotubes   The carbon allotropes known as single-walled carbon nanotubes (SWNTs) are essentially one-atom-thick cylindrical rolls of graphite. SWNTs fluoresce in the range 900 to 1600 nm, a transparent region in biological samples. Near-infrared fluorescence imaging of macrophages has been reported [35], and SWNTs have been suggested as implantable fluorescence reporters for in vivo applications [36].

6.3 PROBE DEVELOPMENT AND APPLICATIONS

6.3.1 Probes Based on Small Molecules

The lethal deployment of toxic agents in recent years [37] has underscored the need for rapid, reliable, and portable detection systems for these compounds. The fluorescent sensor DCP-S4 for chlorophosphates (Fig. 6-2) is an example of a sensor that is operated by modulating a photo-induced electron transfer (PET) effect. Diethyl chlorophosphate (DCP) was used as a model for extremely toxic organophosphorous nerve agents such as sarin and tabun. Toward this goal, sensor DCP-S4 was designed such that the lone pair of a tertiary amine quenches the fluorescence of an adjoining fluorophore via a PET process—the sensor is in an "off" state. Upon exposure to DCP, the alcohol of DCP-S4 is activated (analogous to the reaction with the active-site hydroxyl of acetylcholine esterase [38], the origin of the toxicity of nerve agents), providing a phosphate ester (Fig. 6-2). This phosphate leaves as the amine undergoes intramolecular cyclization to the quaternary azaadamantane salt. This event switches "on" the fluorescence as the lone-pair quenching mechanism is removed. This probe is an example of a type II sensor since the analyte is incorporated irreversibly into the fluorescent product.

The modular design of amine spacer–fluorophore allows a variety of fluorophores whose fluorescence are tuned by PET to be utilized. In this example, 6,7-dimethoxycoumarin-based sensor DCP-S4 undergoes a 20-fold fluorescence enhancement ($\lambda_{em} = 465$ nm) when quantitatively converted to the quaternary salt by treatment with DCP. Similarly, a pyrene-based sensor which exhibits a 22-fold increase in fluorescence ($\lambda_{em} = 378$ nm) is also reported [7].

Carbohydrate Sensing by Boronic Acid Derivatives   In the pursuit of sensors for carbohydrates, boronic acids have been utilized for their ability to covalently but

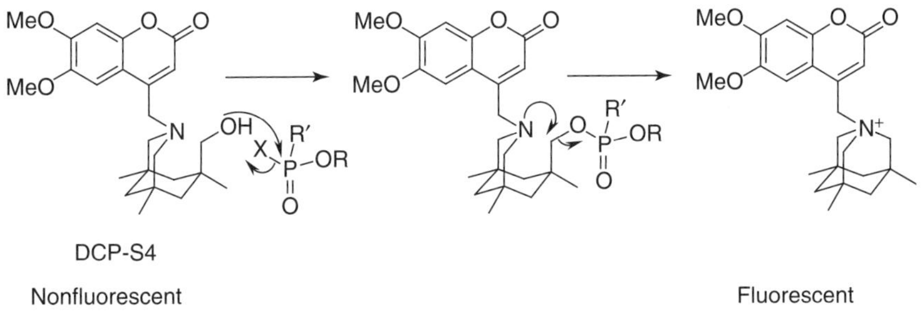

Figure 6-2 PET-based fluorescent sensor for nerve agent mimics.

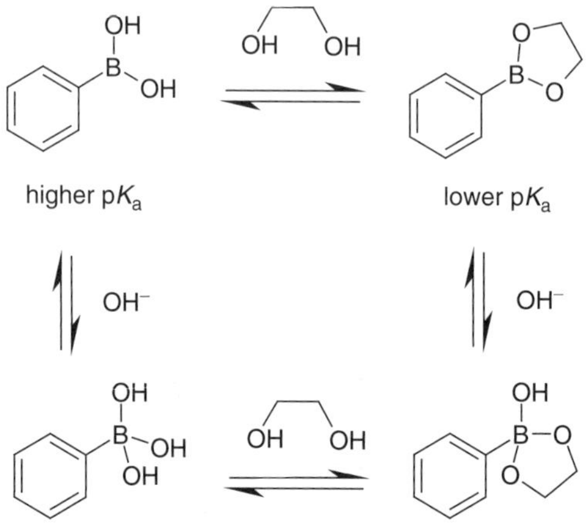

Figure 6-3 Sugar sensing: equilibrium between arylboronic acids and 1,2-diols.

reversibly bind *cis*-1,2- and 1,3-diols as cyclic boronate esters [39,40]. Sensors for carbohydrates are traditionally built around arylboronic acids, the simplest of which, phenylboronic acid, has a higher affinity for D-fructose over D-glucose (the dissociation constants are 0.5 and 10 mM, respectively) [41]. Figure 6-3 shows the equilibrium between diol-bound and unbound arylboron species, and in general it is the changes in pK_a of these species that govern the changes in emission of an attached fluorophore.

Through variation of the aromatic moiety and, moreover, the molecular geometry around the sugar-binding site, selectivity may be tuned. Figure 6-4 shows selected examples of arylboronic acids covalently attached to fluorescent reporters, the output of which are modulated by different mechanisms.

ANDBA is a symmetrical 9,10-disubstituted anthracene bearing two 2-phenylboronic acid units. Each diol-recognizing group is linked to the anthracene via tertiary amines which exert PET quenching upon the anthracene fluorophore. Upon carbohydrate binding (formation of boronate diester), the Lewis acidity of the boron increases and amino lone-pair electron density is transferred away from the anthracene, relieving the fluorescence quenching. The increase in emission ($\lambda_{em} = 420$ nm) can be observed as a function of increasing glucose concentration.

R,R-ANDPBA is a chiral analog of ANDBA, elaborated with two 1R-phenylethyl groups. This compound and its S,S-enantiomer are attempts toward the chiral recognition of sugar acids, especially D- and L-tartaric acid [42]. Operating by the same PET mechanism as ANDBA, the sensors display an increased fluorescence output upon sugar acid binding. Both chiral groups are required for high enantioselective binding. The R,R-enantiomer provides enantioselective reporting of D-tartaric acid, with negligible fluorescent increases upon binding to L-tartaric acid. This stereoselectivity is mirrored by the S,S-enantiomer. Binding constants for these sensor–sugar acid complexes are pH dependent. This means that selectivity may be tuned by alteration of the pH.

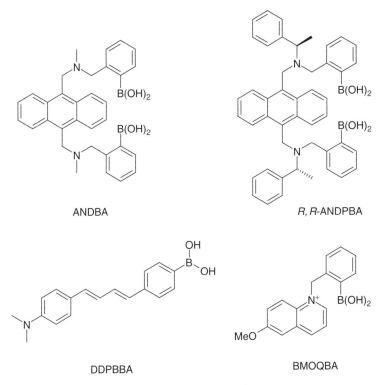

ANDBA

R, R-ANDPBA

DDPBBA

BMOQBA

Figure 6-4 Structures of arylboronic acid–based sugar sensors.

For example, at pH 8.3, R,R-ANDPBA selects D-tartaric acid over D-sorbitol in the ratio 7.2:1. Upon lowering the pH to 5.6, the selectivity becomes irrefutable at a ratio 11,000:1. The pH dependency of these probes offers the potential for switchable sensing of sugar acids in complex mixtures.

1-(p-Boronophenyl)-4-(p-dimethylaminophenyl)buta-1,3-diene (DDPBBA) bears electron-donating (dimethylamino) and electron-withdrawing (boronic acid) substituents at opposite ends of the conjugated system. These substituents transfer charge density across the molecule. In the presence of a diol, and at a certain pH, the charged boronic acid ester form predominates. This switching of an electron-withdrawing group to an electron-donating group interferes with excited-state charge transfer, and spectral changes can be observed. In the case of DDPBBA-binding glucose at millimolar concentrations (phosphate buffer, pH 8.0), a blue-shifted emission increases at $\lambda = 500$ nm.

N-(2-Boronobenzyl)-6-methoxyquinolinium (o-BMOQBA) is a fluorescent cationic quinolinium core bearing a phenylboronic acid [43]. This probe functions in a mechanistically distinct manner compared to other sugar sensors. The key design feature here is that the boronic acid group is an *ortho*-substituent to the N-methylquinolinium. Upon diol binding, the resulting charged adduct is able to form an intramolecular salt bridge with the quaternary nitrogen.

As this permanent charge is partially neutralized, fluorescence decreases. By placing a positive charge on the arylboronic acid, the pK_a of the boronate ester (sugar–boronic acid complex) is 6.1, considerably lower than those of other probe adducts (typically, 7.0 to 8.3). Since the stability of the boronate ester (and hence detection of the sugar) is pH dependent, tuning the pK_a allows the design of probes that are robust in a range of biological tissues. In aqueous buffer, o-BMOQBA responds over physiologically relevant glucose concentrations (3 to 8 mM in the blood of healthy patients, and up to 40 mM in diabetics) [43–45].

FRET sensors for sugars based on a glucose-sensing protein, concanavalin A (Con A), have been prepared [46]. Con A is a plant lectin from the jack bean with four glucose-binding sites [41]. Dual labeling of Con A with fluorescein near a glucose-binding site (Tyr100) and a coumarin (at Lys114) was achieved using site-selective methodologies. Binding of saccharides causes changes in the fluorescence of both coumarin (donor) and fluorescein (acceptor). This ratiometric response could be utilized to image glucose inside HepG2 cells (liver) as well as mannose-rich saccharides on the surface of MCF-7 cells (a breast cancer cell line). Other FRET sensors employing fluorescently labeled substrates with singly labeled Con A as well as other sugar-binding proteins have been reviewed [41]. Genetically encoded sensors for glucose and other carbohydrates are discussed below.

Clinical Applications of Carbohydrate Sensors   The main clinical application for sugar-sensing probes is in monitoring blood glucose levels in diabetes patients. The current enzyme-based method samples glucose by means of "finger prick" tests, up to 10 times per day. Relatively painless, these are undoubtedly useful for predicting diet requirements throughout a day [47]. However, poor compliance by some patients (particularly young and elderly patients) may preclude effective management of this potentially lethal disease. Also, continuous monitoring of glucose levels is often required (in more severe cases, and also while a patient sleeps). While methods based on subcutaneous implants and reverse iontophoresis have recently been brought to market, fluorescence-based methods have been proposed to provide longer-term solutions to noninvasive, continuous, real-time monitoring of blood sugar levels [41].

Glucose concentrations in tears have been shown to track levels in blood, and the eye has been proposed as a noninvasive sampling site [48]. Glucose levels in tears may be in the range 50 to 500 μM and to measure these levels the use of contact lenses doped with fluorescent boronic acids has been investigated [49]. Contact lenses that offer colorimetric changes in response to specific analytes are being developed concurrently as a general approach to small-molecule sensing in the body or environment [44,48]. In studies by Badugu and co-workers, disposable over-the-counter contact lenses were soaked for 24 hours in solutions containing fluorescent boronic acids [48]. The fluorescence signal of intact lenses was measured in a standard quartz cuvette fitted with a bespoke lens holder with consideration of light scattering due to the curvature of the lens. The PET-based probe ANDBA and the charge transfer–operated sensor DDPBBA as well as related compounds were found to yield only negligible responses to glucose when impregnated into contact lenses. This discrepancy with their activity in aqueous solutions was attributed to the lower polarity and lower pH inside the lens. Only the quinolinium-based sensors, including BMOQBA, whose glucose-boronate ester has a lower pK_a (6.6) than those of the other probes tested (typically, 7.0 to 8.3), respond to glucose concentrations in the submillimolar range with a concentration-dependent decrease in fluorescence [43,44]. Since many diabetics wear contact lenses (their condition means that they require sight correction), these results, although preliminary, show that fluorescent molecules are ready to be integrated into innovative clinical solutions [43,49,50].

Subcutaneous implants of devices with a fluorescence output may have potential as reporters of in vivo glucose concentrations [41,51]. Single-walled carbon nanotubes (SWNTs) have been proposed as a fluorescent material around which to construct such a device [41]. Molcules can bind to the outer surface of SWNTs, and such events may modify the fluorescence in some way [36]. In principle, long emission wavelengths allow these responses to penetrate outside the body through the tissue. In a theoretical study to establish design rules for an implantable glucose monitor, a dialysis capillary containing a dextran-coated SWNT and a glucose sensing protein (concanavalin A) was evaluated with respect to optical properties, glucose response, and biofouling [36]. SWNTs have favorable tissue penetration and photostability compared to other fluorophores. In addition, the SWNT-based device yielded a linear response to glucose, which for up to 76 days was not compromised by biofouling. By comparison, the response of an electrochemical device was instantly distorted and produced an error of 20% after nearly 4 days. These model results, although theoretical, show that novel fluorescent nanomaterials may be suitable for use in the next generation of implantable in vivo sensors.

Reactive Oxygen Species Reactive oxygen species (ROSs) are a class of highly pro-oxidant species comprising both radicals and anions. Endogenously produced from molecular oxygen, they are essential to life, playing roles in many biological processes such as signal transduction [52] and smooth muscle relaxation [53]. When overabundant, under conditions collectively known as *oxidative stress*, ROSs are highly damaging, leading to oxidation of enzymes, lipid membranes [54], and DNA [55]. Unregulated oxidative damage to these and other biomolecules is immediately harmful in cases of sunburn [56] and reperfusion injury [57] and is implicit in aging processes [58] and the development of many diseases, including some cancers [59]. For these reasons, the study of ROSs is essential to further understanding of complex cell processes and identification of novel therapies. This places a high value on the development of sensitive and selective fluorescent probes for these small molecules.

Boronate Esters to Detect ROSs Many small-molecule fluorescent ROS probes have been reported in the literature which consist of a reduced form of a fluorophore. Upon redox reaction with a reactive small molecule, the fluorescence signal is switched on. Fluorescent probes based on (among others) fluoresceins, coumarins, and ethidium, to detect a variety of ROSs, have been described. These probes were reviewed comprehensively in 2005 [60].

Since that time, other small-molecule probes have appeared in the literature. In one such study, sensors for hydrogen peroxide (H_2O_2) were proposed from boronate esters [9]. Electron-deficient boranes are well known to react with H_2O_2, and this reaction was chosen to achieve selectivity for one particular ROS over others. The original sensor in this class was based around fluorescein, with two boronate esters incorporated at the 3- and 6-positions, locking the molecule into the nonfluorescent spirolactone form. Two further sensors were synthesized, each also with two boronate esters flanking a profluorescent core. Upon reaction with H_2O_2, these nonfluorescent molecules were designed to release the highly fluorescent 3,6-dihydroxyxanthone, fluorescein, and resorufin (PR1), each with different spectral properties (blue, green, and red emissions, respectively). Figure 6-5 shows the structure of PR1 and its reaction with H_2O_2, together with an example of a second-generation boronate probe (PG1).

In vitro studies revealed that reaction of these type II probes with H_2O_2 produces dramatic increases in fluorescence. PR1 undergoes a greater than 1000-fold fluoresence enhancement when treated with an excess of H_2O_2, with a detection limit as low as 100 nM. This probe was also highly selective for H_2O_2 over other physiologically relevant ROSs: superoxide (O_2^-, greater than $1000 \times$ selective), nitric oxide (NO, $200 \times$ more selective), hydroxyl radical ($OH^\cdot$, $200 \times$ more selective), and singlet oxygen ($^1O_2$, $15 \times$ more selective). The resorufin produced emits red fluorescence, thus avoiding the use of potentially cell-damaging

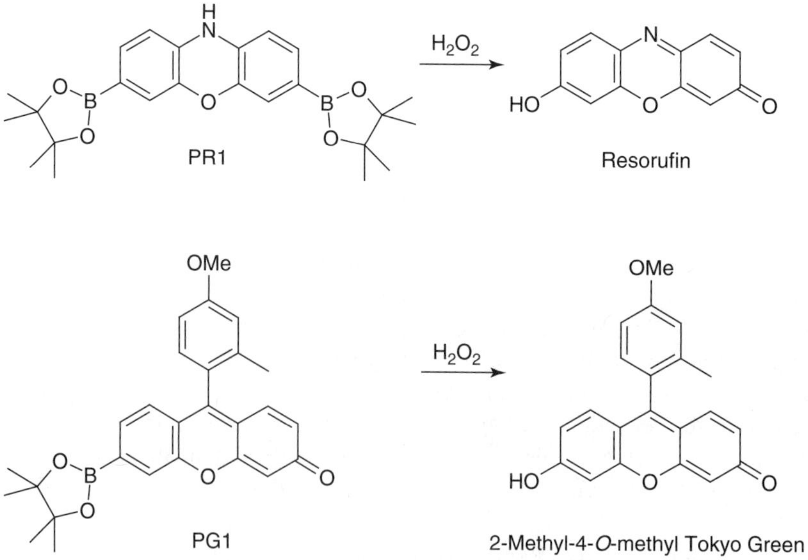

Figure 6-5   Generation of fluorescent products from reactions of PR1 and PG1 with H_2O_2.

shorter-wavelength irradiation and reducing background fluorescence. Another useful property of PR1 is membrane permeability, which provides easy access to H_2O_2 on the inside of cells. The imaging of these dyes in living HEK cells was demonstrated. PR1-loaded cells showed virtually no background fluorescence, and upon administration of 100 μM H_2O_2, red staining occurred. The other boronate probes gave similar responses including the xanthone derivative, which was imaged using two-photon excitation (the advantages of two-photon imaging include minimized photodamage and improved spatial resolution [61]). Exogenously elevated H_2O_2 levels (300 μM treatment) in living hippocampal rat neurons from primary culture were also imaged using these probes.

This first generation of diboronate esters was used successfully to artificial models of extreme oxidative stress. However, these probes were not sufficiently sensitive to detect the low levels of H_2O_2 involved in oxidative signaling [9,10]. Achieving this is an important milestone in deconvoluting complex signaling pathways: for example, the precise role of cysteine oxidation in PTP1B regulation [62,63]. Second-generation boronate esters were developed to provide this level of sensitivity. PG1 is a monoboronated Tokyo Green derivative that yields a brightly fluorescent product ($\lambda_{em} = 510$ nm) upon reaction with H_2O_2 in vitro [10]. This probe is more sensitive than PR1 to H_2O_2, in terms of rate of increase of fluorescence and absolute brightness, while retaining selectivity for H_2O_2 over other ROSs.

Monoboronate esters such as PG1 are also cell permeable, affording the opportunity to images changes in intracellular H_2O_2 concentrations in living cells. To demonstrate the improved sensitivity of the monoboronate esters, living HEK 293 cells were treated with PG1 (5 μM) and imaged after treatment with H_2O_2 (10 μM, earlier experiments were performed at 100 μM). The response was an increase in green fluorescence throughout the cell.

A more rigorous test of the usefulness of these probes was the imaging of physiological changes in H_2O_2 in response to epithelial growth factor (EGF) stimulation. Live A341 cells, chosen for their high expression of epithelial growth factor receptor (EGFR), were stimulated with EGF at a physiologically relevant dose (500 ng/mL). Cells loaded with PG1 gave brighter fluorescence (approximately twofold) relative to pre-stimulation levels. In a further development, inhibitors of EGFR kinase domain, phosphatidylinositol-3-OH kinase and NADPH oxidase were used to attenuate the fluorescence signal, in contrast to a nitric oxide synthase inhibitor, which had no effect. This result was reproduced in living rat hippocampal neurons, the first such visualization of H_2O_2 production in brain cells, exemplifying the use of small-molecule fluorescent probes such as PG1 for mapping molecular pathways to changing levels of specific ROSs.

In addition, a genetically encoded ratiometric sensor for H_2O_2 is now available. This probe, named HyPer, incorporates a YFP variant into the prokaryotic H_2O_2-sensing protein (OxyR) and was used successfully to monitor the formation of H_2O_2 in single mitochondria within a living cell [8].

Nitric Oxide Nitric oxide (NO) is an environment-polluting free radical and a known intracellular second messenger whose action is indicated in many clinical situations [64–66]. Fluorescent probes for NO may be useful tools to further our understanding of cell signaling and guide drug design. The design of a small-molecule sensor of nitric oxide was based on the weakly fluorescent aminofluorescein, and the observation of increased fluorescence upon reaction of the amine (e.g., to an amide) [67]. It was also known that N-nitrosylation (by NO) of 1,2-diamines gives rise to the formation of 1,2,3-triazoles [68]. Thus, when NO reacts with 1,2-diaminofluorescein (DAF), triazole formation increases the quantum yield dramatically, from 0.005 to 0.92 ($\lambda_{em} = 513$ nm) [69]. Of the considerable amount of chemistry performed on a fluorescein template, this reaction is a novel pathway to switching on fluorescence.

Of course, there is an inherent danger in that destructively sensing NO (type II) interferes with the very intracellular signals that one may be attempting to study. However, it has been proposed that 1,2-diaminofluoresceins in fact react with dinitrogen trioxide (N_2O_3), an intermediate formed when NO reacts with O_2. The latter reaction is the rate-determining step in the formation of fluorescent triazoles, which despite reducing the detection limit, offers the possibility that indirectly sensing NO minimizes interference with normal pathways. This fluorogenic reaction is also specific to NO versus higher nitrogen oxides such as nitrite (NO_2^-), nitrate (NO_3^-), and peroxynitrite ($ONOO^-$) and also reactive oxygen species such as superoxide (O_2^-) and H_2O_2.

In the same study, a diacetylated membrane-permeable diaminofluorescein (DAF2-DA, Fig. 6-6) was synthesized. It was shown that this compound (10 μM) easily enters living rat aorta cells and is cleaved by intracellular esterases to produce DAF2, which evenly distributes around the cells and localizes in the nucleus. Addition of an NO donor (NOC13) increased the fluorescence. When an NOS-inhibitor (L-NMMA) was added, fluorescence did not increase.

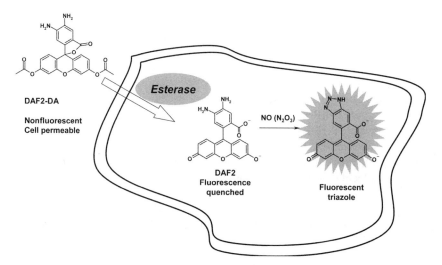

Figure 6-6 DAF2-DA, a cell-permeable diaminofluorescein, is hydrolyzed by intracellular esterases and forms a fluorescent triazole in a type II response to NO.

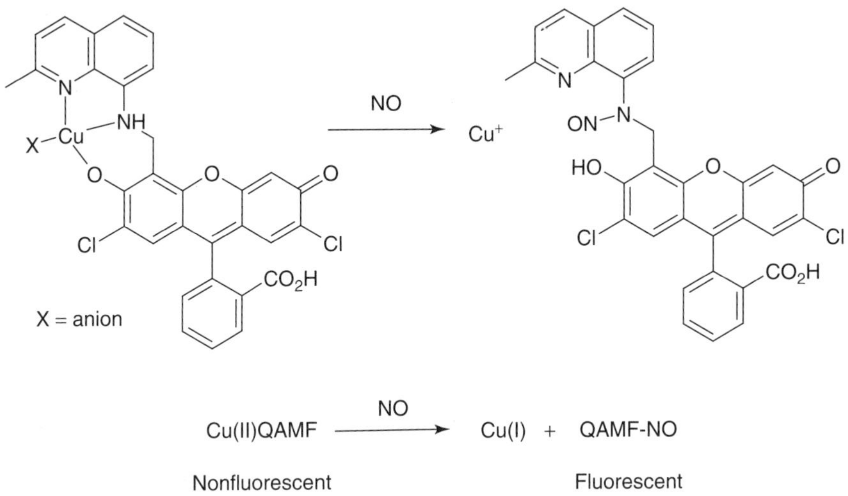

Figure 6-7 CuQAMF, a copper(II) complex of a fluorescein derivative, undergoes reductive nitrosation by nitric oxide. The loss of the copper ion from the complex results in an increase in fluorescence.

More recently, a class of molecules that senses NO itself (not a by-product of its presence) was discovered by Lim and co-workers [70,71]. They have shown that their fluorogenic copper(II) complexes react exclusively with NO to afford up to an 11-fold increase in fluorescence [72]. Figure 6-7 shows how one such complex (CuQAMF) undergoes reaction and fluorescence enhancement. NO is a powerful donor ligand of transition metals [73,74] and also reacts with secondary amines [75]. CuQAMF undergoes nitrosation at the amine and the copper(II) center is reduced to copper(I). This modified ligand, which has a low affinity for copper(I), is released from the complex. As such, the fluorescence is restored. This compound was imaged in live RAW 264.7 macrophages. The incorporation of an inorganic element in the sensing mechanism illustrates how all areas of chemistry may provide chemical biology solutions.

6.3.2 Sensing Fluorophores Generated from Diversity-Directed Approaches

The small-molecule fluorescent sensors described thus far were borne out of rational design and target-driven synthesis. Wam and Chang argue that this approach, although successful in the development of fluorescent probes, is ultimately limited in the volume of results it can achieve considering the efforts invested [76]. In the age of combinatorial chemistry, they propose that libraries of fluorescent compounds may be screened as diversity-directed sensors. Certainly this approach is routine in drug discovery, where the quest for small molecules showing selectivity toward and specific interactions with a given (or range of) target(s) is not dissimilar to the search for novel fluorescent sensors.

In principle, any library of fluorophores generated from combinatorial synthesis may be screened for responses to "biorelevant" analytes. In a screen of

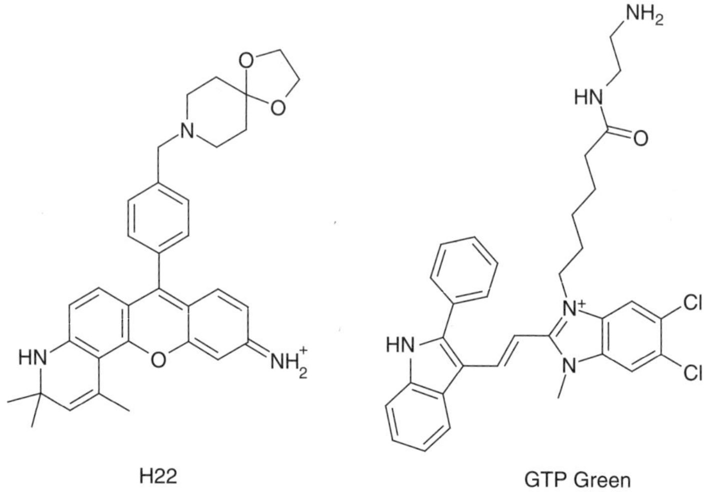

Figure 6-8 Lead sensor candidates, H22 and GTP Green, for glutathione and GTP, respectively, derived from diversity-directed sensor discovery approaches.

rosamine derivatives (built from unsymmetrical xanthones and aromatic Grignard reagents), a candidate sensor for glutathione (GSH) was discovered [77]. This molecule, H22 (Fig. 6-8), responded to 5 mM GSH (typical intracellular concentration) with an 11-fold increase in fluorescence intensity at 550 nm. Selectivity for GSH was observed over other thiol-containing analytes, notably dithiothreitol (DTT) and cysteine. Furthermore, the oxidized form of glutathione, GSSG, elicited no response from H22. These are interesting results since GSH is the most abundant cellular thiol and the redox potential of a cell is a function of the ratio of GSH to GSSH. H22 is cell permeable, and glutathione responses under a range of conditions could be imaged in living 3T3 cells. When H22-loaded cells were treated with α-lipoic acid (enhances GSH levels) and buthionine sulfoxide (BSO, a GSH synthesis inhibitor) the fluorescence levels were intensified and attenuated, respectively. α-Lipoic acid–induced increases could be reversed by addition of *N*-methylmaleimide (a thiol-reactive agent). Artificial manipulations of cell GSH levels are common protocols in cell biology experiments, and by monitoring redox levels in cells, probes such as H22 may aid these studies. The discovery of H22 offers a platform for rational design of sensitive glutathione probes and in general exemplifies the strategy of combinatorial synthesis for obtaining structurally diverse (and therefore potentially selective) fluorescent probes.

In another study, a library of potential fluorescent probes for nucleotides was built from simple condensation reactions of a benzimidazolium core with 96 aromatic aldehydes. Some rationale was introduced into this library—the fixed cationic charge of the benzimidazolium as a receptor for phosphate moieties; the two halves would be joined directly to allow binding events to influence the whole molecule through conjugation (and to minimize molecular weight, a lesson learned from drug discovery); two chlorine atoms afford relatively red-shifted fluorescence for favorable cell imaging properties.

In the initial screen, one compound (given the name GTP Green, Fig. 6-8) displayed fluorescence enhancement ("turn-on") upon GTP binding and was favorably photostable compared to other candidates [76]. Prior to the discovery of GTP Green, the only other GTP sensor (rationally designed) was a "turn-off" sensor [78]. GTP Green displayed a red-shifted 80-fold fluorescence enhancement ($\lambda_{em} = 540$ nm) in the presence of GTP and only $\leq$ two-fold increases with the other nucleotide triphosphates ATP, TTP, and UTP. This compound was also found to be selective for GTP over the mono- and diphosphate homologs.

In conclusion, coupled with a simple fluorescent screening process, the sheer number of compounds generated by combinatorial chemistry may be as powerful a tool as rational design for finding small fluorescent reporter molecules.

6.3.3   Genetically Encoded Probes

Genetically encoded probes belong mostly, but not always, to the group of ligand-binding (type I) sensors. As mentioned above, most of these are fusion constructs with fluorescent proteins [79,80]. Simple sensors feature a recognition domain for the small molecule of interest fused directly to a fluorescent protein. Upon expression in cells the construct is initially found in one cellular compartment. Elevation of the ligand concentration in one cellular compartment then leads to a translocation of the construct to another region of the cell. The change may be monitored both by a loss in fluorescence in the starting compartment and a gain in the detection compartment and offers the possibility of ratiometric analysis. The predominant applications are lipid sensors [81,82]. Changes in the lipid composition of a cellular membrane lead to the translocation of lipid-binding domains from the cytosol to this membrane, and vice versa. In fact, this method is currently the only way to measure changes in lipid concentration in living cells. Prominent examples are the C1 domain of protein kinase C, which recognizes diacylglycerol variants [83,84] and various pleckstrin homology (PH) domains that bind to several phosphoinositides with a varying degree of specificity [82].

A more sophisticated type of sensor employs two fluorophores that exhibit changes in fluorescence resonance energy transfer (FRET) upon binding of the analyte to the sensor unit [80]. When the emissions of the fluorophore pair exhibiting FRET are expressed as an intensity ratio, such measurements are largely independent of sensor concentration. FRET might also be observed when the energy of the donor is quenched by a nonfluorescent dye [85]. Since changes can only be derived from single fluorophore emission and are not ratiometric, this setup is less favorable when widefield or confocal microscopy is used but works very well when changes in the fluorescent lifetime (FLIM) are monitored [86].

Changes in FRET may be caused by a change in the distance of the two fluorophores or by alteration of the two dipole moments. Distance changes are particularly influential on the FRET efficiency around the Foerster radius R_0, and structural information on the sensor protein might permit intelligent guesses regarding the potential functionality of a sensor. Changes in dipole moments are very difficult to predict. Other factors that need to be considered are the sensitivity to ions and pH [87] and the tendency of many fluorescent proteins to form dimers or sometimes tetramers [27]. The dimerization of fluorescent proteins is usually regarded as negative [88]. However, in certain cases the fluorophore–fluorophore interaction might actually help improve sensor performance [89,90].

FRET Sensors Initially, FRET was used to determine distances within proteins [91]. The first useful FRET sensor based on proteins dates from times when fluorescent proteins were not yet cloned. Accordingly, this sensor for cyclic adenosine 3′,5′-monophosphate (cAMP) levels was equipped with a pair of small molecule dyes, fluorescein and rhodamine [92]. The design was based on single labeled catalytic (C) and regulatory (R) subunits forming the holoenzyme of the cAMP-dependent protein kinase A (PKA). Due to its labeling locations, the probe was termed FlCRhR. Upon binding of cAMP the two subunits separated to indefinite distance, resulting in a total loss of FRET which was measured by ratioing the emission intensities of both fluorophores. When the holoenzyme was injected into living cells, agonist-induced increases and decreases in intracellular cAMP levels could be monitored for the first time in real time [92]. Among many other applications, the probe was particularly useful when applied to neurons for measuring diffusion rates of cAMP in single dendrites [93,94]. A similar sensor was prepared for cyclic guanosine 3′,5′-monophosphate (cGMP), based on the cGMP-dependent protein kinase (PKG) [95]. Although structurally similar to PKA, this enzyme consists of a single peptide chain. Therefore, binding of cGMP resulted in a conformational change rather than complete separation of the protein subunits. In this molecule, the affinity of the intramolecular binding of catalytic and regulatory subunit needed to be adjusted to provide a useful dynamic range of the sensor.

In the meantime, genetically encoded sensors based on fluorescent fusion proteins of *Aequorea victoria* (jellyfish) or from anthozoa (corals) families have been taking stage. As mentioned above, this is fostered by very successful efforts to increase the color palette of fluorescent proteins through specific mutations of the fluorophore pocket [23]. In the course of these efforts, other properties of fluorescent proteins, such as ion sensitivity or dimerization tendencies, have been eliminated or standardized. The result is a now significant set of FRET sensors that can be readily expressed in cultured cells to monitor enzyme activities or the abundance of small molecules. In fact, the majority of probes report enzymatic activity. In the following section we describe the few genetically encoded sensors able to detect changes in the concentration of small molecules.

A sensor for cAMP requires a conformational change of the protein upon binding of the ligand. This is provided by the two major cAMP-binding proteins: the regulatory subunit of the cAMP-dependent protein kinase (PKA) or by EPAC (exchange protein directly activated by cAMP). Both proteins were equipped with the standard FRET pair EYFP and ECFP attached to the N- and C-termini, respectively [96]. When expressed in hippocampal neurons or peritoneal macrophages, the sensor responded to agonist-induced changes in cAMP concentration with an approximately 15% change in the ECFP/EYFP emission ratio. The sensor based on a single cAMP-binding site mutant of the regulatory subunit of PKA exhibited slower binding kinetics and a smaller FRET change than those for the EPAC-based sensor [96]. Very recently, similar sensors aimed to measure intracellular cGMP levels were presented. These are based on the cGMP-binding domain B from cGMP-dependent protein kinase I or on the regulatory guanine nucleotide activating factor domain from a cGMP-regulated phosphodiesterase, respectively. The kinase-based sensor responded to elevated cGMP levels with a decrease in FRET, the cPDE-derived sensor with an increase and with largely increased dynamics [97].

A molecule of particular interest is the excitatory neurotransmitter glutamate. Recently, a FRET sensor was introduced that permitted the quantitative measurement of cytosolic and synaptic glutamate levels in real time [98]. The sensor is based on the glutamate/aspartate binding protein ybeJ from *Escherichia coli*. Allosteric binding of glutamate to the mature ybeJ protein led to a drop in the 535/480 nm emission ratio of the flanking fluorophores ECFP and Venus. The original sensor, called FLIPE, had a K_d value of 630 nM. A series of mutants was prepared to report over other physiologically significant glutamate concentration ranges up to millimolar concentrations. In addition, for extracellular applications these sensors were fused to the transmembrane region of the platelet-derived growth factor. These constructs expressed well in PC12 cells and hippocampal neurons. Electrical stimulation of neurons resulted in a drop in emission ratio due to the release of glutamate from the synapses [98].

Similar types of FRET sensors were previously reported for sugars such as maltose, ribose, and glucose [99–102]. All of them are based on bacterial periplasmic-binding proteins. These and especially the glucose sensor FLIPglu should find wide application. Interesting questions regarding the glucose homeostasis of hepatocytes (HepG2 cells) have already been addressed [103]. Most of these applications, especially those in plant cells, have recently been reviewed [104].

A typical type II sensor is HyPer, the above-mentioned sensor for hydrogen peroxide. Upon reaction of H_2O_2 with one of two cysteines of the regulatory domain of OxyR, a disulfide bond is formed, resulting in a dramatic conformational change in the protein. The latter permits an increase in the excitation maximum of the fused circular permutated YFP from 420 to 500 nm. Therefore, the probe can be used ratiometrically by exciting at both wavelength and detecting emission at 516 nm. Oxidants other than H_2O_2 were ineffective. The probe was used successfully in nerve growth factor–stimulated PC12 cells [8].

Translocating Sensors in Living Cells A second approach for monitoring small molecules with genetically encoded sensors is the use of translocation properties of proteins or protein domains inside cells [105]. One advantage compared to FRET sensors is the fact that only a single fusion of the translocating protein with a fluorophore, mostly a fluorescent protein, is required. Many proteins change location upon binding to a ligand. The translocation usually occurs from a membrane to the cytosol, or vice versa. Location changes, for instance of transcription factors, in and from the nucleus are also easily detectable [106,107]. If the locations are sufficiently defined, measuring the ratio of fluorescent intensities of the two locations before and after translocation is possible. As with FRET probes, this procedure is suitable for quantifications in a manner that is, within limits, independent of the sensor concentration.

Most of the sensors published so far report on changes in the lipid composition of cell membranes. Initially, various pleckstrin homology (PH) domains were used to monitor PIP_2 breakdown or the formation of PIP_3 in the plasma membrane [81,108,109]. For PIP_2 the PH domain of phospholipase C $\delta1$ (PLC$\delta1$) was shown to be particularly specific and acting with favorable sensitivity [81]. Interestingly, the results showed that PIP_2 could be detected exclusively in the plasma membrane but not in microsomal membranes, although the abundance of PIP_2 in these membranes was shown previously. This finding suggests that other PIP_2-binding proteins with higher affinity for PIP_2 and specifically residing on microsomal membranes are outcompeting PH domains. For the detection of PIP_3 levels the PH domain of Bruton's tyrosine kinase (BTK) is most commonly used [110]. Other lipids that may be monitored dynamically by translocating domains include diacylglycerol (DAG) via C1 domains or calcium-insensitive PKC isoforms [83,84]. Similarly, phosphatidic acid (PA) is recognizable [111,112]. Both lipids are formed transiently after the stimulation of phospholipase D and C, respectively. In addition, PA is rapidly dephosphorylated to DAG, making the simultaneous use of a sensor for each lipid a particularly valuable goal in the future.

Other translocating reporters are sensitive to changes in calcium concentrations since binding of calcium induces a conformational change in the binding domain that allows the recognition of phospholipids in the plasma membrane. The most commonly used example is the calcium-binding C2 domain of protein kinase C. In some cases the entire PKC enzyme was labeled and used as a sensor [113]. Annexins also translocate to the plasma membrane in a calcium-dependent manner. Here it has been shown that the positively charged calcium ions are involved directly in the interaction with negatively charged phospholipids [114]. Interestingly, annexin A4 translocates much slower to the inner nuclear membrane than to the plasma membrane, even when the calcium concentration is identical in both compartments [115]. This implies that the phospholipid composition of nuclear and plasma membrane are different and that this difference may be measured in the future by using annexin fusion proteins, provided that the specificity of annexin A4 for lipids can be determined in living cells.

In summary, translocating probes are the major tool for monitoring changes in the lipid composition in living cells. It is therefore beneficial to identify more specific lipid-binding domains for other lipids. Caution has to be taken to be aware that results might be blurred by high-affinity binding of endogenous lipid-binding proteins that prevent recognition of the sensor molecule.

6.4 CONCLUSIONS

Although a significant number of sensors are available for measuring, for instance, phosphorylation events, sensors to measure changes in small-molecule concentrations or to determine absolute concentrations of small molecules in living cells are still scarce compared to the number of relevant molecules. This needs to change if we want to understand complex processes in cell biology and if serious attempts in systems biology are ultimately to become successful. To increase the number of sensors available, all types of reporters need to be put forward. Genetically encoded sensors will be easier to make but have the disadvantage of usually affording less signal, an important factor in cell-based screening. Small-molecule sensors do not require transfection and can therefore be used in native cells and tissue provided that they are membrane-permeant. Unfortunately, few small molecule–based sensors are known to be able to measure concentrations of anything but ions.

REFERENCES

1. Tsien, R. Y. (1989). *Methods Cell Biol.*, *30*, 127.
2. Hasan, M. T., Friedrich, R. W., Euler, T., et al. (2004). *PloS Biol.*, *2*, 763.
3. Domaille, D. W., Que, E. L., Chang, C. J. (2008). *Nat. Chem. Biol.*, *4*, 168.
4. Barker, S. L. R., Clark, H. A., Swallen, S. F., Kopelman, R., Tsang, A. W., Swanson, J. A. (1999). *Anal. Chem.*, *71*, 1767.
5. Barker, S. L. R., Zhao, Y. D., Marletta, M. A., Kopelman, R. (1999). *Anal. Chem.*, *71*, 2071.
6. Viguier, R. F. H., Hulme, A. N. (2006). *J. Am. Chem. Soc.*, *128*, 11370.
7. Dale, T. J., Rebek, J., Jr. (2006). *J. Am. Chem. Soc.*, *128*, 4500.
8. Belousov, V. V., Fradkov, A. F., Lukyanov, K. A., et al. (2006). *Nat. Methods*, *3*, 281.
9. Miller, E. W., Albers, A. E., Pralle, A., Isacoff, E. Y., Chang, C. J. (2005). *J. Am. Chem. Soc.*, *127*, 16652.
10. Miller, E. W., Tulyanthan, O., Isacoff, E. Y., Chang, C. J. (2007). *Nat. Chem. Biol.*, *3*, 263.
11. Lavis, L. D., Raines, R. T. (2008). *ACS Chem. Biol.*, *3*, 142.
12. Katerinopoulos, H. E. (2004). *Curr. Pharmaceut. Des.*, *10*, 3835.
13. Woodroofe, C. C., Lim, M. H., Bu, W. M., Lippard, S. J. (2005). *Tetrahedron*, *61*, 3097.
14. Sun, W. C., Gee, K. R., Klaubert, D. H., Haugland, R. P. (1997). *J. Org. Chem.*, *62*, 6469.
15. Zhu, Z. H. (1995). *Dyes Pigm.*, *27*, 77.
16. Loudet, A., Burgess, K. (2007). *Chem. Rev.*, *107*, 4891.
17. Tsukube, H., Yano, K., Ishida, A., Shinoda, S. (2007). *Chem. Lett.*, *36*, 554.
18. Chattopadhyay, N., Mallick, A., Sengupta, S. (2006). *J. Photochem. Photobiol. A*, *177*, 55.

19. Hagen, V., Dekowski, B., Kotzur, N., et al. (2008). *Chem. Eur. J.*, *14*, 1621.

20. Komatsu, K., Urano, Y., Kojima, H., Nagano, T. (2007). *J. Am. Chem. Soc.*, *129*, 13447.

21. Sivakumar, K., Xie, F., Cash, B. M., Long, S., Barnhill, H. N., Wang, Q. (2004). *Org. Lett.*, *6*, 4603.

22. Hsu, T. L., Hanson, S. R., Kishikawa, K., Wang, S. K., Sawa, M., Wong, C. H. (2007). *Proc. Natl. Acad. Sci. USA*, *104*, 2614.

23. Shaner, N. C., Steinbach, P. A., Tsien, R. Y. (2005). *Nat. Methods*, *2*, 905.

24. Brejc, K., Sixma, T. K., Kitts, P. A., et al. (1997). *Proc. Natl. Acad. Sci. USA*, *94*, 2306.

25. Zhang, L. P., Patel, H. N., Lappe, J. W., Wachter, R. M. (2006). *J. Am. Chem. Soc.*, *128*, 4766.

26. Ai, H. W., Shaner, N. C., Cheng, Z. H., Tsien, R. Y., Campbell, R. E. (2007). *Biochemistry*, *46*, 5904.

27. Chudakov, D. M., Lukyanov, S., Lukyanov, K. A. (2005). *Trends Biotechnol.*, *23*, 605.

28. Shaner, N., Campbell, R. E., Steinbach, P. A., G. B. N. P., Palmer, A. E., Tsien, R. Y. (2004). *Nat. Biotechnol.*, *22*, 1567.

29. Shu, X. K., Shaner, N. C., Yarbrough, C. A., Tsien, R. Y., Remington, S. J. (2006). *Biochemistry*, *45*, 9639.

30. Baird, G. S., Zacharias, D. A., Tsien, R. Y. (1999). *Proc. Natl. Acad. Sci. USA*, *96*, 11241.

31. Aryal, B. P., Benson, D. E. (2006). *J. Am. Chem. Soc.*, *128*, 15986.

32. Cordes, D. B., Gamsey, S., Singaram, B. (2006). *Angew. Chem. Int. Ed.*, *45*, 3829.

33. Wang, X. J., Ruedas-Rama, M. J., Hall, E. A. H. (2007). *Anal. Lett.*, *40*, 1497.

34. Medintz, I. L., Uyeda, H. T., Goldman, E. R., Mattoussi, H. (2005). *Nat. Mater.*, *4*, 435.

35. Cherukuri, P., Bachilo, S. M., Litovsky, S. H., Weisman, R. B. (2004). *J. Am. Chem. Soc.*, *126*, 15638.

36. Barone, P. W., Parker, R. S., Strano, M. S. (2005). *Anal. Chem.*, *77*, 7556.

37. Ogawa, Y., Yamamura, Y., Ando, A., et al. (2000). *Nat. Select. Synth. Toxins*, *745*, 333.

38. Fleming, C. D., Edwards, C. C., Kirby, S. D., et al. (2007). *Biochemistry*, *46*, 5063.

39. Springsteen, G., Wang, B. H. (2002). *Tetrahedron*, *58*, 5291.

40. Ni, W. J., Fang, H., Springsteen, G., Wang, B. H. (2004). *J. Org. Chem.*, *69*, 1999.

41. Pickup, J. C., Hussain, F., Evans, N. D., Rolinski, O. J., Birch, D. J. S. (2005). *Biosens. Bioelectron.*, *20*, 2555.

42. Zhao, J., Davidson, M. G., Mahon, M. F., Kociok-Kohn, G., James, T. D. (2004). *J. Am. Chem. Soc.*, *126*, 16179.

43. Badugu, R., Lakowicz, J. R., Geddes, C. D. (2005). *Bioorg. Med. Chem.*, *13*, 113.

44. Badugu, R., Lakowicz, J. R., Geddes, C. D. (2005). *Talanta*, *65*, 762.

45. Badugu, R., Lakowicz, J. R., Geddes, C. D. (2006). *Dyes Pigm.*, *68*, 159.

46. Nakata, E., Koshi, Y., Koga, E., Katayama, Y., Hamachi, I. (2005). *J. Am. Chem. Soc.*, *127*, 13253.

47. Badugu, R., Lakowicz, J. R., Geddes, C. D. (2005) *Curr. Opin. Biotechnol.*, *16*, 100.

48. Badugu, R., Lakowicz, J. R., Geddes, C. D. (2003). *J. Fluoresc.*, *13*, 371.

49. Badugu, R., Lakowicz, J. R., Geddes, C. D. (2004). *J. Fluoresc.*, *14*, 617.

50. Badugu, R., Lakowicz, J. R., Geddes, C. D. (2004). *Analyst*, *129*, 516.

51. Wilson, G. S., Gifford, R. (2005). *Biosens. Bioelectron.*, *20*, 2388.

52. Suzuki, Y. J., Forman, H. J., Sevanian, A. (1997). *Free Radical Biol. Med.*, *22*, 269.

53. Wilcox, C. S. (2005). *Am. J. Physiol. Regul. Integr. Comp. Physiol.*, *289*, R913.

54. Azzi, A., Davies, K. J. A., Kelly, F. (2004). *FEBS Lett.*, *558*, 3.

55. Bruskov, V. I., Malakhova, L. V., Masalimov, Z. K., Chernikov, A. V. (2002). *Nucleic Acids Res.*, *30*, 1354.

56. Russo, P. A. J., Halliday, G. M. (2006). *Br. J. Dermatol.*, *155*, 408.

57. Docherty, J. C., Kuzio, B., Silvester, J. A., Bowes, J., Thiemermann, C. (1999). *Br. J. Pharmacol.*, *127*, 1518.

58. Beckman, K. B., Ames, B. N. (1998). *Physiol. Rev.*, *78*, 547.

59. Waris, G., Ahsan, H. (2006). *J. Carcinogen.*, *5*, 14.

60. Gomes, A., Fernandes, E., Lima, J. L. (2005). *J. Biochem. Biophys. Methods*, *65*, 45.

61. Oheim, M., Michael, D. J., Geisbauer, M., Madsen, D., Chow, R. H. (2006). *Adv. Drug Deliv. Rev.*, *58*, 788.

62. van Montfort, R. L. M., Congreve, M., Tisi, D., Carr, R., Jhoti, H. (2003). *Nature*, *423*, 773.

63. Salmeen, A., Andersen, J. N., Myers, M. P. et al. (2003). *Nature*, *423*, 769.

64. Moncada, S., Higgs, A. (1993). *N. Engl. J. Med.*, *329*, 2002.

65. Nathan, C., Xie, Q. W. (1994). *Cell*, *78*, 915.

66. Broillet, M. C., Firestein, S. (1996). *J. Neurobiol.*, *30*, 49.

67. Munkholm, C., Parkinson, D. R., Walt, D. R. (1990). *J. Am. Chem. Soc.*, *112*, 2608.

68. Nagano, T., Takizawa, H., Hirobe, M. (1995). *Tetrahedron Lett.*, *36*, 8239.

69. Kojima, H., Nakatsubo, N., Kikuchi, K., et al. (1998). *Anal. Chem.*, *70*, 2446.

70. Lim, M. H., Kuang, C., Lippard, S. J. (2006). *ChemBioChem*, *7*, 1571.

71. Lim, M. H., Lippard, S. J. (2006). *Inorg. Chem.*, *45*, 8980.

72. Lim, M. H., Xu, D., Lippard, S. J. (2006). *Nat. Chem. Biol.*, *2*, 375.

73. Hayton, T. W., Legzdins, P., Sharp, W. B. (2002). *Chem. Rev.*, *102*, 935.

74. Hilderbrand, S. A., Lim, M. H., Lippard, S. J. (2004). *J. Am. Chem. Soc.*, *126*, 4972.

75. Tsuge, K., DeRosa, F., Lim, M. D., Ford, P. C. (2004). *J. Am. Chem. Soc.*, *126*, 6564.

76. Wang, S., Chang, Y. T. (2006). *J. Am. Chem. Soc.*, *128*, 10380.

77. Ahn, Y. H., Lee, J. S., Chang, Y. T. (2007). *J. Am. Chem. Soc.*, *129*, 4510.

78. Kwon, J. Y., Singh, N. J., Kim, H. N., Kim, S. K., Kim, K. S., Yoon, J. Y. (2004). *J. Am. Chem. Soc.*, *126*, 8892.

79. Miyawaki, A. (2003). *Dev. Cell*, *4*, 295.

80. Zhang, J., Campbell, R. E., Ting, A. Y., Tsien, R. Y. (2002). *Nat. Rev. Mol. Cell Biol.*, *3*, 906.

81. van der Wal, J., Habets, R., Varnai, P., Balla, T., Jalink, K. (2001). *J. Biol. Chem.*, *276*, 15337.

82. Varnai, P., Lin, X., Lee, S. B., et al. (2002). *J. Biol. Chem.*, *277*, 27412.

83. Oancea, E., Meyer, T. (1998). *Cell*, *95*, 307.

84. Oancea, E., Teruel, M. N., Quest, A. F. G., Meyer, T. (1998). *J. Cell Biol.*, *140*, 485.

85. Rose, T. M., Prestwich, G. D. (2006). *ACS Chem. Biol.*, *1*, 83.

86. Ganesan, S., Ameer-beg, S. M., Ng, T. T. C., Vojnovic, B., Wouters, F. S. (2006). *Proc. Natl. Acad. Sci. USA*, *103*, 4089.

87. Griesbeck, O., Baird, G. S., Campbell, R. E., Zacharias, D. A., Tsien, R. Y. (2001). *J. Biol. Chem.*, *276*, 29188.

88. Zacharias, D. (2002). *Sci. STKE*, PE23.

89. Vinkenborg, J. L., Evers, T. H., Reulen, S. W. A., Meijer, E. W., Merkx, M. (2007). *ChemBioChem*, *8*, 1119.

90. Jost, C. A., Reither, G., Hoffmann, C., Schultz, C. (2008). *ChemBioChem*, *9*, 1379.

91. Stryer, L. (1978). *Annu. Rev. Biochem.*, *47*, 819.

92. Adams, S. R., Harootunian, A. T., Buechler, Y. J., Taylor, S. S., Tsien, R. Y. (1991). *Nature*, *349*, 694.

93. Adams, S. R., Bacskai, B. J., Hochner, B., et al. (1993). *Jpn. J. Physiol.*, *43*, S91.

94. Bacskai, B. J., Hochner, B., Mahautsmith, M., et al. (1993). *Science*, *260*, 222.

95. Honda, A., Adams, S. R., Sawyer, C. L., et al. (2001). *Proc. Natl. Acad. Sci. USA*, *98*, 2437.

96. Nikolaev, V. O., Bunemann, M., Hein, L., Hannawacker, A., Lohse, M. J. (2004). *J. Biol. Chem.*, *279*, 37215.

97. Nikolaev, V., Gambaryan, S., Lohse, M. J. (2006). *Nat. Methods*, *3*, 23.

98. Okumoto, S., Looger, L. L., Micheva, K. D., Reimer, R. J., Smith, S. J., Frommer, W. B. (2005). *Proc. Natl. Acad. Sci. USA*, *102*, 8740.

99. Fehr, M., Frommer, W. B., Lalonde, S. (2002). *Proc. Natl. Acad. Sci. USA*, *99*, 9846.

100. Fehr, M., Lalonde, S., Lager, I., Wolff, M. W., Frommer, W. B. (2003). *J. Biol. Chem.*, *278*, 19127.

101. Lager, I., Fehr, M., Frommer, W. B., Lalonde, S. W. (2003). *FEBS Lett.*, *553*, 85.

102. Fehr, M., Okumoto, S., Deuschle, K., et al. (2005). *Biochem. Soc. Trans.*, *33*, 287.

103. Fehr, M., Takanaga, H., Ehrhardt, D. W., Frommer, W. B. (2005). *Mol. Cell. Biol.*, *25*, 11102.

104. Medintz, I. L. (2006). *Trends Biotechnol.*, *24*, 539.

105. Teruel, M. N., Meyer, T. (2000). *Cell*, *103*, 181.

106. Birbach, A., Bailey, S. T., Ghosh, S., Schmid, J. A. (2004). *J. Cell Sci.*, *117*, 3615.

107. Pranada, A. L., Metz, S., Herrmann, A., Heinrich, P. C., Muller-Newen, G. (2004). *J. Biol. Chem.*, *279*, 15114.

108. Balla, T., Bondeva, T., Varnai, P. (2000). *Trends Pharmacol. Sci.*, *21*, 238.

109. Venkatcswarlu, K., Gunn-Moore, F., Oatey, P. B., Tavare, J. M., Cullen, P. J. (1998). *Biochem. J.*, *335*, 139.

110. Vlahos, C., Matter, W., Hui, K., Brown, R. (1994). *J. Biol. Chem.*, *269*, 5241.

111. Loewen, C. J. R., Gaspar, M. L., Jesch, S. A., et al. (2004). *Science*, *304*, 1644.

112. Rizzo, M. A., Shome, K., Watkins, S. C., Romero, G. (2000). *J. Biol. Chem.*, *275*, 23911.

113. Reither, G., Schaefer, M., Lipp, P. (2006). *J. Cell Biol.*, *174*, 521.

114. Gerke, V., Creutz, C. E., Moss, S. E. (2005). *Nat. Rev. Mol. Cell Biol.*, *6*, 449.

115. Piljic, A., Schultz, C. (2006). *Mol. Biol. Cell*, *17*, 3318.

7

DEVELOPMENT OF SMALL-MOLECULE LIGANDS AND INHIBITORS

LEONID L. CHEPELEV, NIKOLAI L. CHEPELEV, HOOMAN SHADNIA, WILLIAM G. WILLMORE, JAMES S. WRIGHT, AND MICHEL DUMONTIER

Departments of Biology and Chemistry, Carleton University, Ottawa, Ontario, Canada

7.1 INTRODUCTION

Although the complex nature of cellular metabolic and regulatory mechanisms provides ample molecular classes for the alteration of cellular genetic and chemical states, the most widely applied and best understood class is that of small molecules (SMs). It takes a molecular interaction to proliferate life; and where an interaction occurs a SM intervention may be devised to modify it. Although we may not think much of such interventions in our daily activities, they have a strong impact on our lives, both directly and indirectly. Such SM interventions are vital in two fields, medicine and agriculture, where, among other things, they are applied to destroy pathogens, relieve pain, supplement defective metabolic pathways, alter abnormal regulatory networks, and alter growth. Even though immense progress has been made in the application of SM treatments to current problems, a large number of unaddressed questions exist. Additionally, more and more deficiencies of extant SM agents, such as subpopulation-specific side effects

Protein Targeting with Small Molecules: Chemical Biology Techniques and Applications,
Edited by Hiroyuki Osada
Copyright © 2009 John Wiley & Sons, Inc.

of SM drugs and evolving resistance of pathogens to pesticides and antibiotics and other SMs, are being discovered. These circumstances necessitate further development of novel SM ligands and inhibitors and improvement of the existing solutions.

The traditional SM ligand development relies on expensive, large-scale screens with the intent of probing chemical space, and thus involving potentially millions of compounds, to identify an active compound which is subsequently tuned to improve its activity and specificity. In this approach, appropriate library design, selection, and synthesis play a role that is as important as that of the experimental techniques used to survey the interactions of the compounds populating the library. Since we consider the high-throughput quantitative technologies used for SM interaction profiling elsewhere in this work, here we consider only the experimental techniques capable of providing mechanistic information on the mode of SM–target interaction. Unfortunately, due to the cost of the equipment and materials involved in massive chemical library screens, they are carried out primarily by well-funded corporations and institutions. However, the emergence of techniques capable of providing mechanistic information on the mode of binding of native ligands to target proteins, such as x-ray crystallography, along with the development and consistent improvement of computers, has prompted rational ligand design revolution. With this approach, provided that starting experimental data are available and setting aside costs of education, SM ligand development is affordable to the vast majority of citizens of first-world countries.

We believe that it is a combination of the knowledge-based rational design and traditional approaches that provides the most workable route to ligand discovery, design, and development. Since the field of SM ligand and inhibitor development is quite broad and draws on a formidable number of disciplines, in this chapter we provide only a general overview of the theory involved and highlight some of the many interesting directions of the field.

7.2 THEORETICAL BACKGROUND

When dealing with the development of a SM ligand or an inhibitor, it is necessary to consider the processes targeted by the ligand proposed. Some of the most relevant cellular processes include protein–protein interactions in signaling networks, protein–nucleic acid interactions in regulation of transcription and translation, as well as protein–SM interactions that are most relevant to metabolism through enzyme–SM interactions and initial stages of signal transduction through receptor–SM interactions. Although a given ligand may target any of the three interaction classes, it may be argued that the latter class will inevitably be encountered by most extraneous SMs, through either nonspecific binding or xenobiotic metabolism. Furthermore, enzyme inhibitors and receptor blockers have proven to have a rather high level of commercial importance. In the pharmaceutical industry alone, the current market for these compounds has been estimated at approximately $175 billion, with an approximate 7.8%

annual market growth rate [1]. The use of enzyme inhibitors for other commercially important applications, such as pesticides and industrial fungicides, should also not be underestimated. Since the basic mechanisms and relevant factors in enzyme–SM interactions can be applied to the design of ligands for other interaction classes, let us first consider very briefly the theory of enzyme catalysis and mechanisms of enzyme inhibition. For a more in-depth treatment, readers are directed to resources dedicated exclusively to this topic [2,3].

7.2.1 Basic Enzyme Catalysis Theory

The biological catalysts known as *enzymes* are indispensable for life as we know it, since they impart directionality and feasibility to an otherwise chaotic pool of chemical reactions sometimes catalyzed by abiotic factors and often made possible only through energy input from the environment. By bringing substrate(s) of a given reaction together, thus changing the effective molecularity of a reaction, altering the original mechanism of a reaction (e.g., by breaking it up into multiple steps or involving energy carriers such as ATP to "pay" for the thermodynamic "expense" of a reaction), or changing the energies of reactants through the application of intermolecular forces that lead to geometric distortions in the substrate, among other things, enzymes are capable of lowering reaction energy barriers, thus enabling or speeding up otherwise impossible or slow reactions. The precise tactics employed by enzymes to that end are as varied as enzymes themselves and are left to be enumerated by more comprehensive reviews [3]. A simplified energy profile for a simple enzymatic reaction where enzyme E catalyzes the conversion substrate S to product P, which depicts the energy changes of the reaction system as a function of reaction progress, is shown (Fig. 7-1).

Here ES^* is the transition state (TS) for the binding of S to E, EP^* is the TS for the dissociation of P from E, and $(ES/EP)^*$ is the TS for the actual catalytic step, conversion of ES complex to the EP complex. Thus, there exists a certain activation energy barrier for every step of the catalytic mechanism. The step most frequently considered, however, is that of the transformation of the substrate to product, to which an energy barrier of ΔG^* corresponds. In an uncatalyzed reaction, the energy barrier for the conversion of S to P is much greater. Lowering the reaction energy barrier has a direct bearing on the kinetics of the reaction, accelerating it.

Michaelis–Menten (MM) kinetics, although not perfect and in many cases inadequate for a number of reasons, have been used widely to describe the dynamics of catalysis of most enzymes. Databases such as BRENDA [5], containing sets of MM kinetic constants for most enzymes in a wide array of organisms, are widely available. The basic MM equation (Fig. 7-2) is derived for a simple system as shown:

$$\text{E} + \text{S} \underset{k_{-1}}{\overset{k_1}{\rightleftharpoons}} (\text{ES}) \xrightarrow{k_2} \text{E} + \text{P} \qquad (1)$$

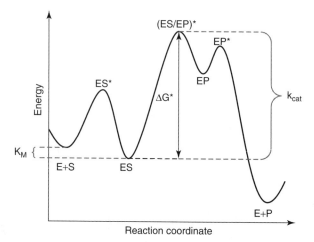

Figure 7-1 Simplified energy profile for an enzymatic reaction. Shown are the kinetic constants which can potentially be calculated from the free-energy changes indicated by the dashed lines [see equations (3) and (4)]. The reaction coordinate can be interpreted as a path on a hyperdimensional potential energy surface that the reaction is most likely to take in order to proceed to completion. (From ref. 4.)

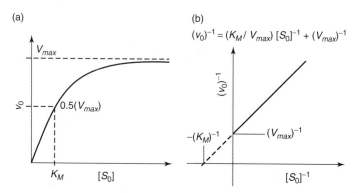

Figure 7-2 Graphic analysis of MM kinetics: (a) an initial velocity–initial substrate concentration plot; (b) a Lineweaver–Burk plot, along with the equation of the resulting line. The two parameters to describe MM kinetics can be obtained from the intercepts.

Here k_1 and k_{-1} are the forward and reverse reaction rate constants for substrate binding, while the process of conversion of the enzyme–substrate complex ES is assumed to be irreversible, thus having only one reaction rate constant, k_2, for the forward reaction. The MM kinetic equation is derived assuming, among other things, that the reaction is operating away from the thermodynamic equilibrium and that the concentration of the ES complex is constant. Skipping the derivation, the final MM equation can be written in the familiar form

$$\frac{dP}{dt} = v = \frac{V_{max}[S]}{K_M + [S]} \tag{2}$$

In other words, the rate of reaction is proportional to the product of the maximal reaction velocity, V_{max}, and the initial substrate concentration, divided by the sum of the MM constant, K_M, and the initial substrate concentration. Under the MM assumptions, equation (3) defines the MM constant, which can be determined computationally given the magnitude of k_1 and k_{-1}, which can be approximated using the free-energy difference between the ES complex and the enzyme and substrate alone (Fig. 7-1), as well as k_2, which can be determined by computing the energy barrier to the biochemical transformation catalyzed by a given enzyme, as in

$$K_M = \frac{[E][S]}{[ES]} = \frac{k_{-1} + k_2}{k_1} \tag{3}$$

On the other hand, V_{max} can be related to the reaction rate constant in equation (1), k_2 (also known as k_{cat}), and the activation energy barrier ΔG^*, through the equation

$$V_{max} = k_2[E_0] = [E_0]\frac{k_B T}{h} e^{-\Delta G^*/RT} \tag{4}$$

where k_B is the Boltzmann constant, T is temperature, R is the gas constant, and h is Planck's constant. The power of equations (3) and (4) lies in the fact that a theoretically computed reaction energy profile can, in principle, provide the kinetic constants for a given reaction, thus completely describing its dynamics. On the other hand, kinetic data can be used to deduce the energies associated with the fundamental reaction steps.

One way to look at the kinetics of a MM-type interaction is by plotting the initial reaction velocity over initial substrate concentrations (Fig. 7-2a). This type of graphic analysis allows for the determination of the relevant kinetic constants directly from the curve: The asymptote of the resultant curve as substrate concentration approaches infinity is of the form $y = V_{max}$, while the substrate concentration at which half the maximal reaction velocity is reached corresponds directly to K_M. Another way to analyze reaction kinetics graphically is by creating a double-reciprocal Lineweaver–Burk plot, where inverse initial reaction velocity is plotted against the inverse of initial substrate concentrations, producing a line that has the inverse of V_{max} as its y-intercept and the negative inverse of K_M as its x-intercept (Fig. 7-2b). This type of plot is often indispensable for the characterization of enzyme kinetic mechanisms, as well as inhibition mechanisms and the relevant kinetic constants through variation of initial concentrations of all the substrates or inhibitors involved. The curves shown in Figure 7-2 are usually constructed by employing kinetic enzyme activity assays, whereby product evolution over time is recorded either directly if a product possesses characteristic absorption or other physicochemical features, or through a coupled assay, where the product is used by a second enzyme that converts it to detectable compounds or takes up detectable compounds as its reaction proceeds.

Although the MM equation is a powerful kinetic form to which the vast majority of enzyme kinetics has been fitted, one should not forget the assumptions and limitations of the model. As a basic example, feedback inhibition, whereby the product of the reaction inhibits the enzyme–substrate cooperativity, multiple-substrate reactions, allosteric modifications, and other deviations from the reaction scheme in equation (1) are treated only adequately by the MM formalism under certain experimental conditions. In other words, enzyme kinetics are often "bent" to conform to the MM formalism for the sake of obtaining a set of parameters easily recognizable by most biochemists. The explicit mathematical and experimental treatment of reaction mechanisms more complex than that shown in equation (1) is highly involved, although a mathematical automated kinetic equation derivation framework for an arbitrary mechanism has been described in the past (e.g., ref. 6).

One of the most important assumptions in MM kinetics is that the reaction in question will proceed in a three-dimensional vessel filled with a well-stirred fluid that obeys Fick's law for diffusion. This is rarely the case in a living cell, where many reactions are localized to membranes (two dimensions) or to small regions somewhere within the cell, creating an effectively one-dimensional environment with little or no diffusion. To circumvent this limitation, fractal kinetics have been developed which allow for the approximation of enzymatic reaction velocities in vivo [7]. Fractal kinetics can utilize MM-type kinetic constants to create a model of events in a spatially restricted environment. Briefly, as the dimensionality of a reaction is reduced from three dimensions to one, the kinetic order of a bimolecular reaction, for example, increases from 2 in a three-dimensional case, to 2.46 in a two-dimensional environment (e.g., membrane), to 3 in a one-dimensional channel, up to 50 for the case where fractal dimensions are less than 1. In simple terms, the kinetic order is the sum of all stoichiometric coefficients of the reactants in a balanced chemical reaction equation. Rearranging the familiar equation for MM kinetics

$$\frac{S}{K_M} = \frac{v}{V_{\max}} \left(1 - \frac{v}{V_{\max}}\right)^{-1} \tag{5}$$

and comparing it to the equation for fractal MM kinetics,

$$\left(\frac{S}{K_f}\right)^{g_s} = \frac{v}{V_{\max}} \left(1 - \frac{v}{V_{\max}}\right)^{-g_e} \tag{6}$$

one becomes aware of the similarity of the two. The only difference is the presence of a fractal Michaelis constant, K_f, and kinetic orders with respect to substrate (g_s) and enzyme (g_e) in the fractal kinetic equation.

Fortunately, the vast majority of the processes observed to occur in our universe are governed by the basic laws of thermodynamics. In everyday terms, this means that there is no free lunch: A process that is favorable in terms of enthalpy and entropy change is expected to proceed, subject to activation energy constraints, whereas it is unlikely that an unfavorable process will occur without energy input. It is therefore expected that the reaction would proceed along the most likely reaction coordinate, resulting in the transition of substrates to products. In some reactions, however, this simple view does not hold. Unexpectedly, the reaction jumps from one point of the energy profile to another without undergoing intermediate steps or visiting the transition state. This phenomenon, referred to as *quantum tunneling*, is observed most frequently in reactions involving proton or electron transfer. According to quantum physics, every particle can be considered to be a particle or a wave, with the wavelength inversely proportional to the square root of double the product of its mass and kinetic energy. Thus, an average electron has a wavelength in the range of tens of angstroms, while the wavelength of a proton with similar kinetic energy can be estimated in the angstrom to subangstrom range. This allows these light particles to be "tunneled" over distances comparable to their wavelengths, resulting in instantaneous product formation. Such effects are extremely important in electron transfer reactions and play an important role in most hydrogen transfer reactions.

7.2.2 Enzyme Inhibitors

One can broadly separate enzyme inhibitors into two classes: reversible and irreversible. The *irreversible inhibitors*, sometimes referred to as *catalytic poisons*, exert their effect primarily by covalent modification of the target enzyme, leading to its complete inactivation. Reversible inhibitors exert their effect through non-covalent interactions such as hydrogen and ionic bonds or van der Waals forces, leading to blockage of the active site or conformational changes in the protein. Such inhibitors may be broadly classified further into three groups: competitive, uncompetitive, and noncompetitive inhibitors, based on the sequence of inhibitor binding to the enzyme (Fig. 7-3).

Competitive inhibitors are so named because they compete for the active site with the native substrate, meaning that only enzyme–inhibitor or enzyme–substrate complex formation is possible (Fig. 7-3a). In this case, the inhibition constant, or enzyme–inhibitor complex dissociation constant, can be defined:

$$K_I = \frac{[E][I]}{[EI]} \tag{7}$$

In this scenario, enzyme affinity for its substrate suffers, and therefore K_M increases to $K_M(1 + [I]/K_I)$, while the maximal reaction velocity is unaffected.

In *uncompetitive inhibition*, the situation is different: The inhibitor is only capable of binding to the enzyme–substrate complex, meaning that both V_{max} and K_M are affected detrimentally (Fig. 7-3b). K_I', the equilibrium constant for the dissociation of the inhibitor from the enzyme–substrate complex,

$$K_I' = \frac{[ES][I]}{[ESI]} \tag{8}$$

is relevant in this case.

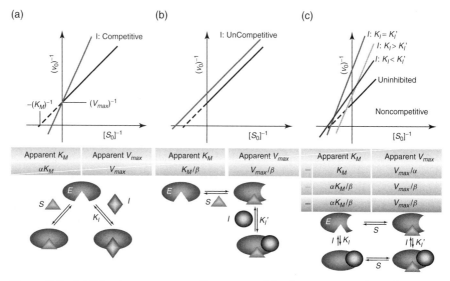

Figure 7-3 Inhibitor types, corresponding apparent kinetic constants, and reaction mechanisms. Here α is defined as $(1 + [I]/K_I)$, while β is short for $(1 + [I]/K_I')$. Note that the rate equation for a given inhibitor type can be obtained simply by substituting the kinetic constants in equation (2) with the apparent constants. *(See insert for color representation of figure.)*

Finally, *noncompetitive inhibitors* present the most complex set of cases of the three basic classes (Fig. 7-3c). Such inhibitors can bind either the enzyme itself, or the enzyme–substrate complex, with different effects depending on the magnitude of K_I relative to K_I'. In *pure* noncompetitive inhibition, where K_I is equal to K_I', the binding of the inhibitor with the enzyme has no bearing on the binding of the substrate, and thus has no effect on the apparent K_M. In *mixed* noncompetitive inhibition, on the other hand, the binding of an inhibitor to the enzyme leads to, for example, enzyme conformational changes that alter its affinity for the substrate, hence decreasing $V_{\max}$.

In inhibitor development, the goal is usually the creation of a compound that in the case of competitive inhibition binds much more tightly to its target enzyme's active site than the ligand it is competing with, or in the case of uncompetitive and noncompetitive inhibition, binds tightly to a site other than the active site, on the enzyme. Alternatively, a competitive inhibitor may not possess binding thermodynamics that are superior to the native ligand, but due to relatively larger half-life (taking longer to degrade), it may still overwhelm the native ligand in the competition for the active site.

7.2.3 Ligand–Receptor Interactions

Receptors constitute an extremely important protein class, responsible for cell growth, hormone response, and neuronal activity, among other things, making them highly promising as therapeutic or industrial SM targets. When receptors are considered, the SM that targets a receptor and inhibits its activity or response is referred to as an *antagonist*, while an *agonist* is designed to trigger the response of the receptor normally associated with binding to its native signaling molecule. An *inverse agonist*, on the other hand, causes the inherent basal activity of the receptor to decline. The descriptions of mechanisms of binding of agonists and antagonists to receptors are highly similar to those of enzyme–substrate and enzyme–inhibitor interactions, respectively. Briefly, *competitive antagonists* compete with the agonist for the receptor site, *noncompetitive antagonists* bind to a site distinct from that of the agonist and inhibit response propagation, while *uncompetitive antagonists* will bind only to the agonist–receptor complex. Of course, a SM can be designed to bind irreversibly to a given receptor.

Assays of antagonist efficacy rely on determination of the response levels or ligand binding kinetics and are complicated for several reasons. For example, if an antagonist or agonist is found to bind tightly and specifically to a given molecule [e.g., by isothermal calorimetry (discussed below)], such binding does not always guarantee changes in the receptor leading to a dispatch of a signal. Tests carried out in vivo or carefully crafted in vitro tests therefore constitute an important aspect of receptor–ligand interaction screening.

7.3 EXPERIMENTAL APPROACHES FOR INHIBITOR DISCOVERY, DESIGN, AND DEVELOPMENT

As we mentioned earlier, drugs that are enzyme and receptor inhibitors constitute a commercially significant class of molecules. Significant similarities exist in the search for a potent drug and an efficient biological ligand or inhibitor; therefore, for the purpose of discussion in this section, the terms *drug*, *inhibitor*, and *ligand* are used interchangeably. Indeed, the process of development of ligands or inhibitors is very similar to drug design, the only difference being that a ligand or inhibitor is not necessarily targeted toward alleviating disease and has not necessarily passed all subsequent clinical trials to be called a drug. The majority of the drugs currently available exert their effect by acting as enzyme inhibitors, receptor antagonists, or modulators of ion channels [8]. Therefore, a great deal of information on the strategies involved in the process of inhibitor discovery and design can be acquired from drug discovery studies. In addition, the term *inhibitor* should be understood in the present context as a modulator of the function of a particular target. There are, in fact, instances when small molecules act as enhancers rather than inhibitors of a particular biological entity [9].

Historically, the roots of modern drug discovery can be traced back to German scientist Paul Ehrlich (1854–1915), who, early in the twentieth century synthesized and tested a library of few hundred compounds in rabbit models. As a result, the first effective treatment against syphilis, Salvarsan, was discovered in 1910 [10]. With a number of chemical companies following Ehrlich's example, a key concept of drug discovery has become the systematic examination of synthetic libraries consisting of hundreds of compounds [11]. This "traditional" or "random" approach is distinguished from the more recent "rational" or de novo approach to inhibitor discovery, where researchers often take advantage of detailed knowledge of the biological system under investigation, such as its physicochemical and structural properties. We will provide a general description of both traditional and rational approaches to ligand and inhibitor development. We also provide examples of where the two approaches are integrated successfully.

7.3.1 Target Selection

The first step in the process of inhibitor discovery is usually the identification of a suitable target. This may involve elucidating possible molecular mechanisms behind a given biological phenomenon. A gene, protein, organelle, pathway, certain cell type, phenotype, or disease can be thought of as potential targets for inhibitor design. At this stage, genes, proteins, or even entire pathways that are involved in the process under investigation are identified with the help of various methodologies, including siRNA (small interfering RNA), gene or protein microarrays, or transgenic animals. For instance, by knocking down a gene expression with siRNA, it may be possible to identify a target whose reduced gene expression leads to specific inhibition of a signaling pathway. Certainly, target identification is not an easy task and one can proceed to further steps of inhibitor discovery and design even if the exact inhibitor target remains to be elucidated. Such was the case of inhibitor design of hypoxia-inducible factor-1 (HIF-1): Use of the siRNA library failed to diminish HIF-1-driven luciferase reporter expression, but out of a small molecule library of approximately 700,000 compounds, 250 significantly decreased the expression of luciferase under HIF-1 control [12]. Further analyses revealed 18 alkyliminophenylacetate compounds that were found to be the inhibitors of mitochondrial reactive oxygen species production, which are thought to be essential for HIF-1 activation under hypoxia.

7.3.2 Library-Based Approaches

If the exact target has been identified, *focused* or *target-oriented library* is utilized for the inhibitor screening. Its creation is aided by some prior biochemical or biophysical characterization of the target and its known ligands. However, when the exact target is unknown, the type of compound library used to identify a potential inhibitor is known as a *generic* or *diversity-oriented library*, consisting of a large number of highly diverse and structurally complex compounds aiming to cover larger areas in chemical diversity space. The goal here is to increase the chance of finding a ligand or an inhibitor by working with a collection of diverse and complex compounds rather than with a group of related compounds comprising a focused library. In general, two types of compound libraries exist: natural product and combinatorial chemistry (combichem) libraries.

Natural Product Libraries   Historically, natural products have been the oldest sources for new drug and inhibitor candidates. It is conceivable that during evolution, natural selection has led to the production of potent ligands and inhibitors yet to be discovered. Furthermore, a proposition was put forward suggesting that all natural products possess receptor-binding capacity to a certain degree [13]. Microorganisms such as molds (e.g., penicillium mold) and bacteria were found to excrete toxic substances prohibiting the growth of other organisms in close proximity. Similarly, certain plants have been known to possess medicinal properties for centuries, with willow bark and leaves being one famous example.

In addition, products of microbial fermentation have been established as rich sources for drug candidates. For instance, the discovery of cholecystokinin antagonist from the fungus *Aspergillus alliaceus* has led to the development of cholecystokinin A and B receptor antagonists by scientists at Merck [14]. Another very famous example of naturally occurring drugs is Taxol, produced by Bristol-Myers Squibb from yew bark. Taxol is effective for the treatment of certain cancer types, as it binds to tubulin and inhibits the process of cell division, which is accelerated in cancer. The importance of natural product libraries has been highlighted by a recent analysis of the sources of the U.S. Food and Drug Administration (FDA)–approved drugs for the period 1981 to 2006 [15]. Surprisingly, of 974 new SM chemical entities, 63% were either nature-derived or nature-inspired compounds, such as analogs of semisynthetic natural products or synthetic compounds built around natural product pharmacophores, which are simply the set of molecular features important to binding.

Owing to the chemical complexity of crude plant extracts or microbial fermentation products, certain steps have to be applied to find the active compound in the natural library. Thus, fractionation of crude natural extracts by polarity, for example, is an important initial step in natural library creation. Library screening can be carried out on the resulting fractions. While multiple small molecules within a given fraction may play a synergistic role lost upon further fractionation, a single inhibitor with a defined chemical structure is sought. Active fraction components have to be profiled to identify the known constituents in the process termed *dereplication*, used to eliminate the possibility of duplicating the effort. This analysis is assisted by bioassay-guided fractionation, a series of extraction rounds aimed at recovering a single, pure active compound whose extraction can be scaled up for further tests, such as secondary tests or structure–activity relationship (SAR) studies (Fig. 7-4).

Natural library screening is more time consuming and resource intensive than high-throughput screening (HTS) of combinatorial chemistry libraries, often taking from several months to one year [16]. It has commonly been recognized that isolation and purification of active compounds from extremely complex matrices is the bottleneck of natural library screening efforts [17]. This could explain the fact that screening natural product extracts has become less popular on an industrial scale [16].

Unexpectedly, another disadvantage of natural compound libraries having to do with the heterogeneous nature of the separated fractions became apparent with

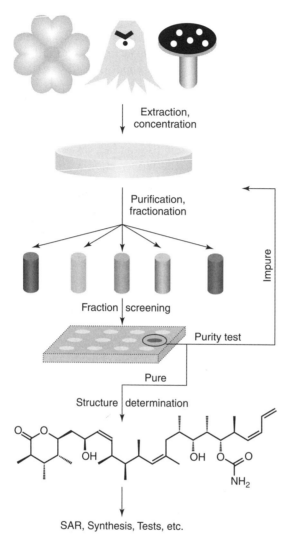

Figure 7-4 General scheme for natural library creation and screening. The chemical structure of natural product (+)- discodermolide is shown.

the advent of HTS. Originally, fractions from the natural compound libraries had been tested in cellular or animal models, in which fraction impurity and complexity was not an issue and the detection of active compounds was possible [17]. In addition to an active inhibitor or ligand of a target under investigation, one fraction from a natural source could contain fluorescent or insoluble impurities interfering with some HTS methods, such as fluorescence polarization. This problem can be circumvented, for example, by using fluorophores with different excitation and emission profiles or by lifetime-discriminated polarization, which reduces compound interference by rejecting signals originating from fluorescent sources with short lifetimes [18]. Whatever the solution to this problem may be for a given HTS system, the presence of this factor complicates the analysis of natural libraries.

Several improvements to the process of natural library creation have been developed. Recently, a high-throughput (HT) method for production and analysis of natural product libraries has been described [16]. Briefly, extracts from 176 plant families and 596 genera from the United States and Africa were collected and fractionated using automated flash chromatography. The resulting fractions were subjected to batch-throughput solid-phase extraction to remove tannins and were filtered to exclude high-molecular-weight compounds. Further separation with high-performance liquid chromatography (HPLC) produced mixtures of about five compounds per well. The resulting fractions are profiled with HT analytical liquid chromatography–evaporative light-scattering detection–mass spectrometry (LC-ELSD-MS). Fractions with sufficient compound quantities were organized into a library from which the compounds can be taken for biological screening. Active fraction constituents are identified using high-resolution nuclear magnetic resonance (NMR) spectrometry. The resulting library consists of 36,000 fractions with up to five compounds per fraction.

Several experts suggested that the shift of pharmaceutical companies from natural toward combinatorial chemistry libraries has not yielded the expected results and that we might be witnessing a renewed interest in the natural compound libraries [16,17,19–21], especially given recent developments accelerating library creation. In addition, closer similarities of libraries derived from natural sources to known inhibitors may be responsible for the renewed interest in the natural libraries as starting points for ligand and inhibitor design.

Recently, the potential of compounds from synthetic and natural product libraries to act as inhibitors of a protein–protein interaction has been evaluated [22]. A library of about 7000 natural SMs from Novartis and the Natural Product Collaborative Drug Discovery Group was screened in an attempt to discover an inhibitor of Tcf/β-catenin interaction, implicated in colorectal and other cancer types. Six potent inhibitors of microbial origin were identified and suggested for further structural refinement for cancer therapy. However, an additional screen of approximately 45,000 synthetic compounds from the Novartis collection yielded no additional inhibitors of this interaction [22].

Since some natural products cannot be harvested in sufficient quantities, organic synthesis may be used to produce and optimize the most promising ligands identified using natural library screening. Such was the case of discodermolide (see Fig. 7-4), an antitumor compound isolated from a marine sponge in 1990 [23]. Low extraction yield and limited natural supply of this potent inducer of microtubule stabilization (acting similarly to Taxol) stimulated several groups to find an efficient route to discodermolide synthesis. Several efficient methods were developed for the synthesis of this complex compound with its 13 stereogenic centers [17].

Synthetic Libraries *Combinatorial chemistry* can be defined as a process in which a large collection of distinct, complex, and diverse compounds is synthesized using combinations of compounds with simpler chemical structures as building blocks. Combinatorial synthesis is most often associated with parallel, solid-phase synthesis [24] carried out with the help of insoluble material (usually, polystyrene beads) on a linker. Attached are several features arranged in linear fashion, including chemical cleavable sites, encoding blocks, and ionizable groups, with the desirable compound being synthesized at the terminus [15]. Beads allow for easy separation of the attached products from reactants in solution. For example, in the split-and-pool strategy, beads are split into reaction vessels to which a unique set of reagents (building blocks) is added (see Fig. 7-5). Further splitting of beads and addition of reagents leads to the creation of diverse compounds immobilized on separate beads. Incorporation of encoding elements (e.g., based on chemical and physical principles such as the use of surrogate chemical compounds and radio chips) helps trace the synthetic history and identity of a sample. Subsequent analytical characterization of such libraries is carried out by utilizing cleavable sites to reveal the identity of both encoding tags and chemical products, often with the help of mass spectrometry.

Initially, the key to synthetic library design was increasing library size and diversity while keeping constituent compounds within certain boundaries, defined by empirical rules such as *Lipinski's rule of 5*, which estimates whether a chemical compound with a certain pharmacological or biological activity has properties that would make it a likely active drug in humans [26]. However, the idea that increasing library size and diversity increases the chance of finding a suitable candidate now seems to be erroneous. If the size of the human genome is about 35,000 genes, the number of targets for SMs is about 10,000 and the size of SM diversity space is anywhere from 10^{14} to 10^{30} molecules, one needs at least 10^8 to 10^9 average-sized (10^5 to 10^6 members) libraries to find SM ligands for all targets [25]. Needless to say, this number is enormous compared to the number of libraries available. The creation of so many compound collections is not conceivable in the near future since all the laboratories worldwide are thought to contain approximately 10^8 molecules [27].

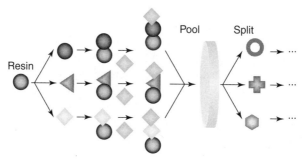

Figure 7-5 Schematic outline of the spit-and-pool approach in combinatorial chemistry.

Similarly, little attention has been paid to library quality control, which affects compound identification. It was anticipated that this process would be accomplished straightforwardly, with more stringent synthesis carried out after promising inhibitors were identified. Unfortunately, this attitude has lead to irreproducible results and disappointing findings that the candidate compound synthesized and profiled by conventional approaches was totally inactive [25]. Unexpectedly, despite the increased number of compounds in combinatorial libraries, the number of successful drug candidates identified from HTS of these libraries was decreasing [19]. This lack of success of combinatorial chemistry libraries is also reflected in the fact that over the 25-year period from 1981 to 2006, only one de novo FDA-approved inhibitor was discovered with the aid of combinatorial chemistry (the antitumor drug Nexavar, from Bayer) [15].

Despite the low level of diversity in many synthetic compound libraries [17], a significant number of drug and inhibitor candidates are still found with HTS. This can be explained by the fact that the synthetic library collections tend to be biased toward compounds that are relatively easy to produce, were generated in previous drug discovery efforts, and are related to known drugs [28]. This bias can be justified by the fact that drug subspace is not distributed randomly across the chemical diversity space and that successful drug and inhibitor candidates can be derived from inhibitors identified previously. Additionally, the often low diversity of targets currently considered may be conducive to lower diversity of screened chemical subspace. For example, sulfonilamides were initially discovered as potent inhibitors of bacterial dihydrofolate synthetase, then related sulfonamide compounds were shown to be carbonic anhydrase inhibitors in human kidney, and finally, structurally related sulfonylureas were established as inhibitors of potassium channels in human pancreas.

The lack of anticipated success of combichem libraries in drug discovery echoed in various recent publications prompted several groups to compare various parameters of compounds from different library types with those of known drugs. It was shown that natural product libraries display more druglike characteristics than do combichem libraries. One of the features that distinguishes between the two library types is the lower number of chiral centers and lack of rigidity (low number of rotatable bonds) in the combichem collections [21]. Structural rigidity is widely regarded as a prerequisite for tight binding of small molecules to relatively flat (compared to the active sites of enzymes) protein–protein interaction sites [24].

Nonetheless, the power of combinatorial chemistry in the structural optimization of a promising inhibitor–ligand candidate through SAR studies is exceptional. As a result, we are witnessing a shift in combinatorial chemistry libraries from large and diverse to small and focused collections containing from hundreds to several thousands of compounds. It is also desirable that combinatorial libraries become enriched with "druglike" elements, such as a higher degree of chirality and rigidity, seen in libraries of natural products. Several successful attempts to address the shortcomings of combinatorial chemistry libraries by developing synthetic libraries of natural-like compounds have been reported

[29,30]. We anticipate that the current shift in combinatorial chemistry libraries away from numerically large to more focused, target-oriented libraries will continue in the near future and that these libraries will be helpful in developing ligands with improved binding characteristics. It is also plausible that such a shift may open efficient inhibitor and ligand discovery opportunities to academic laboratories.

7.3.3 Fragment-Based Design

While natural and synthetic compound libraries are used to identify promising ligands from a static pool, fragment-based ligand design is more flexible and conceptually diverse. This is a new technique introduced in the past decade, whose goal is to identify structurally simple ligands ("fragments") and to use several fragments as building blocks to design a more efficient ligand (Fig. 7-6). The fragments binding to a target are identified in the first round of screening. The strength of binding of "successful" fragments can be moderate or even weak. Typically, the molecular size of fragments is below 300 Da [31] and the number of fragments screened initially is on the order of 1000 to 10,000 compounds [32]. Next, several fragments found to bind to a target of interest are linked together to produce a collection of larger compounds, some of

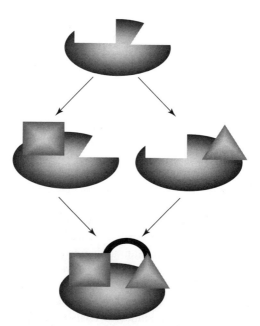

Figure 7-6 Fragment-based design: When two optimized and independent ligands (square and triangle) are joined, the binding of the resulting compound to its target protein (red) is superior to that of separate fragments. *(See insert for color representation of figure.)*

which often display exceptionally strong binding characteristics. Promising candidates are then modified and optimized further to produce an inhibitor with desirable parameters. One of the requirements of this approach is that the fragments tested be water soluble at higher concentrations, as they tend to be weak binders.

A number of advantages for fragment-based design include the following:

1. Better coverage of chemical space with a lower number of compounds, since a large library of compounds can be reduced to a small number of fragments in the same way that a lengthy novel can be reduced to the few letters of the alphabet used repeatedly.
2. A fragment has greater flexibility and better access to its receptor site on the target than do the more rigid and larger compounds of the natural and synthetic libraries.
3. The synthesis and maintenance of fragment libraries are more facile than for other libraries [33].

On the other hand, significant disadvantages to fragment-based inhibitor design can arise from the fact that the optimal linkage of successful fragments in three dimensions is not obvious, and due to the limited number of folds utilized by proteins, receptor sites to which fragments are being docked can be quite similar among a number of targets. As a result, the binding of a fragment or even an inhibitor assembled from several fragments can be nonspecific. Finally, the use of fewer fragments can represent a large collection of compounds (a database of 2.6 million compounds was reduced to 70,000 fragments [31]), but that still constitutes a small proportion of chemical diversity space.

Fragment-based design was used successfully in a number of studies reviewed elsewhere [34]. For instance, consider one recent work on identifying a specific inhibitor of neuronal nitric oxide synthase, nNOS, using fragment hopping—a variation on fragment-based design [33]. Dipeptide inhibitors of nNOS developed previously were used as a starting point. Five focused fragment libraries, such as a "basic fragment library" and a "side chain library," were utilized to calculate the binding energy of fragments docked to the nNOS active site using the crystal structure of the enzyme in silico. Minute differences in the active sites of the three NOS enzymes, identification of minimal pharmacophore elements needed for nNOS inhibition, and common pharmacological parameters were used to create a very potent and specific nNOS inhibitor whose potency and mechanism of binding was validated using enzyme assays and x-ray crystallography. This work nicely illustrates the power of combined research utilizing structural, biological, computational, and organic chemistry methods to design a potent enzyme inhibitor. It is foreseeable that these types of studies will become very common and fruitful in the future.

7.4   EXPERIMENTAL ELUCIDATION OF LIGAND AND INHIBITOR ACTIVITY

Once the most suitable library has been chosen, rapid screening of the compound collection is carried out. During the screening process, three typical questions are being addressed, in order of increasing complexity: (1) Does a library member bind or modulate the activity of the target; (2) what is the mode of action; and (3) what is the strength of binding or relative potency of a given ligand to alter the target's function? Although numerous methods have recently been introduced to provide a rapid answer to the first question, and it is not uncommon for researchers to have a proposed mechanism of the ligand's action, including its binding site and chemical interactions with the target, answering the third question helps to select and develop the most potent ligand. In other words, although a number of ligands could be discovered for a given target, exerting their effect in a more or less similar manner (e.g., binding to the active site and forming hydrogen bonds with selected amino acids), the most potent ligand with the greatest affinity for its target will usually be selected for future development. Here we introduce briefly several popular screening methods, together with their usefulness in elucidation of the mechanism of the ligand's action, paying special attention to technologies capable of either absolute or relative ranking of inhibitors. Some of the HTS methods developed to accommodate the rapid screening of large libraries of compounds are discussed in greater detail elsewhere in the book.

7.4.1   Selected Mass Spectrometry–Based Methods

Mass spectrometry (MS) is a very popular analytical tool for compound identification in inhibitor design. MS devices determine the mass of an ionized compound with high resolution and sensitivity by detecting ions according to their mass-to-charge (m/z) ratios. Since most compounds will have unique m/z spectra, MS allows for the identification of analytes. Not only does MS provide an ideal platform for enzyme assays identifying the conversion of substrates to products, but it also permits the identification and quality assessment of compounds in natural and synthetic libraries. Early enzyme assays, linked to gas chromatography (GC-MS), often suffered from the requirement of analyte derivatization to increase compound volatility. Fortunately, electrospray ionization (ESI) and matrix-assisted laser desorption ionization (MALDI) techniques introduced in the late 1980s allowed direct measurement of compounds, leading to the exponential growth in MS applications for enzyme assays [35]. A number of MS-based approaches have recently been reported; their comprehensive overview is offered elsewhere [35].

Frontal affinity chromatography–mass spectrometry (FAC-MS) is very popular for ligand and inhibitor discovery and design [36]. This two-dimensional detection method physically separates compounds by affinity chromatography and determines their identities by MS. Introduced in 1999 [37], FAC-MS is capable of ranking several compounds present in solution in terms of their binding affinity to a protein immobilized on a thin capillary tube. As the solution flows continuously through the column, equilibrium between bound and nonbound ligands is reached, and individual ligands bind to their target and are retained on the column, eluting later than do nonbinding ligands. The binding strength of each ligand is manifested in its *breakthrough volume*, defined as the effluent volume coming through the column at which the input ligand concentration equals its output concentration [36]. Breakthrough volume can be determined by observing the ion intensity over time with MS. The extent to which the breakthrough volume differs from the void volume of the column relates directly to the binding affinity of each ligand (K_d) [38] and allows unambiguous ranking of each inhibitor's binding strength.

Advantages of FAC-MS include high sensitivity, cost efficiency due to its reuse of the target, quantification of the ligand's affinity in terms of the complex dissociation constant K_d, and the ability to analyze complex mixtures [37]. FAC-MS is suitable for HTS; for instance, in one automated method, two columns are used and regenerated repeatedly, allowing the HTS of up to 10,000 compounds per day [36]. Recent examples of FAC-MS applications include successful discovery of an inhibitor of EphB2 receptor tyrosine kinase domain among 468 compounds selected by virtual screening out of over 50,000 unique compounds derived from a chemical diversity library [39].

7.4.2 Nuclear Magnetic Resonance Applications

Nuclear magnetic resonance (NMR) technologies are based on the fact that atomic nuclei of elements with either odd mass or atomic number, such as ^{1}H, ^{13}C, and ^{15}N, possess nuclear magnetic moments, behaving like miniature spinning objects. Upon application of an external magnetic field, such nuclei start precessing about their spin axes with an angular frequency (Larmor frequency) proportional to the external magnetic field's strength, creating an oscillating electric field that can be coupled with externally applied radio-frequency radiation of the same magnitude to cause spin change (resonance). Due to the circulation of the valence electrons around a nucleus, a counter magnetic field opposing the applied magnetic field is created. The differences in chemical environment or electron density of each precessing nucleus in a molecule are manifested in different resonance frequencies, expressed as chemical shifts. This allows NMR to distinguish between different nuclei in relatively complex molecules. By tracing the changes in nuclear chemical environments during a binding event, structural and kinetic information of protein–ligand interactions can be obtained. The many observables in NMR include chemical shifts, differential line broadening, and transferred nuclear Overhauser effect (NOE). More in-depth, theoretical treatment of NMR is presented elsewhere [40].

Perhaps the most famous examples of using NMR in inhibitor discovery is the approach termed *SAR by NMR*, introduced as early as 1996 [41], a fragment-based design type of approach, as discussed above. It involves the monitoring of protein–ligand binding with ^{1}H or ^{15}N chemical shift of amide groups of a ^{15}N-labeled protein using two-dimensional ^{15}N-heteronuclear single-quantum correlation (^{15}N-HSQC). Once a ligand with sufficient affinity for every binding site on a protein is found in a stepwise fashion, two or more fragments are linked and optimized to produce a strong binding ligand.

SAR can refer to the process in which the structure of a promising binder is derivatized to come up with a ligand with higher activity than the starting compound. SAR by NMR was used successfully to discover high-affinity ligands for FK506-binding protein (FKBP) [41]. Rendering the capability to analyze mixtures of 10 compounds and to screen about 1000 compounds per day, the method necessitates the use of high concentrations of low-molecular-weight proteins. More recent modifications of SAR by NMR increased the throughput enormously, testing mixtures of 100 compounds with significantly lower (about 50 µM) concentrations of ^{15}N-labeled proteins using cryogenic NMR technology [42]. Specific recent applications of this approach include the development of novel ligands such as ABT-737, an extraordinarily potent SM inhibitor of Bcl-2 (overexpressed in certain tumors) [61], as well as several novel creatine kinase inhibitors [62]. A modification of this approach, potentially applicable to protein–protein and protein–nucleic acid interaction inhibitor development, has been shown to generate high-affinity, selective ligands rapidly and efficiently for p38α [63].

Similarly, ^{1}H–^{15}N-HSQC NMR was useful in discovery of the FRB domain of mTOR [43]. Following in silico virtual screening of 30,000 compounds that yielded 56 potential ligands, the primary screen monitored spectral changes in the ^{1}H NMR aliphatic regions of ^{15}N-labeled FRB. Four compounds inducing significant changes in the NMR spectrum were subjected to titrations involving [^{15}N]Gly-labeled FRB, varying ligand concentrations, and two-dimensional ^{1}H–^{15}N-HSQC. A simple plot of changes in chemical shifts arising due to the ligand binding to G2040 residue in both ^{15}N and ^{1}H dimensions versus ligand concentrations yielded ligand-binding constants (K_d).

Another very popular NMR-based method is saturation transfer difference spectroscopy (STD). It is based on the magnetization transfer (^{1}H saturation) from protein to its ligand through space. The *off-resonance spectra*, in which the irradiation applied is off-resonance with respect to the protein and its ligand(s), is subtracted from the *on-resonance spectra*, in which the protein and its ligand(s) are selectively irradiated, to yield the spectrum representing the resonances of interacting ligand [40]. Protein resonance is usually invisible due to low protein concentrations, while the ligand peak intensity is observed to decrease as a result of binding. Unlike NMR by SAR and HSQC, which require small labeled proteins, STD is applicable for large-molecular-weight unlabeled proteins. It is also possible to derive structural information from STD experiments. For example, the functional groups of ligands located in close proximity to the protein during ligand binding tend to produce stronger saturation of their NMR signal intensity [44].

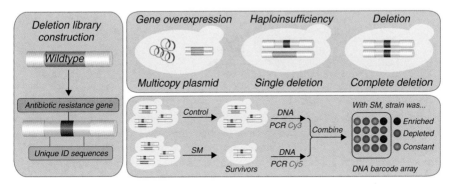

Figure 2.5 Typical SM-induced growth alteration screening procedure.

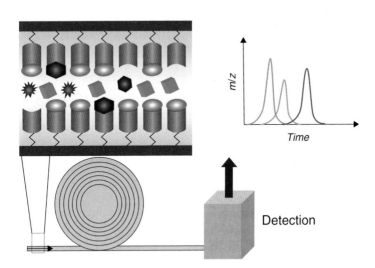

Figure 2.7 FAC-MS: SM affinity is proportional to retention time.

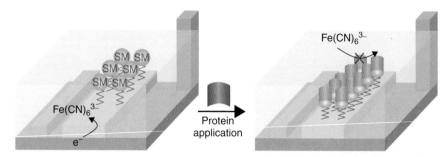

Figure 2.9 Typical EIS device. Auxiliary, working, and reference electrodes are shown in gold, blue, and green, respectively. In a solution (gray transparent cube), a redox compound such as $Fe(CN)_6^{3-}$ will "complete the circuit" between the auxiliary and working electrodes, but upon application of a protein that would bind to a SM immobilized on the working electrode, the access of the redox solute may be severely limited, resulting in impedance changes.

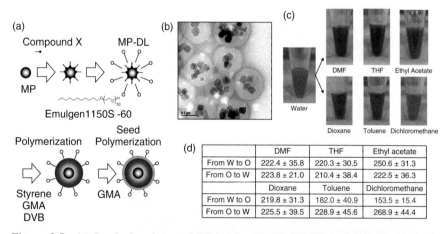

Figure 3.5 (a) Synthetic scheme of FG beads; (b) FE-TEM image of the isolated FG beads; (c) photo image of FGNE beads dispersed in DMF, THF, ethyl acetate, 1,4-dioxane, toluene, and dichloromethane (these dispersions contain 0.4 mg of FGNE beads); (d) dynamic light-scattering (DLS) analyses of FGNE beads in DMF, THF, ethyl acetate, 1,4-dioxane, toluene, and dichloromethane and water-resuspended beads from each organic solvent.

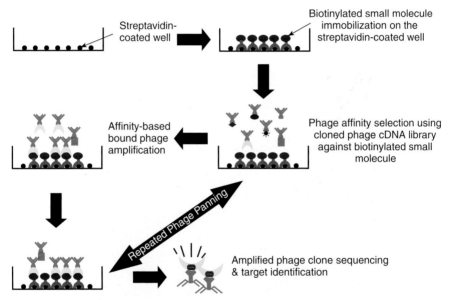

Figure 5.1 Overall phage display biopanning scheme. Identification of a target protein of a small molecule.

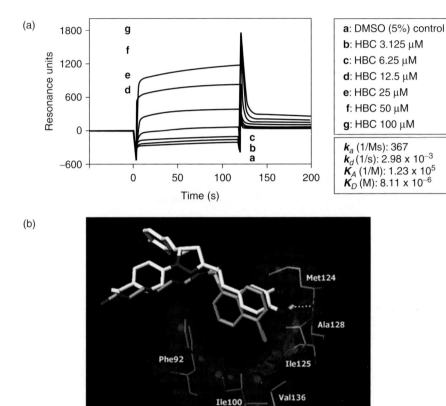

(a)

a: DMSO (5%) control
b: HBC 3.125 μM
c: HBC 6.25 μM
d: HBC 12.5 μM
e: HBC 25 μM
f: HBC 50 μM
g: HBC 100 μM

k_a (1/Ms): 367
k_d (1/s): 2.98 x 10^{-3}
K_A (1/M): 1.23 x 10^5
K_D (M): 8.11 x 10^{-6}

(b)

Figure 5.3 Validation of CaM as a target protein of HBC. (a) Surface plasmon resonance analysis of interaction between HBC and Ca^{2+}/CaM. Purified Ca^{2+}/CaM was immobilized on a CM5 sensor chip and various concentrations of HBC were loaded into the sensor cell. Binding sensor grams were obtained from the BIAcore evaluation software. Kinetic parameters of k_a, k_d, K_A, and K_D are shown. (b) Docking model of HBC in a complex with the C-terminal Ca^{2+}/CaM domain. The docking mode of HBC (gray carbon) and W7 (orange carbon) obtained from FlexX. The Connolly molecular surface of the active site is shown in purple with amino acid residues occupying the active site. Hydrogen atoms are not shown for clarity. The yellow dotted line indicates the hydrogen-bonding interaction ($d = 1.244$ Å). (From ref. 15.)

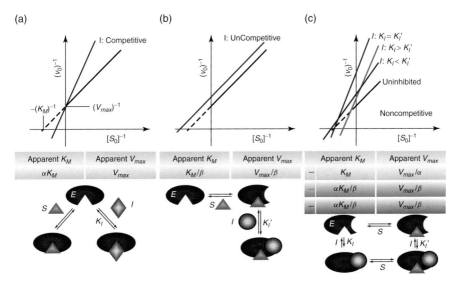

Figure 7.3 Inhibitor types, corresponding apparent kinetic constants, and reaction mechanisms. Here α is defined as $(1 + [I]/K_1)$, while β is short for $(1 + [I]/K_1')$. Note that the rate equation for a given inhibitor type can be obtained simply by substituting the kinetic constants in equation (2) with the apparent constants.

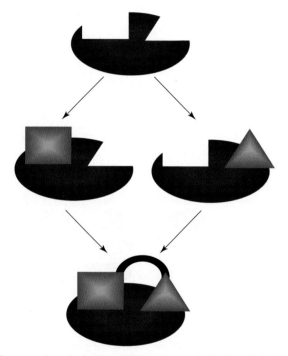

Figure 7.6 Fragment-based design: When two optimized and independent ligands (square and triangle) are joined, the binding of the resulting compound to its target protein (red) is superior to that of separate fragments.

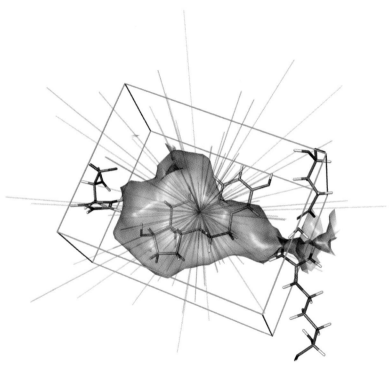

Figure 7.7 U Dock 1.1, in a search for an ideal ligand conformation. Note that this pose will probably be rejected since it clashes with the surface of the active site and protrudes into the protein.

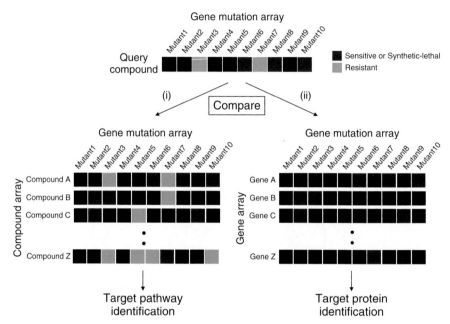

Figure 11.2 From phenotype observation to target pathway and target protein identi-fication. Profiling of chemical genetic interactions is of help for identification of target pathway and target protein. Once phenotypes of gene mutants for a query compound are defined, modes of action of the compound can be expected from statistical analysis using phenotype compendia or functional analysis using public databases. For example, the mutants 6, 8, and 10 are sensitive (shown in red) to the query compound and the mutants 3 and 7 are resistant (shown in green) to the query compound. (i) Comparison with the compendia of chemical genetic interaction profiles reveals that the query compound has a similar target pathway or modes of action with those of the compound A. (ii) Comparison with the synthetic lethal profiles shows that the gene A product is predicted to be a target protein of the query compound. This strategy is particularly useful for reverse genomics approaches using yeast.

7.4.3 Isothermal Titration Calorimetry

Isothermal titration calorimetry (ITC) is a powerful technique capable of measuring the thermodynamic parameters involved in protein–ligand interactions. It takes advantage of the fact that molecular binding is accompanied by heat capacity change (ΔC_p), observable with ITC. This makes it the only technique capable of direct determination of standard Gibbs free energy (ΔG^0), standard enthalpy (ΔH^0), and standard entropy (ΔS^0) associated with protein–ligand binding. In a typical ITC experiment, titrations are carried out using fixed protein and variable ligand concentration. Depending on the nature of the interaction, heat is either generated (exothermic) or consumed (endothermic process). Since the calorimeter is designed to maintain constant-temperature (isothermal) conditions, the extent of the power needed to be supplied to the system is dependent on the amount of heat generated or consumed. After a series of ligand injections into the vessel with fixed amount of protein, the saturation of protein with its ligand is eventually reached, as indicated by a lack of further changes in the calorimeter's heating power. The raw data in this experiment represent changes in heating power over time, which is integrated to produce the plot of heat increments for each injection as a function of protein/ligand molar ratio, from which the thermodynamic parameters ΔH^0 and K_d are derived. These are substituted into the equation

$$\Delta G^0 = -RT \ln(K_d) = \Delta H^0 - T \Delta S^0 \tag{9}$$

to calculate ΔG^0 and ΔS^0 [45].

An interesting illustration of the ITC approach to inhibitor screening is the investigation of the binding of *Escherichia coli* ketopantoate reductase (KPR) with its cofactor NADPH [46]. Constituent fragments and analogs of NADPH were subjected to ITC to reveal the structural determinants to cofactor binding. Site-directed mutagenesis was used to determine the importance of specific residues to ligand binding. By observing what fragment of NADPH had an altered affinity (measured by binding free energy) as a result of mutation, it was possible to elucidate the binding sites of each fragment. This revealed hot spots in the KPR–NADPH interaction which can be targeted in further inhibitor design, which will probably be conducted using a fragment-based approach. Enthalpic and entropic contributions to the KPR–NADPH binding were measured. Interestingly, ATP binding to an N98A mutant resulted in no detectable change in ΔG^0, whereas the significant changes in ΔH^0 and ΔS^0 were of the opposite magnitude. This has revealed that while a perturbation to a system may result in no observable effect in ΔG^0, which is a parameter measured directly or indirectly in most methods, thus yielding no significant information on the effect of perturbation, the underlying changes in enthalpy and entropy can provide a more complete view of the mechanism by which the perturbation acts on the system in question.

ITC is also applicable to dissect protein–protein interactions. A recent study [47] looked at the energy landscape of Ras/Raf protein–protein interaction with double mutants of the two proteins and ITC. Again, it was noted that small changes in the standard Gibbs free energy can have significant underlying enthalpy and entropy changes. ITC was also used to establish tetrapeptides as successful candidates for peptidomimetic inhibitor design of BRCT [48].

The major drawback of ITC is that it requires large quantities (milligrams) of pure proteins and can only be used for low-throughput analysis. Currently, a limited number of compounds can be screened for protein binding due to low throughput; however, integrated-circuit microcalorimeters that are being developed may soon provide the ability to run ITC in microplate format, enabling researchers to analyze more samples with lower amounts of protein in a shorter period of time [49].

7.5   RATIONAL LIGAND AND INHIBITOR DESIGN

Once a critical mass of mechanistic and structural information has been obtained, possibly using some of the techniques described above, a rational drug design endeavor can be initiated. As opposed to traditional ligand design and development, rational ligand design relies on the fundamental physicochemical principles underlying protein–SM interaction, structure of the target (in target-based design), or the consensus features of a set of compounds previously identified to be potent (in ligand-based design), to construct a rather limited set of compounds for further screening and refinement. This approach is ideal for an academic setting, where funds for large-scale screening campaigns may simply be unavailable, yet where an abundance of mechanistic expertise exists. In this section we review briefly some of typical rational ligand design approaches and important theoretical considerations.

7.5.1   In Silico Ligand Design Principles: Forces Important in Binding

Irrespective of inhibitory mechanism or ligand mode of interaction, the free energy change associated with a binding event can be decomposed into several contributions. The most familiar decomposition involves enthalpic and entropic contributions [e.g. equation (9)]. Thus, for illustration purposes, if a given SM is observed to increase the entropy or degree of disorder of a given solution, its removal from the solution into an enzyme-bound form may be unfavorable. Another decomposition may be performed along the lines of the intermolecular forces contributing to the stabilization or destabilization of the interaction.

The two major forces one should be aware of in ligand design are van der Waals interactions and electrostatic forces, which include hydrogen bonding and ionic interactions. Van der Waals forces arise from very small and transient polarization of the electron cloud around two neighboring nuclei and are more pronounced for nuclei that are a part of an aromatic moiety or nuclei termed as "soft." The potential energy in such an interaction can be fitted with a Lennard-Jones potential function, for example, and drops rather quickly with increasing internuclear separation. In nature, van der Waals forces are often called upon for transient and reversible adhesion.

The electrostatic forces involve two oppositely charged objects that attract each other with a force proportional to their charge and inversely proportional to the square of the distance between the two. This force is responsible for hydrogen bonding, which occurs between a hydrogen atom bound to an electronegative atom and a different electronegative atom. This force is also responsible for ionic bridges between two ionized and oppositely charged functional groups. It is the combination of these two major forces that contributes most to the overall binding free energy.

Storing a certain amount of potential energy in an object such as a spring allows one to let the spring release energy later in the form of movement, heat, and sound. The dissolution of a hydrophobic compound in aqueous media is similar to storing energy in a spring, since such a compound would disrupt the extensive hydrogen-bonding network of water and offer no hydrogen-bond donors or acceptors in return. If that is the case, the hydrophobic compound would tend to be excluded from aqueous media and fractionate into nonpolar environments, if possible. Proteins often contain hydrophobic environments in their active sites, meaning that the interaction of protein with ligand would be favorable relative to the interaction of ligand with aqueous solution. Thus, hydrophobic interactions can also play a significant role in the stabilization of protein–SM interactions.

It is these parameters and considerations that one should always keep in mind when designing novel ligands or considering the mode of binding of native substrates. The endpoint in the in silico ligand design is a ligand that possesses the lowest possible Gibbs free energy of interaction and the highest possible specificity for its particular target. This can be achieved by minimizing the number of steric clashes between the prospective ligand and the target enzyme while maximizing the number of electrostatic interactions. Unfortunately, current in silico ligand design and evaluation methods, with the exception of quantitative structure–activity relationship (QSAR)–based algorithms, cannot predict ligand toxicity. The treatment of backbone rearrangement change upon ligand binding also poses a prominent challenge.

7.5.2 Tools of the Trade

The calculation or approximation of binding free energies is a complex task that can be carried out in a number of different ways. The process involves calculation of the free energy of formation of the solvated ligand, target, and target–ligand complex and usually starts with geometry optimization of the compounds in question. Steepest descent-, Monte Carlo-, or genetic algorithm–based methods can be used to determine the equilibrium geometry of the molecules in question. A conformer search is usually performed, and conformers within a certain energy range of the lowest-energy conformer are selected for further study. Target geometry is not usually optimized rigorously (although it should be), while a full conformational analysis should be done on the ligand.

To determine the optimal configuration of the target–ligand interaction, every relevant conformation of a ligand is placed within the active site in different orientations, to arrive at a set of poses in a process referred to as *docking*. The resulting ligand poses are usually evaluated using a rapid scoring function, which can be derived empirically. Poses can also be evaluated based on rigorous interaction energy calculations (e.g., involving solution of the Poisson–Boltzmann equation). The poses are ranked and the most favorable ones are selected.

Figure 7-7 shows an experiment using U Dock 1.1 [50]. Except for the three hydrogen-bonding amino acids of the target protein (human estrogen receptor alpha, PDB ID: 1GWR), the others are hidden. The active site cavity is rendered in transparent gray. The ligand is randomly displaced inside a theoretical docking box (green) using random displacement vectors (shown in pink). The program will repeat this step until it finds the ideal orientation. After generating initial poses, most programs usually optimize their geometry to produce local minima complexes. Theoretically, the lowest-energy complex should represent the optimum binding mode for the ligand–receptor pair. Unfortunately, there is a strong indication that the scoring functions used to evaluate the optimal ligand pose are inadequate and may lead to erroneous results [51,52].

Theoretically rigorous computation is expected to improve the accuracy of docking scores; however, this improvement may come at a prohibitive cost.

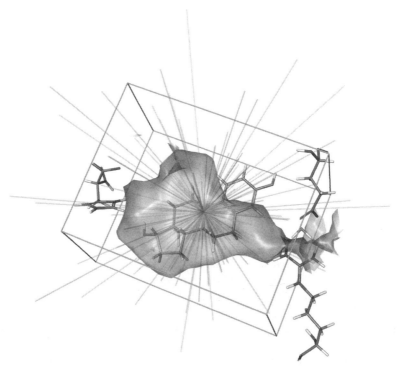

Figure 7-7 U Dock 1.1, in a search for an ideal ligand conformation. Note that this pose will probably be rejected since it clashes with the surface of the active site and protrudes into the protein. *(See insert for color representation of figure.)*

One can enumerate four different levels of theory at which the energetics of ligand–target interaction can be determined or estimated.

1. Molecular mechanics, where atoms and bonds are treated as hard spheres and springs, respectively. The spring constants are derived empirically, from spectroscopy, for example. In this approach, molecular mechanics force fields (MMFFs) are used, which are simply collections of potential energy functions. As the system in question is distorted from equilibrium values, or strained, a potential energy penalty is awarded, in a Hooke's law type of fashion. Coulombic interactions may also lower or increase the levels of strain. Thus, the energy of a given chemical system derived using MMFF-based methods is merely a reflection of strain, not of Gibbs free energy of formation. However, such methods are quite inexpensive and can be applied to macromolecules for geometry optimization.

2. Semiempirical methods incorporating elements of the Schrödinger time-independent equation, along with experimental parameters for electron correlation. The difficulty in all quantum chemical approaches is the fact that systems with more than two electrons cannot be treated adequately. Although these methods provide an improved level of accuracy and can be used to calculate Gibbs free energies of formation, albeit not very accurately, they are somewhat more computationally intensive than MMFF-based methods and are only beginning to gain ground in small macromolecule-scale calculations.

3. Density functional theory–based methods operating on the experimentally proven correlation of ground-state electron density and the ground-state wavefunction, used to determine the system energy. Such methods are moderately computationally expensive and are applied mostly to compounds of only a few hundred daltons.

4. Ab initio methods incorporating a completely theoretical approach to energy calculation and, with the current computational power, applicable only to rather small molecules.

Many software packages exist for carrying out ligand–target interaction studies. One of the most popular is the free AutoDock [53], which comes with a graphical interface called AutoDock Tools. This combination of software packages, along with the support literature produced by the AutoDock user community, can provide a good starting point for any researcher wishing to undertake target-based ligand design.

7.5.3 Structure-Based Methods: QSAR

QSAR-based methods are quite simple in general operational principles, yet can become surprisingly complex in the statistical methods involved in the implementation of this approach. Briefly, a typical QSAR study uses ligand structures and the corresponding empirically determined test outcomes to uncover the physicochemical properties with most impact on the parameter tested. Such parameters can include toxicity, activity, or any other property that is difficult or presently impossible to link to compound structures directly and mechanistically. First, an arbitrary set of calculated ligand properties, as comprehensive as possible but usually including parameters such as pK_a, surface area, number of hydrogen donors, lowest unoccupied molecular orbital energy, molecular weight, octanol/water partition coefficient, and many others, is generated. It is important that all the empirical data are derived in a consistent fashion. For example, data from experiments on mouse enzymes should not be lumped with human enzyme data. Fitting these descriptors to the empirically derived parameter in question, using any of a wide array of statistical techniques, from linear regression to sophisticated factor selection methods, then allows for estimation of the relative importance of each molecular descriptor.

The power of such approaches lies in the fact that once a relationship is obtained, it can (in most cases) provide a glimpse of mechanistic information on the process in question as well as allowing for formulation of predictions of the parameter which was previously obtained only empirically, solely from the compound structure. Unfortunately, QSAR can also be abused. For instance, if a nonrepresentative (either too small or too uniform) set of compounds is used, one may expect elevated QSAR failure rates, especially for compounds of a type distinct from that of the training set [54,55]. Additionally, even minor alterations of compound structure may lead to a change in the mechanism of activity of a given compound, rendering a given relationship useless for a broader set of compounds [56]. Cross-validation and careful delimitation of the working domain of a QSAR model with the use of molecular descriptors may alleviate these problems to some extent, however.

7.5.4   Target-Based Methods: A General Overview

We have already discussed a portion of target-based methods when we considered the "tools of the trade" of ligand discovery. Let us consider the anatomy of a typical target-based ligand design project in more detail by breaking it down into constituent steps (summarized in Fig. 7-8).

1. Target-based approaches start with the structure of the target. The undreamed of collection of tens of thousands of protein structures whose growth accelerates every year, is available from the Research Collaboratory for Structural Bioinformatics (RCSB). Structures obtained here may contain erroneous information, solvent, and particles irrelevant to docking, such as gold atoms and solutes used to crystallize a protein of interest so that its structure can be determined using x-ray crystallography. Due to the nature of x-ray crystallography, hydrogen atoms are not included in solved protein structures and need to be added manually, followed by geometry optimization to avoid atom–atom clashes.

2. If the target structure is unavailable from a database, it may be approximated with protein homology modeling, where a close match of protein

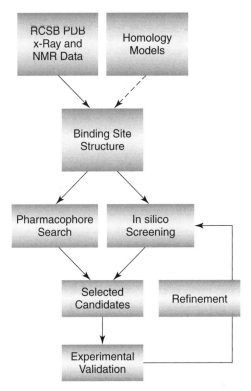

Figure 7-8 Anatomy of a typical target-based design project.

sequence is found in a database of proteins with solved structures and the queried structure is threaded through the known structure, followed by a quick geometry optimization. Homology modeling has shown itself as a reliable technique to obtain a meaningful protein structure [57].

3. In cases where the exact location of the active or binding site is unknown, computational methods capable of identifying it, such as Cavbase, can be utilized [58]. Cavbase assumes that recognition of similar ligands on different proteins is mediated by similar physicochemical and functional properties of the residues defining the ligand binding pocket or cavity. Therefore, knowledge of the ligands of a target protein with available three-dimensional structure and an unmapped active site may help locate the ligand binding on the target protein. This is carried out by attempting to find similarities between the binding pockets available on the protein in question and those of known binding sites for a given ligand found on other proteins. This approach can serve as a starting point in rational inhibitor design by comparing similar binding pockets of a target protein with the inhibitor binding pockets characterized and identifying the structural features that could be involved in the ligand binding to the target protein.

4. Once the target structure is prepared, one may want to identify the phar-macophore, or the set of ligand structural features implicated directly in binding and biological activities through facilities in many software packages. One way to determine these features is to consider the nature of the binding site, and in particular, its shape and charge distribution. By mimicking the shape and charge of the active site, it is possible to design highly specific and tightly binding ligands. One may also perform a high-throughput docking campaign, the cost of which is nowhere near that of its experimental companion, to identify structures that can be improved.

5. The screened and refined structures can subsequently be subjected to more rigorous theoretical tests at higher levels of theory, and the resulting ther-modynamic parameters may be used to run kinetic models to determine the effect of the addition of a given ligand on the overall system.

Finally, the most promising candidates that survive the rigorous screening are synthesized and tested. Herein lies one of the most dangerous parts of target-based design, since in the physical world, it may not be possible to synthesize the ligand so easily created in the virtual world. This procedure is iterative, meaning that experimental results and other screening steps will provide feedback that can be used to improve the promising ligands continuously and prune the number of possible ligands until a satisfactory compound is found.

7.5.5 Selected Applications of Rational Ligand Design

We have already seen the successful application of virtual screening and ratio-nal ligand design in inhibitor development. However, let us consider two recent examples of inhibitor development that drew on in silico, in vitro, and cellu-lar assays. A recent study on the human β-secretase BACE-1 [59], an aspartic protease responsible for the β-amyloid precursor protein cleavage implicated in Alzheimer's disease, nicely demonstrates the power of computational methods in inhibitor design. Two commercially available libraries containing about 300,000 compounds were employed in high-throughput fragment-based docking that is conceptually similar to the fragment-based design discussed above. Briefly, each library compound was broken down into fragments that were placed into the BACE-1 active site available from the RCSB Protein Data Bank in about 1 mil-lion possible orientations followed by the binding energy evaluation considering van der Waals interactions and electrostatic energy. Linear interaction energy with continuum electrostatics (LIECE) was applied to calculate Gibbs free energy of potential candidates' binding. Further tests of 88 prospective candidates yielded 10 potent inhibitors of BACE-1 function in enzyme assay and whole-cell formats. Up to now, very few inhibitors of BACE-1 have been identified with conventional approaches [59].

In another study aimed at the discovery of HIV-1 Nef-SH3 domain of the p59fyn protein tyrosine kinase protein–protein interaction inhibitor [60], we have seen another powerful application of rational design. Cell-based assays and in silico docking with FlexX software were employed to screen the National Cancer Institute Diversity library of nearly 2000 compounds. Interestingly, out of a handful of inhibitors identified, one compound was revealed by both approaches, validating the docking results. Next, using the structural features of the newly discovered inhibitor, 70 structurally similar compounds were identified out of 435,000 compounds from Chembridge EXPRESS-Pick database. Ten promising inhibitors of Nef-SH3 interaction were revealed by cell-based assays, ITC was used to rank the inhibitors by their K_d values, and the interaction surface between the most potent inhibitor and Nef was mapped with NMR.

7.6 CONCLUSIONS

In the last decade we have seen the development of technology with enormous potential for both the creation of large and chemically diverse libraries and their rapid and quantitative screening. Unfortunately, HTS campaigns on large combichem libraries were met with somewhat limited success. At the same time, there are certain complications with natural library-based screening campaigns, efficient solutions to which are only now being introduced. Fortunately, however, the concomitant rise of rational ligand and inhibitor design will help alleviate some of the deficiencies of the past via virtual library screening and directed ligand design. We may be witnessing a shift from large-scale experimental library screening campaigns to smaller, higher-quality, more targeted efforts. Finally, we emphasize especially the power of combined traditional and rational approaches in designing ligands and inhibitors for the future, which promise to be safer, cheaper, more functional, and more effective.

REFERENCES

1. Cyran, R. (2002). *New Developments in Therapeutic Enzyme Inhibitors and Blockers*. Report BIO050B. BCC Research Inc., Wellesley, MA.

2. Smith, H. J., Simons, C. (2005). *Enzymes and Their Inhibition: Drug Development*. CRC Press, Boca Raton, FL.

3. Copeland, R. A. (2005). *Evaluation of Enzyme Inhibitors in Drug Discovery*. Wiley, Hoboken, NJ.

4. Segel, I. H. (1993). *Enzyme Kinetics: Behavior and Analysis of Rapid Equilibrium and Steady-State Enzyme Systems*. Wiley-Interscience. New York.

5. Barthelmes, J., Ebeling, C., Chang, A., Schomburg, I., Schomburg, D. (2007). BRENDA, AMENDA and FRENDA: the enzyme information system in 2007. *Nucleic Acids Res.*, *35*, D511–D514.

6. Yildirim, N., Ankaralioglu, N., Yildrim, D., Akcay, F. (2003). Application of Gröbner bases theory to derive rate equations for enzyme catalysed reactions with two or more substrates or products. *Appl. Math. Comput.*, *137*, 67–76.

7. Savageau, M. A. (1995). Michaelis–Menten mechanism reconsidered: implications of fractal kinetics. *J. Theor. Biol.*, *176*, 115–124.

8. Pagliaro, L., Felding, J., Audouze, K., et al. (2004). Emerging classes of protein–protein interaction inhibitors and new tools for their development. *Curr. Opin. Chem. Biol.*, *8*, 442–449.

9. Sarkar, S., Perlstein, E. O., Imarisio, S., et al. (2007). Small molecules enhance autophagy and reduce toxicity in Huntington's disease models. *Nat. Chem. Biol.*, *3*, 331–338.

10. Burger, A. (1954). Paul Ehrlich. *Chem. Eng. News*, *32*, 4173–4176.

11. Ganellin, C. R. (1993). *Medicinal Chemistry*, 2nd ed. Academic Press, San Diego, CA, p. 127.

12. Lin, X., David, C. A., Donnelly, J. B., et al. (2008). A chemical genomics screen highlights the essential role of mitochondria in HIF-1 regulation. *Proc. Natl. Acad. Sci. USA*, *105*, 174–179.

13. Verdine, G. L. (1996). The combinatorial chemistry of nature. *Nature*, *384*, 11–13.

14. Chang, R. S., Lotti, V. J., Monaghan, R. L., et al. (1985). A potent nonpeptide cholecystokinin antagonist selective for peripheral tissues isolated from *Aspergillus alliaceus*. *Science*, *230*, 177–179.

15. Newman, D., Cragg, G. (2007). Natural products as drug over the past 25 years. *J. Nat. Prod.*, *70*, 461–477.

16. Eldridge, G. R., Vervoort, H. C., Lee, C. M., et al. (2002). High-throughput method for the production and analysis of large natural product libraries for drug discovery. *Anal. Chem.*, *74*, 3963–3971.

17. Koehn, F. E., Carter, G. T. (2005). The evolving role of natural products in drug discovery. *Nat. Rev. Drug Discov.*, *4*, 206–220.

18. Fowler, A., Swift, D., Longman, E., et al. (2002). An evaluation of fluorescence polarization and lifetime discriminated polarization for high throughput screening of serine/threonine kinases. *Anal. Biochem.*, *308*, 223–231.

19. Muegge, I. (2003). Selection criteria for drug-like compounds. *Med. Res. Rev.*, *23*, 302–321.

20. Abel, U., Koch, C., Speitling, M., Hansske, F. G. (2002). Modern methods to produce natural-product libraries. *Curr. Opin. Chem. Biol.*, *6*, 453–458.

21. Feher, M., Schmidt, J. M. (2003). Property distributions: differences between drugs, natural products, and molecules from combinatorial chemistry. *J. Chem. Inf. Comput. Sci.*, *43*, 218–227.

22. Lepourcelet, M., Chen, Y. N., France, D. S., et al. (2004). Small-molecule antagonists of the oncogenic Tcf/beta-catenin protein complex. *Cancer Cell*, *1*, 91–102.

23. Gunasekera, A. P., Gunasekera, M., Longley, R. E., Schulte, G. K. (1990). Discodermolide: a new bioactive polyhydroxylated lactone from the marine sponge *Discodermia dissoluta*. *J. Org. Chem.*, *55*, 4912–4915.

24. Schreiber, S. L. (2000). Target-oriented and diversity-oriented organic synthesis in drug discovery. *Science*, *287*, 1964–1969.

25. Geysen, H. M., Schoenen, F., Wagner, D., Wagner, R. (2003). Combinatorial compound libraries for drug discovery: and ongoing challenge. *Nat. Rev. Drug Discov.*, *2*, 222–230.

26. Lipinski, C. A., Lombardo, F., Dominy, B. W., Feeney, P. J. (2001). Experimental and computational approaches to estimate solubility and permeability in drug discovery and development settings. *Adv. Drug Deliv. Rev.*, *46*, 3–26.

27. Hann, M. M., Oprea, T. I. (2004). Pursuing the leadlikeness concept in pharmaceutical research. *Curr. Opin. Chem. Biol.*, *8*, 255–263.

28. Croston, G. E. (2002). Functional cell-based uHTS in chemical genomic drug discovery. *Trends Biotechnol.*, *20*, 110–115.

29. Reayi, A., Arya, P. (2005). Natural product–like chemical space: search for chemical dissectors of macromolecular interactions. *Curr. Opin. Chem. Biol.*, *9*, 240–247.

30. Peng, L. F., Kim, S. S., Matchacheep, S., et al. (2007). Identification of novel epoxide inhibitors of hepatitis C virus replication using a high-throughput screen. *Antimicrob. Agents Chemother.*, *51*, 3756–3759.

31. Siegal, G., Eiso, A. B., Schultz, J. (2007). Integration of fragment screening and library design. *Drug Discov. Today*, *12*, 1032–1039.

32. Hajduk, P. J., Greer, J. (2007). A decade of fragment-based drug design: strategic advances and lessons learned. *Nat. Rev. Drug Discov.*, *6*, 211–219.

33. Ji, H., Stanton, B. Z., Igarashi, J., et al. (2008). Minimal pharmacophoric elements and fragment hopping, an approach directed at molecular diversity and isozyme selectivity: design of selective neuronal nitric oxide synthase inhibitors. *J. Am. Chem. Soc.*, *130*, 3900–3914.

34. Erlanson, D. A. (2006) Fragment-based lead discovery: a chemical update. *Curr. Opin. Biotechnol.*, *17*, 643–652.

35. Gries, K. D. (2007). Mass spectrometry for enzyme assays and inhibitor screening: an emerging application in pharmaceutical research. *Mass Spectrom. Rev.*, *26*, 324–339.

36. Ng,W., Dai, J. R., Slon-Usakiewicz, J. J, Redden, P. R., Pasternak, A., Reid, N. (2007). Automated multiple ligand screening by frontal affinity chromatography-mass spectrometry (FAC-MS). *J. Biomol. Screen.*, *12*, 167–174.

37. Schriemer, D. C., Bundle, D. R., Li, L., Hindsgaul, O. (1999). Micro-scale frontal affinity chromatography with mass spectrometric detection: a new method for the screening of compound libraries. *Angew. Chem. Int. Ed.*, *37*, 3383–3387.

38. Koehn, F. E. (2008). High impact technologies for natural product screening. *Prog. Drug Res.*, *65*, 176–210.

39. Toledo-Sherman, L., Deretey, E., Slon-Usakiewicz, J. J., et al. (2005). Frontal affinity chromatography with MS detection of EphB2 tyrosine kinase receptor: 2. Identification of small-molecule inhibitors via coupling with virtual screening. *J. Med. Chem.*, *48*, 3221–3230.

40. Lepre, C. A., Moore, J. M., Peng, J. W. (2004). Theory and applications of NMR-based screening in pharmaceutical research. *Chem. Rev.*, *104*, 3641–3676.

41. Shuker, S. B., Hajduk, P. J., Meadows, R. P., Fesik, S. W. (1996). Discovering high-affinity ligands for proteins: SAR by NMR. *Science*, *274*, 1531–1534.

42. Hajduk, P. J., Gerfin, T., Boehlen, J. M., Häberli, M., Marek, D., Fesik, S. W. (1999). High-throughput nuclear magnetic resonance-based screening. *J. Med. Chem.*, *42*, 2315–2317.

43. Leone, M., Crowell, K. J., Chen, J., et al. (2006). The FRB domain of mTOR: NMR solution structure and inhibitor design. *Biochemistry*, *45*, 10294–10302.

44. Haselhorst, T., Blanchard, H., Frank, M., et al. (2007). STD NMR spectroscopy and molecular modeling investigation of the binding of *N*-acetylneuraminic acid derivatives to rhesus rotavirus VP8∗ core. *Glycobiology*, *1*, 68–81.

45. Velazquez-Campoy, A., Leavitt, S. A., Freire, E. (2004). Characterization of protein–protein interactions by isothermal titration calorimetry. *Methods Mol. Biol.*, *261*, 35–54.

46. Ciulli, A., Williams, G., Smith, A. G., Blundell, T. L., Abell, C. (2006). Probing hot spots at protein–ligand binding sites: a fragment-based approach using biophysical methods. *J. Med. Chem.*, *49*, 4992–5000.

47. Kiel, C., Serrano, L., Herrmann, C. (2004). A detailed thermodynamic analysis of ras/effector complex interfaces. *J. Mol. Biol.*, *340*, 1039–1058.

48. Lokesh, G. L., Muralidhara, B. K., Negi, S. S., Natarajan, A. (2007). Thermodynamics of phosphopeptide tethering to BRCT: the structural minima for inhibitor design. *J. Am. Chem. Soc.*, *129*, 10658–10659.

49. Perozzo, R., Folkers, G., Scapozza, L. (2004). Thermodynamics of protein–ligand interactions: history, presence, and future aspects. *J. Recept. Signal Transduct. Res.*, *24*, 1–52.

50. Navidpour, L., Shadnia, H., Shafaroodi, H., Amini, M., Dehpour, R. A., Shafiee, A. (2007). Design, synthesis, and biological evaluation of substituted 2-alkylthio-1,5-diarylimidazoles as selective COX-2 inhibitors. *Bioorg. Med. Chem.*, *15*, 1976–1982.

51. Labute, P. (2006). High strain energies of bound ligands: What is going on? *232nd ACS National Meeting, San Francisco, CA.* American Chemical Society, Washington, DC.

52. Park, H., Lee, J., Lee, S. (2006). Critical assessment of the automated AutoDock as a new docking tool for virtual screening. *Proteins: Struct. Funct. Bioinf.*, *65*, 549–554.

53. Morris, G. M., Goodsell, D. S., Halliday, R. S., et al. (1998). Automated docking using a Lamarckian genetic algorithm and an empirical binding free energy function. *J. Comput. Chem.*, *19*, 1639–1662.

54. Martens, H. A., Dardenne, P. (1998). Validation and verification of regression in small data sets. *Chemom. Intell. Lab. Syst.*, *44*, 99–121.

55. Stone, M. (1977). An asymptotic equivalence of choice of model by cross-validation and Akaike's criterion. *J. R. Statist. Soc. B*, *38*, 44–47.

56. Shadnia, H., Wright, J. S. (2008). Understanding the toxicity of phenols: using QSAR and enthalpy changes to discriminate between possible mechanisms. *Chem. Res. Toxicol.*, *21*, 1197–1204.

57. Honma, T., Hayashi, K., Aoyama, T., et al. (2001). Structure-based generation of a new class of potent Cdk4 inhibitors: new de novo design strategy and library design. *J. Med. Chem.*, *44*, 4615–4627.

58. Schmitt, S., Kuhn, D., Klebe, G. (2002). A new method to detect related function among proteins independent of sequence and fold homology. *J. Mol. Biol.*, *323*, 387–406.

59. Huang, D., Lüthi, U., Kolb, P., Cecchini, M., Barberis, A., Caflisch, A. (2006). In silico discovery of beta-secretase inhibitors. *J. Am. Chem. Soc.*, *128*, 5436–5443.

60. Betzi, S., Restouin, A., Opi, S., et al. (2007). Protein–protein interaction inhibition (2P2I) combining high throughput and virtual screening: application to the HIV-1 Nef protein. *Proc. Nat. Acad. Sci. USA*, *104*, 19256–19261.

61. Oltersdorf, T., Elmore, S. W., Shoemaker, A. R., et al. (2005). An inhibitor of Bcl-2 family proteins induces regression of solid tumours. *Nature*, *435*, 677–681.

62. Bretonnet, A. S., Jochim, A., Walker, O., et al. (2007). NMR screening applied to the fragment-based generation of inhibitors of creatine kinase exploiting a new interaction proximate to the ATP binding site. *J. Med. Chem.*, *50*, 1865–1875.

63. Chen, J., Zhang, Z., Stebbins, J. L., et al. (2007). A fragment-based approach for the discovery of isoform-specific p38alpha inhibitors. *Am. Chem. Soc. Chem. Biol.*, *2*, 329–336.

8

INTERACTION OF A BIOLOGICAL RESPONSE MODIFIER WITH PROTEINS

YUICHI HASHIMOTO

Institute of Molecular and Cellular Biosciences, The University of Tokyo, Tokyo, Japan

8.1 INTRODUCTION

The widespread availability of antibiotics, drugs based on species-selective toxicity, since the middle of the twentieth century has shifted the nature of our lethal diseases from infectious and acute to noninfectious and chronic. For example, the highest mortality rates in Japan before the 1950s were due to infectious diseases such as tuberculosis, but these fell dramatically in the 1950s, and instead, the death rate from cancer has been rising. Until recently, almost all drugs used clinically have been based on the concept of selective toxicity. However, biological response modifiers (BRMs: compounds that regulate cell behavior/gene expression or internal circumstances) offer a different approach [1–3].

BRMs can be divided into two classes. The first includes agents that act directly on cells at the gene expression level to regulate cellular processes. Typical BRMs of this class are nuclear receptor (NR) ligands [4–9]. NRs are ligand-dependent transcription factors which regulate the expression of responsive genes and thereby affect diverse processes, including cell growth,

Protein Targeting with Small Molecules: Chemical Biology Techniques and Applications,
Edited by Hiroyuki Osada
Copyright © 2009 John Wiley & Sons, Inc.

development, differentiation, and metabolism [10]. Based on the elucidated human genome sequence, 48 NRs are thought to exist in humans [10]. So far, the ligands of only about half of them, including steroid hormone receptors [estrogen receptors (ERs)/estradiol, progesterone receptor (PR)/progesterone, androgen receptor (AR)/testosterone, glucocorticoid receptor (GR)/cortisone, and mineral corticoid receptor (MR)/aldosterone], retinoid receptors [retinoic acid receptors (RARs)/*all-trans*-retinoic acid (ATRA) and retinoid X receptors (RXRs)/9-*cis* retinoic acid], thyroxine hormone receptors (TRs)/thyroxine hormones, vitamin D receptor (VDR)/1,25-dihydroxyvitamin D_3 (DHD3), peroxisome proliferator–activated receptors (PPARs)/fatty acids, liver X receptors (LXRs)/oxysterols, farnesoid receptor (FXR)/bile acids, and steroid xenobiotic receptor (SXR)/steroids, have been identified.

The second class of BRMs consists of agents that modulate physiological processes to restore a normal state. One such agent is thalidomide, which was developed in the 1950s as a nontoxic sedative/hypnotic drug, but was withdrawn from the market in the early 1960s because of its serious teratogenicity [1–3,7,8]. However, thalidomide has been established to be useful for the treatment of Hansen's disease (leprosy), and the drug was formally approved for this purpose (treatment of erythema nodusum in Hansen's disease) by the U.S. Food and Drug Administration (FDA) in 1998, under critical control. Many reports have appeared on the therapeutic usefulness of thalidomide in various diseases, such as multiple myeloma (MM), various cancers (including prostate tumor and colon cancer), rheumatoid arthritis, graft-versus-host diseases (GVHD), acquired immunodeficiency syndrome (AIDS), and others [1,7,8]. In 2006, the FDA approved thalidomide (in combination with dexamethasone) for the treatment of MM. Official approval for the use of thalidomide to treat MM has also been applied for in Japan. Although pharmacological applications of thalidomide have been widely investigated, the molecular basis of its actions remains to be fully clarified.

In this chapter, research by the author's group on structural development studies of NR ligands and thalidomide-related molecules is reviewed.

8.2   RAR LIGANDS (RETINOIDS)

RAR is a receptor for *all-trans*-retinoic acid (ATRA: **1**), an active form of vitamin A (except for its function in vision), as well as its bioisosters [called retinoids; including ATRA (**1**)]. The ability of retinoids to modulate the growth and differentiation of a wide variety of cells has been well documented, and ATRA (**1**) plays a role in the control of embryonic development and cell differentiation [4,5]. Retinoids have received much attention from a clinical standpoint, because they are useful in the treatment of vitamin A deficiency, proliferative dermatological diseases, leukemia, and several types of tumors. ATRA (**1**) has been established as the first-choice medicament for the treatment of acute promyelocytic leukemia (APL) [11,12]. Subsequently, many retinoids have been synthesized for potential clinical application [4–8,12]. As regards the target of retinoids, there are three subtypes of RAR: RARα, RARβ, and RARγ. Each of them is a retinoid-dependent transcription factor, which acts as a heterodimer with another member of the NRs, retinoid X receptor (RXR). A close relationship between aberrancy of RARs and malignancy of cells has been well documented [5,13,14]: Truncated RARα fused with another gene (PML or PLZF) as a result of chromosome translocation [t(15;17) and t(11;17), respectively] is found in APL, loss of RARβ is seen in some types of lung tumors, expression of a dominant negative isoform of RARβ in some breast cancers, and so on.

Many researchers have been engaged in structural development studies of retinoids aiming at superior RAR subtype selectivity and amelioration of the clinical disadvantages of ATRA (**1**) and other conventional retinoids with a hydrocarbon skeleton. The major disadvantage of ATRA (**1**)/conventional retinoids is their high lipophilicity and very slow elimination from the body, which cause long-lasting toxicity (hypervitaminosis A). One initial strategy for overcoming these disadvantages is introduction of a heteroatom(s) to obtain structural mimics of ATRA (**1**). A typical example is Am80 (tamibarotene: **2**) (Fig. 8-1) [4–8,12,15]. Another potent synthetic retinoid is TAC101 (**3**) (Fig. 8-1), which possesses two trimethylsilyl groups [4–8,12,16]. One of the unique features of Am80 (**2**) and TAC-101 (**3**) is the lack of binding affinity toward cellular retinoic acid–binding protein (CRABP) [17,18]. Therefore, they are active in patients with CRABP overexpression-based ATRA resistance, which often appears during long-lasting ATRA therapy. Am80 (**2**) possesses potent cell differentiation-inducing activity, and has been approved for the treatment of APL (launched in June 2005 in Japan), while TAC101 (**3**) possesses potent anti-angiogenic activity and anti-hepatometastasis activity, and was successful in a phase II clinical study for the treatment of solid tumors in the United States (a phase III study is starting). Both Am80 (**2**) and TAC101 (**3**) are RARα/β-selective retinoids [4,5,12,18]. Further structural development based on computer-assisted docking (CAD) studies resulted in benzoic acid (**4–9**) and non–benzoic acid types of retinoids (**10–12**) (Fig. 8-1).

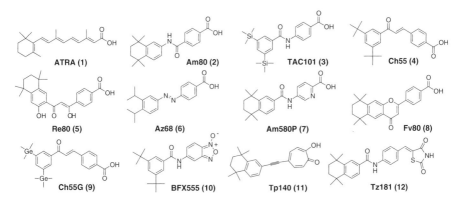

Figure 8-1   Structures of ATRA (**1**) and some synthetic retinoids (**2–12**).

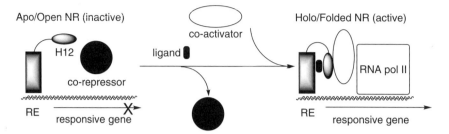

Figure 8-2   Schematic of the transcriptional activation of NR.

CAD studies also afforded valuable information concerning the molecular design of NR antagonists. In the ligand-dependent activation of NRs, a general and important structural feature has been elucidated based on x-ray crystal structure analysis of the ligand-binding domain (LBD) of several NRs with (*holo*-form) or without (*apo*-form) the ligand [i.e., the ligand-induced folding of helix 12 (H12), which is one of the substructures located in the LBD] (Fig. 8-2). In the *apo*-form, H12 takes an open conformation, while in the *holo*-form, it functions as a lid covering the ligand-binding pocket (closed conformation) (Fig. 8-2) [6,19]. This major conformational change induced by ligand binding is thought to be the key structural feature in the activation of NRs. Thus, a compound that binds with the ligand-binding pocket but interferes with the folding of H12 should be an antagonist of the corresponding NR (H12-folding inhibition hypothesis for molecular design of NR antagonists) [6,9]. Thus, the first retinoid antagonist, TD550 (**13**), was created by introduction of a bulky diamantyl group (Fig. 8-3) [6,20]. As expected, CAD studies using the H12-closed conformation of the LBD of RARα suggested that ATRA (**1**) and Am80 (**2**) fit precisely into the ligand-binding pocket, while TD550 (**13**) cannot fit into the pocket [i.e., collision between the diamantyl moiety of TD550 (**13**) and H12 of the LBD was suggested to occur] (Fig. 8-3). Other RAR antagonists (**14–16**) and RXR antagonists (**17–19**) were prepared using a similar strategy (Fig. 8-4) [7,8,12].

8.3   NR ANTAGONISTS DESIGNED BASED ON THE HELIX 12-FOLDING INHIBITION HYPOTHESIS

8.3.1   AR Antagonists

AR is a receptor of androgens, typically testosterone and/or its active form, 5α-dihydrotestosterone, which are endogenous ligands essential for the development and maintenance of the male reproductive system and secondary male sex characteristics [21]. Androgens play diverse physiological and pathophysiological roles in both males and females [21,22]. Among the pathophysiological effects elicited by androgens, a role as endogenous tumor promoters, especially for prostate tumor, is well known. This action is considered to be mediated by

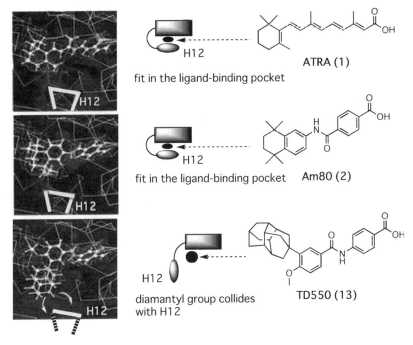

Figure 8-3 CAD studies of retinoids with the LBD of RARα.

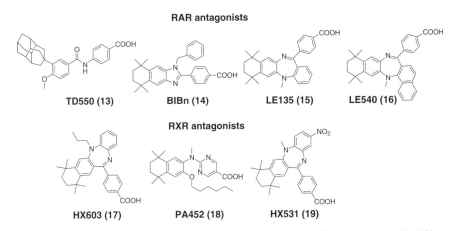

Figure 8-4 Structures of RAR antagonists (**13–16**) and RXR antagonists (**17–19**).

androgen binding to AR. Thus, AR antagonists are expected to be effective for treatment of androgen-dependent tumors, especially for prostate tumor.

Various types of nonsteroidal AR antagonists have been developed so far, including phthalimide-type AR antagonists (**20–22**) (Fig. 8-5) [1–3,6–8,23]. Unlike the RAR/RXR antagonists above (Fig. 8-4), these AR antagonists

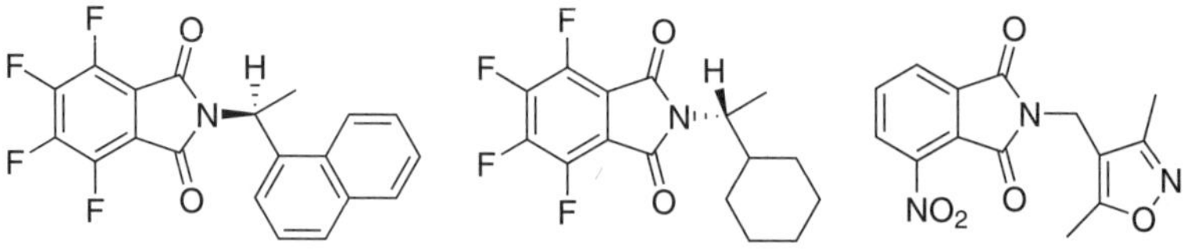

Figure 8-5   Structures of phthalimide-type AR antagonists (**20–22**).

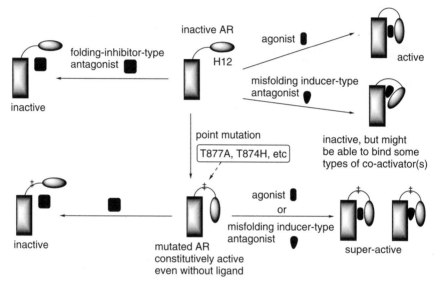

Figure 8-6   Classification of NR antagonists and mechanism of AR antagonist resistance based on point mutation of AR.

(Fig. 8-5) were thought not to be H12-folding inhibitors, judging from CAD studies. In addition, there are many NR antagonists that appear through CAD studies to fit into the ligand-binding pocket, and which are structurally similar in size to the corresponding agonist. A possible interpretation is that these antagonists induce misfolding of H12, making the receptor unable to bind with co-activator(s) (Figs. 8-1 and 8-6) [6,24,25]. Such antagonists are expected to act as partial antagonists/agonists, because NRs with misfolded H12 might still bind some types of co-activator(s). The phthalimide-type AR antagonists (**20–22**) were considered to elicit AR-antagonistic activity by inducing misfolding of H12 [6,23,24].

The major obstacle in the treatment of prostate tumor with AR antagonists is the sudden appearance of AR antagonist-resistant cells. One major molecular mechanism of such resistance is point mutation of AR, as found in the human prostate cancer cell line LNCaP. The AR of LNCaP cells possesses a point mutation T877A, and is considered to take an H12-folded conformation that

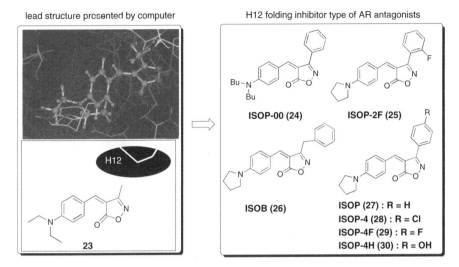

Figure 8-7   Molecular design of H12-folding-inhibitor type of AR antagonists.

is constitutively active even in the absence of the cognate ligand, androgen (Fig. 8-6). Of course, the mutated AR can bind androgens and misfolding inducer-type AR antagonists, which stabilize the active conformation of the AR, leading to superactivation of mutated AR. In fact, our phthalimide-type AR antagonists were ineffective as antagonists toward LNCaP cells, and acted as AR agonists. Therefore, to overcome the problem of such resistance based on AR mutation, H12-folding-inhibitor types of AR antagonists, which bind to the mutated ARs and induce unfolding (or inhibit folding) of H12, would be useful (Fig. 8-6).

CAD studies have yielded only misfolding inducer-type antagonists because the *holo*-form LBD (H12 folded conformation) has been used as a template for docking calculations. Therefore, candidate molecules into which a bulky substituent can be introduced at the region where H12 interacts should be chosen from the group of compounds identified by CAD studies. One such compound is the isoxazolone derivative **23** (Fig. 8-7) [6,24]. Using this skeleton as a scaffold, various substituents were introduced at the position at which H12 interacts, and various H12-folding-inhibitor-type AR antagonists, including ISOP-4 (**28**) and its derivatives (**24–27,29,30**), were obtained (Fig. 8-7) [6,24,25]. These isoxazolone-type AR antagonists (**24–30**) possess high binding affinity for normal and T877A-mutated ARs and have been shown to be active as AR antagonists in both normal and LNCaP cells [25].

8.3.2   VDR Antagonists

VDR is a receptor of 1,25-dihydroxyvitamin D_3 (DHD3: **31**), which is an active form of vitamin D and plays critical roles in a variety of biological activities, including regulation of calcium homeostasis, bone mineralization and control of cellular growth, differentiation, and apoptosis. VDR antagonists can be expected

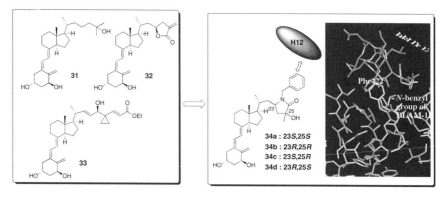

Figure 8-8 Structures of DHD3 (**31**), TEI-9647 (**32**), ZK168281 (**33**), molecular design of DLAM-1P (**34**) based on H12-folding-inhibition hypothesis, and CAD study results of 23*S*, 25*S*-DLAM-1P (**34a**) with the VDR LBD.

to be useful for the treatment of Paget's disease [26,27]. So far, more than 3000 derivatives of DHD3 (**31**) have been synthesized, but only a few types of compounds, including TEI-9647 (**32**) and ZK168281 (**33**) and its analogs (Fig. 8-8), have been reported as VDR antagonists [26,27].

On the basis of CAD studies using the *holo*-form LBD of VDR, a novel VDR antagonist possessing a nitrogen atom in its structure [i.e., DLAM-1P (**34a–d**)] has been designed (Fig. 8-8) [6,28,29]. The nitrogen atom was chosen as a key component to install a bulky side-chain group, which is expected to inhibit the folding of H12. Among the four configurational isomers of DLAM-1P (**34a–d**), the (23*S*,25*S*)-isomer (**34a**) shows the highest affinity for VDR, and only this isomer shows VDR-antagonistic activity with similar potency to that of TEI-9647 (**32**) for inhibition of DHD3 (**31**)-induced cell differentiation of HL-60 cells and in a VDR transcriptional activation reporter gene assay system [6,28,29]. In the VDR transcriptional activation reporter gene assay, TEI-9647 (**32**) shows slight activation at a very high concentration (10 μM), while (23*S*,25*S*)-DLAM-1P (**34a**) does not show any VDR activation. This result suggests that (23*S*,25*S*)-DLAM-1P (**34a**) is a full VDR antagonist i.e., an H12 folding inhibitor). In agreement with this, CAD studies of (23*S*,25*S*)-DLAM-1P (**34a**) using the *holo*-form LBD of VDR suggested collision of the *N*−benzyl group of (23*S*,25*S*)-DLAM-1P (**34a**) with phenylalanine 422 in the H12 region of VDR's LBD (Fig. 8-8) [29].

8.4 NUCLEAR RECEPTOR LIGANDS WITH A 3,3-DIPHENYLPENTANE SKELETON

8.4.1 VDR/AR Ligands

Almost all of the known VDR ligands prepared so far, including DLAM-1P, possess a secosteroidal skeleton. Recently, Boehm et al. reported VDR modulators

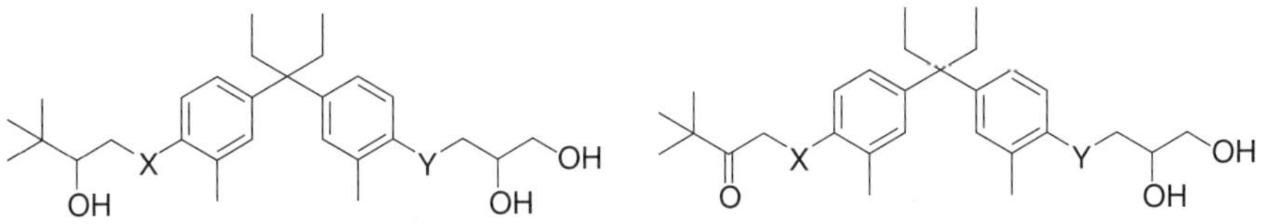

DPP-1123 (35a): X=Y=NH
DPP-1023 (35b): X=NH, Y=O
DPP-0123 (35c): X=O, Y=NH
LG190178 (35d): X=Y=O

DPP-1113 (36a): X=Y=NH
DPP-1013 (36b): X=NH, Y=O
DPP-0113 (36c): X=O, Y=NH
DPP-0013 (36d): X=Y=O

Figure 8-9   Structures of 3,3-diphenylpentane-type dual ligands for VDR (agonists) and AR (antagonists).

of a novel structural type [i.e., compounds with a bisphenol/3,3-diphenylpentane skeleton, including LG190178 (**35d**)] (Fig. 8-9) [30,31]. Based on this skeleton, aza analogs of LG190178 (**35d**) (i.e., compounds **35a–c** and their pivaloyl derivatives, **36a–d**) (Fig. 8-9) have been prepared [9,32,33]. The biological activities were evaluated by receptor-binding assay and human leukemia cell line HL-60 monocytic cell differentiation–inducing assay. HL-60 cell differentiation–inducing activity is thought to correlate well with the binding/activation of VDR by ligands.

Concerning the triol derivatives (**35a–d**), there seems to be a reasonably good correlation between affinity for VDR and HL-60 cell differentiation–inducing activity, with the (R,S)-isomers being the eutomers and the (S,R)-isomers showing the lowest activities in both VDR-binding and HL-60 cell differentiation-inducing assays. (R,S)-DPP-1023 [(R,S)-**35b**] shows the most potent activity among the derivatives for both VDR-binding and HL-60 cell differentiation, with the latter activity being more potent ($EC_{50} = 4.1$ nM) than that of DHD3 (**31**) ($EC_{50} = 9$ nM) [9,33].

On the other hand, although pivaloyl derivatives (**36a–d**) show moderate to potent HL-60 cell differentiation–inducing activity, they lack VDR-binding activity. This discrepancy has been interpreted in terms of fast metabolism of the pivaloyl derivatives (**36a–d**) in HL-60 cells [i.e., the pivaloyl compounds (**36a–d**) have been found to be metabolically reduced in HL-60 cells to the corresponding VDR-binding triol derivatives (**35a–d**)] [9,33]. The importance of the metabolically formed hydroxyl group of **35a–d** in VDR binding was also suggested by CAD studies (i.e., this group is involved in hydrogen bonding to histidine 397 of the VDR-binding pocket) [33]. Therefore, the pivaloyl derivatives (**36a–d**) can be regarded as pro-drugs that require metabolic activation to act as VDR agonists.

The 3,3-diphenylpentane derivatives (**35a–d** and **36a–d**) mentioned above also possess AR-binding affinity [9,32,33]. Concerning the AR-binding activities of the triol derivatives (**35a–d**), the (S,S)-isomers are the eutomers, in contrast to the case of VDR-binding activity, where the (R,S)-isomers are the eutomers. (S,S)-DPP-0123 [(S,S)-**35c**] shows the most potent activity, with a K_i value of 400 nM. Nevertheless, all optical isomers of LG190178 (**35d**) show similar

AR-binding affinity, with K_i values of 1000 to 1100 nM, suggesting that the exchanged nitrogen atom(s) interacts with some residue(s) in the ligand-binding pocket of AR.

In contrast to the case of VDR-binding activity, all the pivaloyl analogs (**36a−d**) show moderate to potent AR-binding activity with K_i values of 720 to 7400 nM, suggesting that metabolic activation (reduction of the pivaloyl group) is not necessary to elicit their AR-binding activity. Among the compounds, (S)-DPP-0113 [(S)-**36c**] shows the most potent AR-binding activity ($K_i = 720$ nM): this compound is more potent than hydroxyflutamide ($K_i = 940$ nM), an active form of the clinically useful antiandrogenic drug flutamide.

None of the compounds mentioned above (**35a−d** and **36a−d**) shows growth-promoting activity on the androgen-dependent cell line SC-3, suggesting that they are not androgen agonists but androgen antagonists [32,33]. In fact, the (S,S)-isomers of **35a−d** (i.e., the eutomers for AR-binding activity) are more potent growth inhibitors of testosterone-induced SC-3 cells than hydroxyflutamide, which is in accordance with their higher affinity for AR than that of hydroxyflutamide. (S,S)-DPP-0123 [(S,S)-**35c**] shows very potent activity, with an IC_{50} value of 4.7 nM, being almost 40 times more potent than hydroxyflutamide [33]. The pivaloyl derivatives (**36a−d**) can be regarded as intrinsically AR-selective ligands, because their reduction has been shown to be necessary to bind VDR (Section 8.3).

8.4.2   FXR and PPAR Ligands

The results above indicate that the 3,3-diphenylpentane skeleton is an effective steroid skeleton substitute, and the NR selectivity of its derivatives can be attributed to the structure(s) of the side chain(s). In fact, on the basis of this consideration, 3,3-diphenylpentane-based specific ligands for other NRs, including FXR and PPAR, have been created successfully [9,34].

FXR is a well-characterized member of the "metabolic" subfamily of NRs and is a transcriptional sensor for bile acids [35]. Its ligands, including chenodeoxycholic acid (CDCA: **37**) (Fig. 8-10), act as signaling molecules and participate in an intricate network of interactions that ultimately govern lipid, steroid, and cholesterol homeostasis and are involved in processes such as glucose utilization, inflammation, and carcinogenesis [35]. Maloney et al. reported GW4064 (**38**) (Fig. 8-8) as a potent synthetic ligand for FXR [36]. Based on the structure of GW4064 (**38**) and the H12 folding inhibition hypothesis (Section 8.3), novel FXR antagonists (**39,40**) have been created (Fig. 8-10). Considering the structures of CDCA (**37**) and GW4064 (**38**), and the potential of 3,3-diphenylpentane as a steroid skeleton substitute, 3,3-diphenylpentane derivatives have been designed as FXR ligand candidates [i.e., DPPF-01 (**41**) and DPPF-13 (**42**)] (Fig. 8.10). The R group of DPPF-01 (**41**), containing a carboxylic acid group, was chosen to mimic the carboxylic acid group found in CDCA (**37**) and GW4064 (**38**). The R group of DPPF-13 (**42**), containing a diol moiety, was chosen to mimic side chains found in several oxycholesterols. In the transcriptional activation reporter

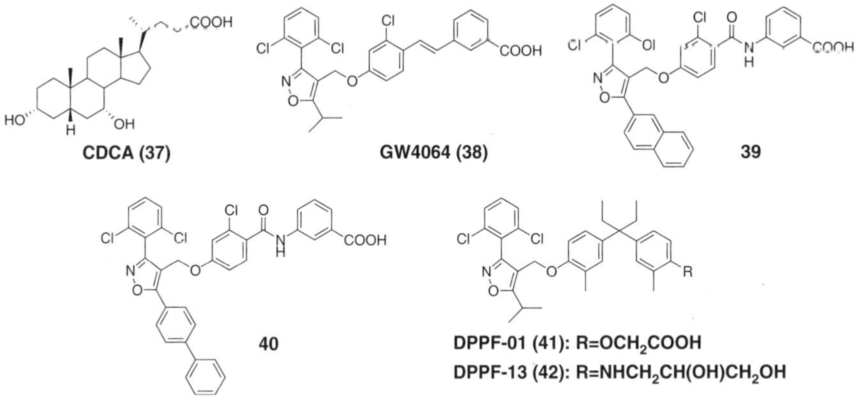

Figure 8-10 Structures of CDCA (**37**), GW4064 (**38**), and designed FXR ligands [antagonists (**39 and 40**) and agonists (**41 and 42**)].

gene assay system, the agonistic activity of DPPF-01 (**41**) for transcriptional activation of FXR is far more potent than that of the physiological ligand, CDCA (**37**). The EC_{50} values are 3.4 µM for DPPF-01 (**41**) and 11.7 µM for CDCA (**37**) under the experimental conditions employed. DPPF-13 (**42**) also shows agonistic activity toward FXR, although it is less potent than CDCA (**37**) [9,34].

PPARα is an NR whose physiological ligands are considered to be endogenous fatty acids, and it is well known as the target molecule of fenofibrate (**43**) (Fig. 8-11), a drug used to treat dyslipidemia and type 2 diabetes, whose active form is considered to be its hydrolyzed analog, fenofibric acid (FA: **44**) (Fig. 8-11). Various synthetic PPARα ligands have been reported, including phenylpropionic acid derivatives derived from KCL (**45**) (Fig. 8-11) [37–39]. PPARα has a large Y-shaped ligand-binding pocket of approximately 1300 to 1400 $Å^3$ [40]. Based on the structures of FA (**44**) and KCL (**45**), as well as the shape of the ligand-binding pocket of PPARα and the potential usefulness of the 3,3-diphenylmethane skeleton as a scaffold for PPARα ligands, the 3,3-diphenylpentane derivative DPPK-01 (**46**) and a diphenylcyclohexane derivative, DPHK-01 (**47**), have been designed, both of which possess a butyric acid moiety (Fig. 8-11). The butyric acid moiety was chosen to mimic the carboxylic acid side chain of KCL (**45**), and a cyclohexyl moiety was adopted to provide a rigid Y-shape of the molecule, as the carbonyl group of FA (**44**) does.

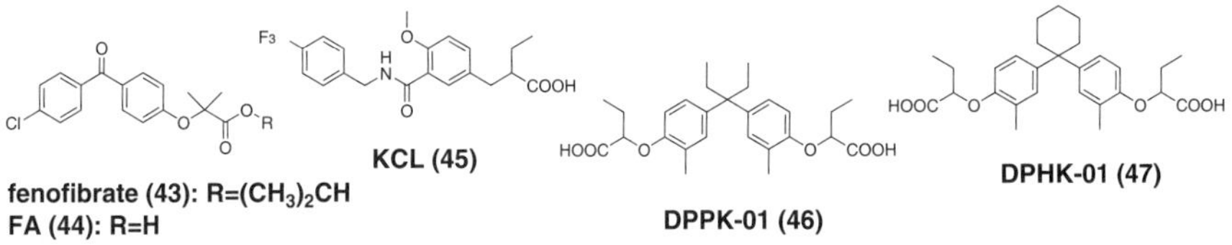

Figure 8-11 Structures of fenofibrate (**43**), FA (**44**), KCL (**45**), and designed PPAR agonists (**46 and 47**).

In the transcriptional activation reporter gene assay, DPHK-01 (**47**) shows more potent agonistic activity for transcriptional activation of PPARα than does FA (**44**). The EC_{50} values were reported to be 3.5 μM for DPHK-01 (**47**) and 9.2 μM for FA (**44**) under the experimental conditions used. The less rigid derivative, DPPK-01 (**46**), shows only very weak agonistic activity toward PPARα [9,34].

These results suggest the usefulness of the 3,3-diphenylmethane skeleton as a multitemplate for NR ligands. Furthermore, the skeleton might also be useful as a pharmacophore for preparing steroidal medicaments other than NR ligands and medicaments whose target molecule(s) are involved in steroid biosynthesis, metabolism, and homeostasis. In fact, 3,3-diphenylmethane-type inhibitors of 5α-reductase (an enzyme that activates testosterone) have been prepared successfully [41].

8.5 THALIDOMIDE-RELATED MOLECULES

8.5.1 Tumor Necrosis Factor (TNF)-α Production Regulators Derived from Thalidomide

Although pharmacological applications of thalidomide (**48**) have been investigated widely, the molecular basis of its actions has not yet been clarified [42]. The beneficial pharmacological effects elicited by thalidomide (**48**) include (A) anti-cachexia activity (cachexia is a major direct cause of cancer death), (B) anti-tumor-promoting activity, (C) anti-angiogenic activity, (D) anti-cell invasion (anti-metastasis) activity, (E) anti-viral activity, and (F) hypoglycemic effect (Fig. 8-12). Although thalidomide (**48**) affects the production of various cytokines, the prevailing hypothesis had been that all of the beneficial effects of thalidomide (**48**) are elicited through regulation of TNF-α production. TNF-α is one of the inflammatory cytokines which is produced mainly by macrophages and T cells (and also by other cell types, including adipocytes and fibroblasts) in response to various stimuli. The activities elicited by TNF-α extend beyond the well-characterized pleiotropic pro-inflammatory properties to include diverse signals for cellular differentiation, proliferation, and death. The growing understanding of the pathophysiological role of TNF-α in induction of cachexia and in various diseases, including AIDS, tumors, and diabetes, has led to the development of strategies to intercept the deleterious effects of TNF-α. However, the TNF-α production-regulating activity of thalidomide (**48**) has been revealed to be bidirectional, depending on both cell types and cell stimulators [1–3,7,8,43,44]. Studies on structural development of thalidomide (**48**) have resulted in very potent bidirectional TNF-α production regulators [e.g., PP-33 (**49**) and FPP-33 (**50**)] and complete separation of the bidirectionality [pure inhibitors (e.g., R-FPTP (**51**) and R-FPTN (**52**)] and pure enhancers [e.g., S-FP13P (**53**)] (Fig. 8-13) [1–3,7,8,44,45]. Some of the TNF-α production regulators prolong the life span of mice with cachexia induced by lipopolysaccharide injection and show more potent anti-angiogenic activity than that of thalidomide (**48**) at a much lower dose than thalidomide (**48**) [2]. However, the

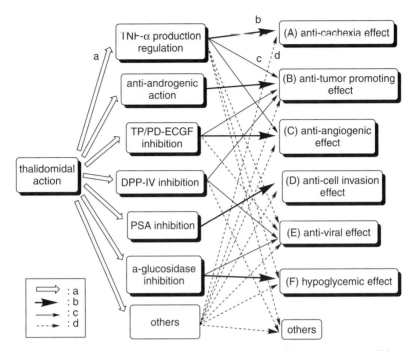

Figure 8-12 Pharmacological effects elicited by thalidomide and their possible target phenomena/molecules: (a) established or possible target; (b) major contribution; (c) partial contribution; (d) minor or unknown contribution.

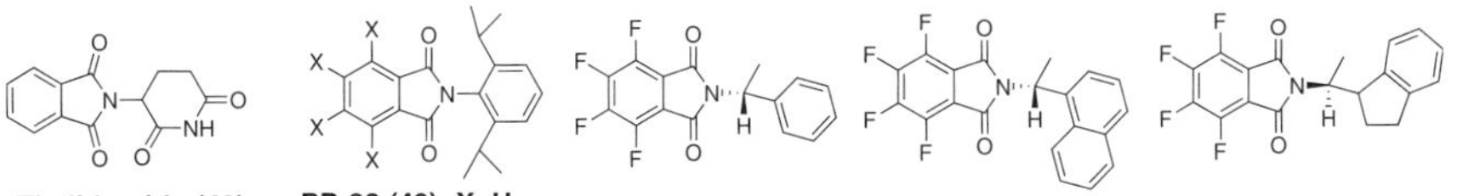

Thalidomide (48) PP-33 (49): X=H R-FPTP (51) R-FPTN (52) S-FP13P (53)
                  FPP-33 (50): X=F

Figure 8-13 Structures of and TNF-α production regulators (**49**–**53**) derived from thalidomide (**48**).

structure–activity relationship studies indicated that the pharmacological effects of thalidomide (**48**) cannot be attributed to its TNF-α production-regulating activity alone [1–3,7,8]. Therefore, structural modifications of thalidomide (**48**) based on different target molecules/phenomena (other than TNF-α) which are considered to be related to the above-mentioned six pharmacological effects (A–F) elicited by thalidomide (**48**) have been studied.

8.5.2 Anti-angiogenic Agents Derived from Thalidomide

For anti-angiogenic activity, thymidine phosphorylase (TP)/platelet-derived endothelial cell growth factor (PD-ECGF) was considered as a putative

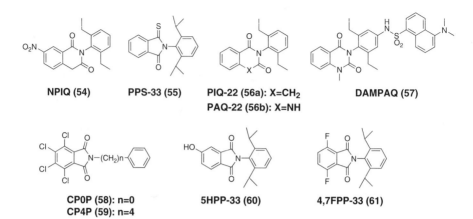

NPIQ (54)   **PPS-33 (55)**   **PIQ-22 (56a): X=CH$_2$**   **DAMPAQ (57)**
PAQ-22 (56b): X=NH

CP0P (58): n=0   **5HPP-33 (60)**   **4,7FPP-33 (61)**
CP4P (59): n=4

Figure 8-14 Enzyme inhibitors (**54–59**) and tubulin polymerization inhibitors (**60** and **61**) derived from thalidomide (**48**).

target molecule (Fig. 8-12). Structural development studies targeting TP/PD-ECGF-inhibitory activity yielded several homophthalimide analogs, including NPIQ (**54**) (Fig. 8-14), which shows more potent inhibitory activity than that of the classical inhibitor, 5-nitrouracil [2,46]. NPIQ (**54**), and related inhibitors are considered to be lead compounds for the development of novel type(s) of TP/PD-ECGF inhibitors.

A preliminary study indicated that the above-mentioned TNF-α production regulators show moderate anti-tumor-promoting activity. This is reasonable, because TNF-α is reported to be one of the endogenous tumor promoters. To develop more potent anti-tumor-promoting agents, researchers have focused on another endogenous tumor promoter [i.e., fibroblast growth factor 10 (FGF-10)]. FGF-10 is reported to act as a tumor promoter, especially in prostate cancer, and its production is induced by the steroid hormone androgen. Considering the effectiveness of thalidomide (**48**) in the treatment of prostate cancer, structural development of thalidomide (**48**) aiming at nonsteroidal androgen antagonists has been performed [1–3,7,8]. The phthalimide-type AR antagonists (Fig. 8-5) were derived from thalidomide (**48**).

8.5.3 Aminopeptidase Inhibitors Derived from Thalidomide

From the viewpoint of anticell invasion/adhesion activity, aminopeptidase-inhibitory activity has been investigated. The structural development studies of thalidomide (**48**) afforded specific dipeptidyl peptidase IV (DPP-IV) inhibitors [e.g., PPS-33 (**55**)] and specific puromycin-sensitive aminopeptidase (PSA) inhibitors [e.g., PIQ-22 (**56a**) PAQ-22 (**56b**)] (Fig. 8-14) [1–3,7,8,46–48]. DPP-IV appears to be involved in various pathophysiological effects, including tumor cell adhesion and the entry of human immunodeficiency virus (HIV) into CD4$^+$ T cells, and therefore DPP-IV inhibitors are expected to be immunomodulators and to have potential pharmacological/clinical applications. Although the physiological role of PSA has not yet been clarified in detail, specific and potent inhibitors, PIQ-22 (**56a**) and PAQ-22 (**56b**), which are more potent than bestatin or actinonin, show much more potent tumor cell invasion-inhibiting activity than bestatin or actinonin [1–3,7,8,46–48]. This suggests that PSA could be a novel target molecule for the development of antimetastatic agents. PIQ-22 (**56a**) and PAQ-22 (**56b**) are completely inactive toward other aminopeptidases, including aminopeptidase N (APN), which has almost the same substrate selectivity as PSA, and leucine aminopeptidase (LAP), against which bestatin and actinonin are potently active. Lineweaver–Burk plot analysis indicates that PIQ-22 (**56a**) and PAQ-22 (**56b**) are noncompetitive inhibitors of PSA, while puromycin and bestatin are competitive inhibitors [3,46–48]. This mode of action might explain the high specificity of PIQ-22 (**56a**) and PAQ-22 (**56b**) for PSA. Generally, aminopeptidase family members possess similar substrate selectivity, with similar structures of the substrate-binding pocket. Therefore, competitive inhibitors generally cross-inhibit aminopeptidases, as bestatin does. Because PIQ-22 (**56a**) and PAQ-22 (**56b**) are noncompetitive inhibitors, it is supposed that PIQ-22 (**56a**) and PAQ-22 (**56b**) bind at a specific site of PSA other than its substrate-binding site. These PSA-specific, potent, nonpeptide, small-molecular inhibitors should be useful as probes to investigate in detail the physiological function of PSA and as lead compounds to develop superior antimetastatic agents. Fluorescent probes [e.g., DAMPAQ-22 (**57**)] that visualize PSA in living cells have also been developed [49].

8.5.4 α-Glucosidase Inhibitors Derived from Thalidomide

Of the six pharmacological effects of thalidomide (**48**) shown in Fig. 8-12, only the anticachexia effect (A) can definitely be interpreted in terms of TNF-α production-regulating activity (TNF-α has been known as one of major factors whose overexpression induces cachexia). The antitumor promotion effect (B) can also be interpreted partly in terms of the same activity but is more likely to be due mainly to AR-antagonistic activity, especially in the case of prostate cancer. The antiangiogenic effect (C) can be interpreted partly in terms of TNF-α production-regulating activity and partly in terms of TP/PD-ECGF-inhibitory activity. The latter activity might also play a role in the antiviral effect (F). The antiviral effect, especially against immunodeficiency virus (HIV), might be explained partly by TNF-α production-regulating activity. The anticell invasion effect (D) can be interpreted in terms of PSA-inhibiting activity. As for the remaining effects, (E) and in part (F), α-glucosidase-inhibiting activity is suspected to be important. α-Glucosidase is an enzyme that catalyzes the final step in the digestion of carbohydrate. Inhibitors of this enzyme may retard the uptake of dietary carbohydrates and suppress postprandial hyperglycemia, and could be useful in the treatment of diabetes, obesity, and certain forms of hyperlipoproteinemia. They also have potential as antiviral agents controlling viral infectivity through interference with the normal biosynthesis of N-linked oligosaccharides by glycosidation of viral coat/envelope glycoproteins, and are being investigated for the treatment of both cancer and AIDS. A well-established classical α-glucosidase inhibitor is 1-deoxynojirimycin (dNM). Some derivatives of dNM have been shown to be effective against AIDS and types B and C of viral hepatitis. Structural development studies based on α-glucosidase-inhibitory activity resulted in potent noncompetitive inhibitors [e.g., CP0P (**58**)] and potent competitive inhibitors [e.g., CP4P (**59**)] (Fig. 8-14) [2,50]. Comparison of the IC_{50} values indicates that CP0P and CP4P are about 13 and 16 times more potent than dNM, respectively.

8.5.5   Other Enzyme Inhibitors Derived from Thalidomide

The studies described above have indicated that the effectiveness and potential of thalidomide (**48**) for the treatment of various diseases cannot be attributed solely to its TNF-α production-regulating activity. Or rather, thalidomide (**48**) should be recognized as a multitarget drug, at least acting on AR, TP/PD-ECGF, DPP-IV, PSA, and α-glucosidase (Fig. 8-12). As described in this article, specific and potent compounds for each of these target molecules/phenomena can be prepared by appropriate modification of the thalidomide structure. This means that thalidomide (**48**) intrinsically possesses pharmacophores with a wide range of activities in its small molecular skeleton. Extraction of the phthalimide and homophthalimide structures of thalidomide (**48**) and use of these skeletons have made it possible to obtain specific and potent TNF-α production regulators, including bidirectional ones (e.g., **49** and **50**), pure inhibitors (e.g., **51** and **52**) and pure enhancers (e.g., **53**), TP/PD-ECGF inhibitors (e.g., **54**), androgen antagonists (e.g., **20–22**), DPP-IV inhibitors (e.g., **55**), PSA inhibitors (e.g., **56,57**), and α-glucosidase inhibitors (e.g., **58** and **59**) (Figs. 8-13 and 8-14). The same strategy is expected to allow the development of many agents with thalidomidal action(s) (i.e., hypnotic, antimalarial, and others). There are also other biological effects of thalidomide (**48**), besides those listed in Fig. 8-12. Inhibition of cyclo-oxygenases [51,52], μ-calpain [46], NO synthase [53], and histone deacetylase (HDAC) [54] are candidate actions, and based on these biological activities, specific and potent compounds have been developed [46,51–54]. Thalidomide (**48**) itself has relatively low potency, or is inactive, toward some of the target molecules listed in Figure 8-12. There are at least two possible interpretations of this. One is that the overall effects of thalidomide (**48**) on the target molecules are additive. Simultaneous actions on plural target molecules by thalidomide (**48**), even though each is not so potent, might appear as clinically useful effects. The other interpretation involves metabolism of thalidomide (**48**). Thalidomide (**48**) is both chemically and metabolically labile, and various metabolites, including hydroxylated metabolites [55–57], are known to be produced invivo. Therefore, one or more metabolites might possess very potent activity on some or a specific target molecule(s) among those listed above. In fact, teratogenicity of thalidomide (**48**) has been reported to be attributed to a metabolite, rather than to thalidomide itself. Also, some thalidomide metabolites are known to possess potent cell differentiation induction-enhancing activity [55] and tubulin polymerization-inhibiting activity [56,58,59]. Based on the latter activity and the structure of the thalidomide metabolite, potent tubulin polymerization inhibitors [e.g., 5HPP-33 (**60**) and 4,7FPP-33 (**61**) (Fig. 8-14)] have been developed [56,58,59]. 5HPP-33 (**60**) has also been reported by other researchers to be a tubulin de-polymerization inhibitor [60], and this issue has not been resolved yet.

8.6 CONCLUSIONS

Structural development studies of BRMs (NR ligands and thalidomide-related molecules) were reviewed in this chapter. As regards retinoids, Am80 (tamibarotene: **2**) is already in clinical use, and TAC-101 (**3**) is now starting phase III studies. Computer-assisted molecular design has resulted in novel retinoids with completely different skeletons. This expansion of the structural variation of retinoids should ultimately afford superior drugs for differentiation therapy. Concerning structural development studies of NR antagonists based on the H12-folding-inhibition hypothesis, the target can be said to be movement (dramatype) of the receptor molecules rather than the receptor molecules themselves (genotype/phenotype) [61]. In this sense, the strategy can be called *dramatype design*.

Concerning the structural development of thalidomide (**48**), our strategy can be considered to be universal template-based deductive drug design, like that adopted in the structural development of NR ligands based on the 3,3-diphenylpentane skeleton as a steroid substitute. In the structural development of thalidomide (**48**) described in this chapter, the following pointss should be emphasized. First, six pharmacological and biological effects of thalidomide (**48**) were identified. Then a hypothesis as to the molecular target or target phenomenon which is potentially relevant to each pharmacological/biological effect was made. It is important to note that it does not matter whether thalidomide itself really binds to the hypothetical molecular target. The aim is simply to reproduce the relevant pharmacological/biological effect specifically by using newly prepared compounds. The third step was the creation of potent and specific compounds. Compounds thus prepared, of course, might merely mimic thalidomide's pharmacological/biological effects, but might have no relation to thalidomide (**48**) at the molecular mechanistic level. Nevertheless, we believe that by preparing compounds (thalidomide substitutes) that mimic the pharmacological/biological effects elicited by thalidomide (**48**) [even if the molecular mechanism is different from that of thalidomide (**48**)], and using combinations of the prepared compounds, it will be possible to reproduce or reconstruct the spectrum of pharmacological/biological effects of thalidomide (**48**). Thus, clinically useful effects, which are the same as those elicited by thalidomide (**48**), will be able to be achieved by using the "thalidomide substitutes," avoiding the risk of teratogenicity, which is the major obstacle in thalidomide therapy.

REFERENCES

1. Hashimoto, Y. (1998). *Curr. Med. Chem.*, *5*, 163–178.
2. Hashimoto, Y. (2002). *Bioorg. Med. Chem.*, *10*, 461–479.
3. Hashimoto, Y., Tanatani, A., Nagasawa, K., Miyachi, H. (2004). *Drugs Future*, *29*, 383–391.
4. Hashimoto, Y. (1991). *Cell Struct. Funct.*, *16*, 113–123.
5. Hashimoto, Y., Shudo, K. (1991). *Cell Biol. Rev.*, *25*, 209–230.
6. Hashimoto, Y., Miyachi, H. (2005). *Bioorg. Med. Chem.*, *13*, 5080–5093.
7. Hashimoto, Y. (2002). *Mini-Rev. Med. Chem.*, *2*, 543–551.
8. Hashimoto, Y. (2003). *Cancer Chemother. Pharmacol.*, *52*, S16–S23.
9. Hosoda, S., Hashimoto, Y. (2007). *Pure Appl. Chem.*, *79*, 615–626.
10. Chawla, A., Pepa, J. J., Evans, R. M., Mangelsdorf, D. J. (2001). *Science*, *294*, 1866–1870.
11. Chomienne, C., Ballerini, P., Bailitrand, N., et al. (1990). *Blood*, *76*, 1710–1717.
12. Kagechika, H. (2002). *Curr. Med. Chem.*, *9*, 591–608.
13. Houle, B., Rochette-Egly, C., Bradley, W. E. C. (1993). *Proc. Natl. Acad. Sci. USA*, *90*, 985–989.
14. Liu, Y., Lee, M., Wang, H., et al. (1996). *Mol. Cell. Biol.*, *16*, 1138–1149.
15. Kagechika, H., Kawachi, E., Hashimoto, Y., Himi, T., Shudo, K. (1988). *J. Med. Chem.*, *31*, 2182–2192.
16. Yamakawa, T., Kagechika, H., Kawachi, E., Hashimoto, Y., Shudo, K. (1990). *J. Med. Chem.*, *33*, 1430–1437.
17. Hashimoto, Y., Kagechika, H., Kawachi, E., Shudo, K. (1988). *Jpn. J. Cancer Res. (Gann)*, *79*, 473–483.
18. Hashimoto, Y., Kagechika, H., Shudo, K. (1990). *Biochem. Biophys. Res. Commun.*, *166*, 1300–1307.
19. Wurz, J., Bourguet, W., Renaud, J., et al. (1996). *Nat. Struct. Biol.*, *3*, 87–94.
20. Kaneko, S., Kagechika, H., Kawachi, E., Hashimoto, Y., Shudo, K. (1992). *Med. Chem. Res.*, *2*, 361–367.
21. Mooradian, A. D., Morley, J. E., Korenman, S. G. (1987). *Endocr. Rev.*, *8*, 1–28.
22. Zhi, L., Martinborough, E. (2001). *Annu. Rep. Med. Chem.*, *36*, 169–180.
23. Miyachi, H., Azuma, A., Kitamoto, T., et al. (1997). *Bioorg. Med. Chem. Lett.*, *7*, 1483–1488.
24. Ishioka, T., Kubo, A., Koiso, Y., Nagasawa, K., Itai, A., Hashimoto, Y. (2002). *Bioorg. Med. Chem.*, *10*, 1555–1566.
25. Ishioka, T., Tanatani, A., Nagasawa, K., Hashimoto, Y. (2003). *Bioorg. Med. Chem. Lett.*, *13*, 2655–2658.
26. Miura, D., Manabe, K., Ozono, K., et al. (1999). *J. Biol. Chem.*, *274*, 16392–16399.
27. Herdick, M., Steinmeyer, A., Carlsberg, C. (2000). *Chem. Biol.*, *7*, 885–894.
28. Kato, Y., Nakano, Y., Sano, H., et al. (2004). *Bioorg. Med. Chem. Lett.*, *14*, 2579–2583.
29. Nakano, Y., Kato, Y., Imai, K., et al. (2006). *J. Med. Chem.*, *49*, 2398–2406.
30. Boem, M. F., Fitzgerald, P., Zou, A., et al. (1999). *Chem. Biol.*, *6*, 265–275.

31. Swann, S. L., Bergh, L., Farach-Carson, M. C., Ocasio, C. A., Koh, J. T. (2002). *J. Am. Chem. Soc.*, *124*, 13795–13805.
32. Hosoda, S., Tanatani, A., Wakabayashi, K., et al. (2005). *Bioorg. Med. Chem. Lett.*, *15*, 4327–4331.
33. Hosoda, S., Tanatani, A., Wakabayashi, K., et al. (2006). *Bioorg. Med. Chem.*, *14*, 5489–5502.
34. Kainuma, M., Kasuga, J., Hosoda, S., et al. (2006). *Bioorg. Med. Chem. Lett.*, *16*, 3213–3218.
35. Pellicciari, R., Constantino, G., Fiorucci, S. J. (2005). *J. Med. Chem.*, *48*, 5383–5404.
36. Maloney, P. R., Parks, D. J., Haffner, C. D., et al. (2000). *J. Med. Chem.*, *43*, 2971–2974.
37. Nomura, M., Tanase, T., Ide, T., et al. (2003). *J. Med. Chem.*, *46*, 3581–3599.
38. Kasuga, J., Makishima, M., Hashimoto, Y., Miyachi, H. (2006). *Bioorg. Med. Chem. Lett.*, *16*, 554–558.
39. Kasuga, J., Hashimoto, Y., Miyachi, H. (2006). *Bioorg. Med. Chem.*, *16*, 771–774.
40. Nolte, R. T., Wisely, G. B., Westin, S., et al. (1998). *Nature*, *395*, 137–143.
41. Hosoda, S., Hashimoto, Y. (2008). *Bioorg. Med. Chem. Lett.*, *17*, 5414–5418.
42. Bartlett, J. B., Dredge, K., Dalgleish, A. G. (2004). *Nat. Rev. Cancer*, *4*, 314–322.
43. Miyachi, H., Azuma, A., Hioki, E., Iwasaki, S., Kobayashi, Y., Hashimoto, Y. (1996). *Biochem. Biophys. Res. Commun.*, *224*, 426–430.
44. Miyachi, H., Azuma, A., Ogasawara, A., et al. (1997). *J. Med. Chem.*, *40*, 2858–2865.
45. Miyachi, H., Azuma, A., Hioki, E., Iwasaki, S., Kobayashi, Y., Hashimoto, Y. (1996). *Biochem. Biophys. Res. Commun.*, *226*, 439–444.
46. Kakuta, H., Takahashi, H., Sou, S., Kita, T., Nagasawa, K., Hashimoto, Y. (2001). *Recent Res. Dev. Med. Chem.*, *1*, 189–211.
47. Komoda, M., Kakuta, H., Takahashi, H., et al. (2001). *Bioorg. Med. Chem.*, *9*, 121–131.
48. Kakuta, H., Tanatani, A., Nagasawa, K., Hashimoto, Y. (2003). *Chem. Pharm. Bull.*, *51*, 1273–1282.
49. Kakuta, H., Koiso, Y., Nagasawa, K., Hashimoto, Y. (2003). *Bioorg. Med. Chem. Lett.*, *13*, 83–86.
50. Sou, S., Mayumi, S., Takahashi, H., et al. (2000). *Bioorg. Med. Chem. Lett.*, *10*, 1081–1084.
51. Noguchi, T., Shimazawa, R., Nagasawa, K., Hashimoto, Y. (2002). *Bioorg. Med. Chem. Lett.*, *12*, 1043–1046.
52. Sano, H., Noguchi, T., Tanatani, A., Hashimoto, Y., Miyachi, H. (2005). *Bioorg. Med. Chem.*, *13*, 3079–3091.
53. Noguchi, T., Sano, H., Shimazawa, R., Tanatani, A., Miyachi, H., Hashimoto, Y. (2004). *Bioorg. Med. Chem. Lett.*, *14*, 4141–4145.
54. Shinji, C., Maeda, S., Imai, K., Yoshida, M., Hashimoto, Y., Miyachi, H. (2006). *Bioorg. Med. Chem.*, *14*, 7625–7651.
55. Noguchi, T., Shinji, C., Kobayashi, H., Makishima, M., Miyachi, H., Hashimoto, Y. (2005). *Biol. Pharm. Bull.*, *28*, 563–564.

56. Inatsuki, S., Noguchi, T., Miyachi, H., et al. (2005). *Bioorg. Med. Chem. Lett.*, *15*, 321–325.

57. Nakamura, T., Noguchi, T., Kobayashi, H., Miyachi, H., Hashimoto, Y. (2006). *Chem. Pharm. Bull.*, *54*, 1709–1714.

58. Yanagawa, T., Noguchi, T., Miyachi, H., Kobayashi, H., Hashimoto, Y. (2006). *Bioorg. Med. Chem. Lett.*, *16*, 4748–4751.

59. Kizaki, M., Hashimoto, Y. (2008). *Curr. Med. Chem.*, *15*, 754–765.

60. Li, P.-K., Pandit, B., Sackett, D. L., et al. (2006). *Mol. Cancer Ther.*, *5*, 450–456.

61. Hino, O. (2004). *Int. J. Oncol.*, *9*, 257–261.

9

CHEMICAL BIOLOGY OF CELL MOTILITY INHIBITORS

Tatsuro Kawamura, Mitsuhiro Kitagawa, and Masaya Imoto

Department of Biosciences and Informatics, Faculty of Science and Technology, Keio University, Yokohama, Kanagawa, Japan

9.1 INTRODUCTION: BIOLOGICAL BACKGROUND

9.1.1 Directed Cell Migration in Physiology and Pathology

Directed cell migration (Fig. 9-1) is a fundamental cellular process during many phases of development and adult life. For example, it is essential for gastrulation, the process by which the endoderm and mesoderm take up their correct topological positions in the embryo, the formation and wiring of the nervous system, and the development of organs such as the heart and intestinal tract. Furthermore, in the renewal of skin and intestine, fresh epithelial cells migrate up from the basal layer and the crypts, respectively. During skeletal muscle regeneration caused by injury, muscle satellite cells proliferate toward the site of muscle injury. Migration is also a prominent component of tissue repair and immune surveillance, in which leukocytes from the circulation migrate into the surrounding tissue to destroy invading microorganisms and infected cells and to clear debris. On the other hand, migration also contributes to several important pathological processes, including vascular diseases, osteoporosis, chronic inflammatory disease such as rheumatoid arthritis and multiple sclerosis, and metastasis of tumor cells [1].

Protein Targeting with Small Molecules: Chemical Biology Techniques and Applications,
Edited by Hiroyuki Osada
Copyright © 2009 John Wiley & Sons, Inc.

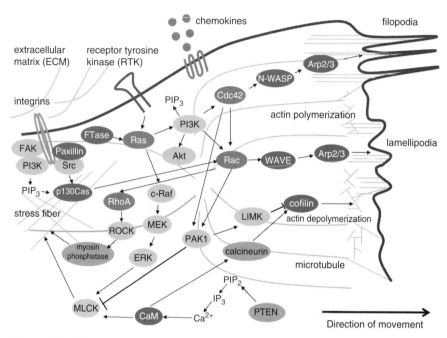

Figure 9-1 Signal transduction cascades involved in cell migration. Cell migration is regulated by various molecules, such as small GTPases, kinases, phosphatases, and other molecules (ellipses) which regulate actin dynamics in response to extracellular stimuli.

9.1.2 Molecular Mechanisms Underlying Cell Migration

Cells can either migrate as individuals or move in the context of tissues. Movement is controlled by internal and external signals, which activate complex signal transduction cascades, resulting in highly dynamic and localized remodeling of the cytoskeleton, cell–cell, and cell–substrate interactions. Although the detailed molecular mechanisms underlying cell migration differ among cells, the actin cytoskeleton and its regulatory proteins are crucial for cell migration in most cells. Cell migration is generally induced by three ordered steps: cell polarization, front-membrane protrusion formation and adhesion to the substrate, and retraction of the rear part of the cell [2].

Cell Polarization   Cell polarization, the difference in molecular processes from the front to the rear, is established and maintained in response to extracellular stimuli. It is mediated by Rho family GTPases, phosphoinositide 3-kinases (PI3Ks), microtubules, and vesicular transport [3]. Cdc42 is a key regulator of cell polarity in eukaryotic organisms and is activated toward the front of migrating cells. Cdc42 can affect polarity by localizing the microtubule-organizing center (MTOC) and Golgi apparatus in front of the nucleus, oriented toward the leading edge. Cdc42-induced MTOC orientation may contribute to polarized migration by facilitating microtubule growth into the lamella and microtubule-mediated delivery of Golgi-derived vesicles to the leading edge, providing membrane and associated proteins needed for forward protrusion. The effects of Cdc42 on the MTOC position appear to be exerted mainly through a pathway involving the Cdc42 effector PAR6, which exists in a complex with PAR3 and an atypical protein kinase C (aPKC). Recent evidence suggests that it could occur as a result of local capture of microtubules at the leading edge via APC, a protein that binds tubulin and localizes to the end of microtubules, via CLIP170 and IQGAP and/or via the microtubule-based dynein/dynactin motor protein complex. PAK1 can mediate Cdc42 activation downstream of heterotrimeric GTP-binding protein-coupled receptors (GPCRs), which are activated by many chemoattractants. These interactions define a positive feedback loop between Cdc42 and PAK1, resulting in high Cdc42 activity at the leading edge. The phosphoinositides PtdIns(3,4,5)P_3 (PIP$_3$) and PtdIns(3,4)P_2 (PIP$_2$) are essential signaling molecules that become rapidly polarized in cells that are exposed to a gradient of chemoattractant. This chemoattractant gradient amplification process involves both localized accumulation and activation of PI3Ks, which generate PIP$_3$/PIP$_2$ at the leading edge of cells in response to a chemoattractant, and the phosphatase PTEN, which removes them at the side and the rear. Rac is also a key regulator of maintaining directional protrusion, which is restricted to the front of the cell and is activated by PI3K product-activated Rac GEFs. On the other hand, Rho is localized at the rear and stabilizes microtubules.

Front-Membrane Protrusion Formation and Adhesion to Substrate The front-membrane protrusion formation process is regulated primarily by the Rho family of GTPases [2,4]. During membrane protrusion, rapid actin polymerization is induced at the leading edge, resulting in the formation of filopodia and lamellipodia; filopodium formation is mediated by the Cdc42 and lamellipodium formation by Rac. Rac and Cdc42 activate WAVE and WASP proteins, respectively, resulting in Arp2/3 complex activation. WAVE/WASP proteins may regulate the activity of Rac and Cdc42 by binding to GAPs and GEFs, and could thereby generate positive or negative feedback loops to regulate the extent of Rac/Cdc42-induced actin polymerization. WAVE/WASP proteins can also be regulated by other stimuli apart from Rac and Cdc42, including Src family kinases, the adaptor proteins Nck and WIP, and phosphoinositides. Several actin-binding proteins also regulate the rate and organization of actin polymerization in protrusions by affecting the pool of available monomers and free ends. For example, profilin prevents self-nucleation by binding to actin monomers and also serves to selectively target monomers to barbed ends. Disassembly of older filaments, which is needed to generate actin monomers for polymerization at the front end, is assisted by proteins of the ADF/cofilin family, which sever filaments and promote actin dissociation from the pointed end. Other proteins play supporting roles in the dendritic network: cortactin stabilizes branches, whereas filamin A and α-actinin stabilize the entire network by cross-linking filaments.

After protrusion formation, stabilization by attaching to the extracellular matrix (ECM) or other cells is an essential step in migration. Although many different receptors are involved in the migration of different cell types, the integrins are a major family of migration-promoting receptors. Ligand-binding integrins initiate intracellular signals such as protein tyrosine phosphorylation, activation of small GTPases, and changes in phospholipid biosynthesis that regulate the formation and strengthening of adhesion sites, the organization and dynamics of the cytoskeleton, and cell polarity during migration. Integrin affinity is increased by activated intermediates such as the GTPase Rap1 or PKC. Conversely, activation of Raf-1 kinase often suppresses integrin activation. The cytoskeletal linker protein talin promotes integrin activation. Integrin is also regulated by post-translational modification. For example, integrin α_4 phosphorylation blocks the binding of paxillin, and this step is required to maintain stable lamellipodia of migrating cells. Small adhesions known as *focal complexes*, which depend on Rac and Cdc42, can be observed at the leading edge, whereas focal adhesions are severely tight and are typically observed either in nonmigratory cells or in cells that move very slowly.

Retraction of the Rear Part of the Cell   Finally, adhesion disassembly and retraction at the rear are also important steps in cell migration. Adhesion assembly is observed both at the leading edge and at the rear. Both protein kinases and phosphatases appear to be important for the regulation of adhesion turnover and stability [1,5]. For example, cells lacking the tyrosine kinase FAK or Src have more and larger adhesions and migrate poorly. The interaction of FAK with Src and the adapter proteins p130Cas and Crk, which in turn activate Rac-specific GEFs, appears to regulate adhesion turnover. Adhesion turnover in migrating cells is also regulated by a complex of Rac-associated proteins and by the mitogen-activated protein kinase ERK. In addition, intracellular calcium levels are implicated in the disassembly of adhesions at the rear. The tension generated in migrating cells by strong adhesions at the rear can be sufficient to open stretch-activated calcium channels. Potential targets for calcium are the calcium-regulated phosphatase calcineurin and the calcium-activated protease calpain, which is also activated by ERK and has the potential to cleave several focal adhesion proteins, including integrins, talin, vinculin, and FAK.

In rear retraction, the force transmitted to sites of adhesion derives from the interaction of myosin II with actin filaments that attach to these sites. Myosin II activity is regulated by myosin light-chain (MLC) phosphorylation, which is positively regulated by MLC kinase (MLCK) or Rho kinase (ROCK) and negatively regulated by MLC phosphatase, which itself is phosphorylated and inhibited by ROCK. Whereas MLCK is regulated by intracellular calcium concentration as well as by phosphorylation by a number of kinases, ROCK is regulated by binding Rho-GTP. MLC phosphorylation activates myosin, resulting in increased contractility and transmission of tension to sites of adhesion.

9.2   CHALLENGES FROM A CHEMICAL VIEWPOINT

Cell migration requires the complex coordination of numerous cellular processes of cell polarization, actin cytoskeletal reorganization, membrane cycling, and adhesion. Those processes are triggered by extracellular stimuli, which cause intracellular signal transduction sequentially through highly regulated molecular networks. To gain insight into those complex mechanisms, a remarkable approach of chemical biology, using low molecular inhibitors, has emerged. This report focuses on the discovery of these inhibitors, summarizing the various efforts to develop screening methods for inhibitors and analysis methods for cell migration.

The challenge of developing antagonists for membrane receptors is one standpoint to reveal key extracellular stimuli for cell migration. Since huge numbers of stimuli for a wide variety of migratory cell types are known, the specificity of the designed antagonist is apparently the core of the challenge. For example, about 50 chemokines corresponding to >20 chemokine receptors are characterized in the inflammatory reaction of leukocytes [6], and there are many reports of antagonists against them for inflammatory leukocytes [7], each of which often has >1000-fold specificity. For those, high-throughput assays have been developed in which people or automated machines evaluate chemokine binding to their receptors using radiolabeled chemokines and by monitoring Ca^{2+} mobilization triggered by chemokine stimuli. Diverse growth factors are also key triggers—VEGF for epithelial cell migration, KGF for keratinocyte wound healing, HGF for regeneration of tissues—and almost every growth factor is associated to some degree with developmental cell motility and cancer cell migration [8]. Growth factor receptors possess tyrosine kinase activity, and several reports have evaluated the specificity of kinase inhibitors by enzyme-linked immunosorbent assay (ELISA) using phosphorylated kinase-specific antibody. Indeed, commercial profiling services are now available that utilize hundreds of kinases. Recently, many studies have reported these profiling methods to distinguish the specificity of kinase inhibitors.

Another challenge focuses on intracellular signal transduction–regulating cellular processes. Actin cytoskeleton, membrane cycling, and adhesion are the processes being studied but no universal method seems to promise success in obtaining inhibitors against proteins in these processes. However, several innovative and unique methods have been developed; one unique example is the actin polymerization assay. In this assay, pyrene-tagged actin monomer is an interesting element. When an actin filament is generated, a tag pyrene molecule approaches, and a special fluorescence called *pyrene excimer fluorescence* can be monitored. Using this method, several small-molecular compounds have been discovered and evaluated. In this assay system, the protein for inducing polymerization can vary. Extracts of bacteria or eukaryote cells containing actin cytoskeletal proteins are used mainly as inducers of actin polymerization. Peterson et al. used *Xenopus* egg larvae, which contain Cdc42, N-WASP, and Arp2/3 complex, and found an N-WASP inhibitor [9]. *Sacharomyces cerevisiae* and *Listeria* monocytogenes are also used by other research teams to evaluate the effect of compounds on actin polymerization [10]. In silico screening is another innovative method, in which computers calculate the virtual affinity of compounds to target proteins.

Once an inhibitor is obtained by these in vitro, in silico screening methods, it is necessary to evaluate its effect on cell systems. The Boyden chamber assay and the wound healing assay are the major simplified model assays of migration. The Boyden chamber assay has two compartments separated by a membrane filter. Cells seeded in the upper compartment migrate through the filter pores to the lower compartment. Cells are treated with/without the inhibitor, and migration is evaluated by counting the cells in the lower compartment. Additionally, researchers individually conduct gene-knockdown analysis and/or overexpression analysis to prove that the inhibitor target is implicated in cell migration.

On the other hand, phenotype screening is promising for inhibitory activity against cell migration. The wound healing assay is a unique high-throughput method. In this method, cell monolayers are scratched with a pin, compounds are added, cells are incubated for several hours to recover from the scratch by cell migration, and the size of the remaining scratched wound is evaluated. When the compound inhibits cell migration, the scratched area remains. This is very convenient to evaluate activity to inhibit cell migration; however, as Sutherland introduced in *Drug Discovery Today*, billions of compounds probably inhibit cell migration in the wound healing assay [11]. This is because a multitude of molecules, from transcription factors to membrane trafficking proteins, could be involved in this model assay. Although this assay cannot elucidate the inhibitory mechanism of the compound, several different research teams have discovered inhibitors against proteins that are related to the GTPase family. Yarrow et al. showed the importance of a secondary screening assay to narrow the potential targets of inhibitors and found that inhibitors affect the Rho pathway [10]. Lee et al. used differential analysis screening for the integrin inhibitor [12]. They generated cells highly expressing integrin from normal cells by gene transfection, and screened for an inhibitor which affected cell migration expressing integrin, but not original cells.

Many proteins are implicated in membrane protrusion in cell migration, which can be observed under microscopy. Recently, a cell-imaging assay was proposed for high-throughput phenotype screening by Eggert et al. [13]. In this assay, compounds are added to cells, proteins of interest are rendered fluorescent by a combination of molecular probes or antibodies or a green fluorescence protein tag, and microscopy captures the images. Those images contain an enormous amount of information about cell shape, protein expression, and protein localization, called *high-content screening*. However, this phenotypic screening is not used as commonly because it still has experimental limitations. Data mining is one problematic issue. Eggert et al. mentioned collecting a large amount of image data (data are in terabytes), and optical solutions are required. Interestingly, they also stated that open-source image analysis packages, such as CELLProfiler and ImageJ, are useful tools for profiling the images. The most significant limitation of phenotypic screening methods is the difficulty of establishing single-cell-based assay systems that reflect the complex biology. Moreover, target identification of small molecules is also a challenge. On this matter, Eggert et al. suggested a strategy in which phenotype images of a small-molecule library are compared with those of an RNAi library and links between the targets of compounds and the knocked-down genes [13].

In summary, numerous screening methods have been established for small molecules. These inhibitors are used extensively and may enable the dissection of the complex cellular processes of cell migration. Several success stories are described, but there are even more innovative methods. The findings of inhibitors and target identification are introduced in the next section.

9.3 MOLECULAR MECHANISM OF INHIBITORS

Cell migration requires the coordination of numerous cellular processes, including polarization, actin cytoskeletal reorganization, membrane cycling, and adhesion. One approach to dissecting these complex cellular events has come through the use of small molecules that can temporally and spatially regulate individual proteins and processes. Numerous small molecules known to inhibit cell migration have been used extensively; however, our ability to perturb specific protein function is limited by the compounds' availability. Therefore, many attempts to screen for new inhibitors of specific components of a biochemical network, with unique screening sources and by innovative approaches, have been instigated. There are two approaches to screening such inhibitors. One approach is using pure proteins as target molecules and screening for inhibitors in activity assays (reverse chemical genetic approach). The other approach is screening for compounds that affect phenotype (forward chemical genetic approach). In this section, several challenges to discovering or designing small molecules that regulate cell migration and insights obtained from studies with them are summarized.

9.3.1 Small Molecules That Affect Extracellular Stimuli

Chemokines Chemokines comprise a superfamily of small proteins that demonstrate a wide variety of biological and pathological functions [7]. Many reports have shown that chemokines and their receptors regulate cell differentiation, proliferation, and migration. In this section, chemoattractant activity and small-molecular antagonists are focused on and described concisely as follows.

The chemokine family is classified into four subfamilies on the basis of their amino acid sequence near the N-terminus: CXCL, CCL, XCL, and CX3CL (L represents ligand). Receptors are also categorized by their ligand affinity into four groups: CXCR, CCR, XCR, and CX3CR (R represents receptor). Until now, approximately 50 chemokines and nearly 20 receptors have been identified, and understandably, it is quite complex to distinguish their biological functions [6]; however, since many inflammatory diseases are known to be due to chemokine-mediated migration of leukocytes, chemokine receptors are considered attractive drug targets [7]. A selection of reports describing chemokine antagonists is presented (see Fig. 9-2).

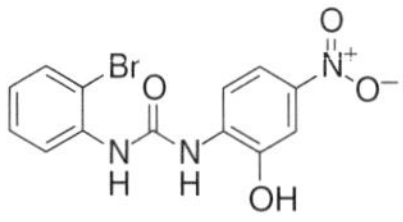

SB225002 (1)
(CXCR1,2)

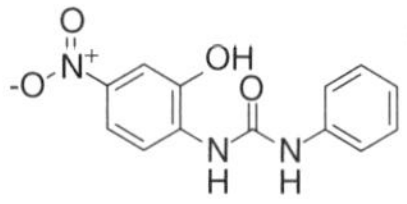

SK&F83589 (2)
(CXCR1,2)

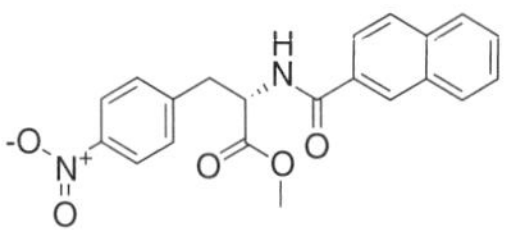

SB328437 (3)
(CCR3)

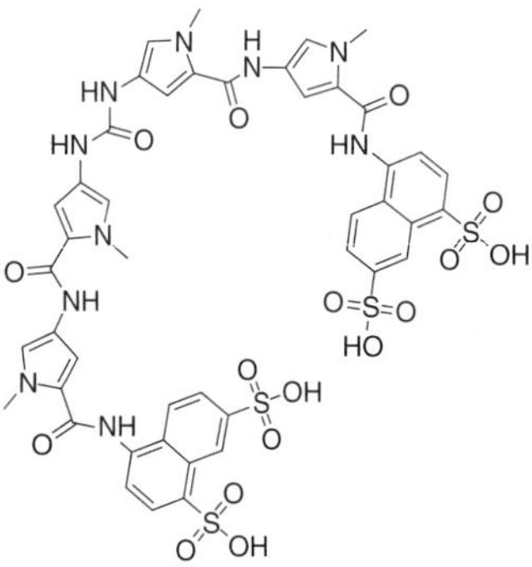

NSC651016 (4)
(CXCR4)

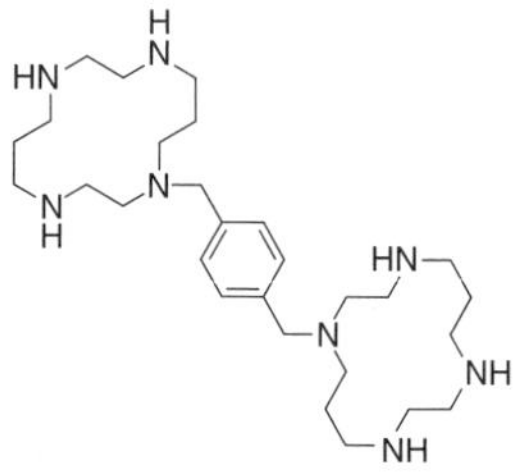

AMD3100 (5)
(CXCR4)

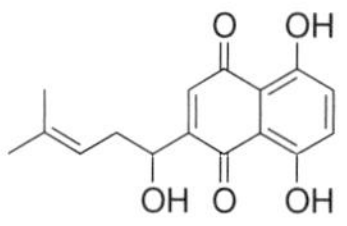

shinkonin (6)
(CCR1)

Figure 9-2 Structures of compounds that inhibit chemokine receptors. Molecules in parentheses are targets of small molecules.

The discovery of SB225002 (**1**) is one of the first success stories, developed as a nonpeptide antagonist against CXCR1 or CXCR2 for anti-inflammatory agents, which was described by White et al. in 1998 [14]. CXCR1 and CXCR2 had already been shown to be present on the surface of human neutrophils, and excess recruitment of neutrophils to the site of preinflammation was thought to be a critical cause of several inflammatory diseases, such as adult respiratory distress syndrome, chronic bronchitis, and asthma. This recruitment was shown to be due to the activation of CXCR1 and/or CXCR2 in an IL-8 (CXCL8)-dependent manner. The authors developed high-throughput screening for antagonists that inhibit [125I]IL-8 binding to the membrane of cells expressing CXCR1 or CXCR2 in vitro and found SK&F 83589 (**2**) and its higher affinity derivative SB225002 (**1**) as specific inhibitors of CXCR2. Furthermore, they showed that CXCR2 inhibitors were effective in their animal model. They concluded that neutrophil migration was caused by agonist binding to CXCR2 in vivo, and that specific antagonists have the potential to identify the important role of CXCR2 in inflammatory diseases and may help in therapeutics.

Apart from neutrophil migration, eosinophils are also a potential cause of several inflammatory diseases. The same research team, including J. R. White, also developed an antagonist for CCR3 receptor, the activation of which by their agonists seemed critical for eosinophil migration [15]. In their in-house library, SB-328437 (**3**) was the most selective inhibitor against CCR3 binding with eotaxin-1 (CCL11), eotaxin-2 (CCL24), and MCP-4 (CCL13). Its specificity against CCR3 was >2500-fold higher than that against other receptors.

Other reports of antagonist discovery are summarized as follows: NSC651016 (**4**) and AMD3100 (**5**) antagonize CXCR4, an antagonistic compound against CCR2B with MCP-1 (CCL2), 3-peperidinyl-1-cyclopentane carboxamide as a specific, potent antagonist against CCR2, and so on [16–18]. Given those reports, it appears a quite remarkable strategy to set a molecular target in searching for anti-inflammatory agents, a reverse chemical genetic approach; however, there is also a forward chemical genetic approach. Chen et al. identified CCR1 as the target of shikonin (**6**), which was an isolated component of the anti-inflammatory Chinese herbal medicine Zicao [purple gromwell, the dried root of *Lithospermum erythrorhizon* Sieb. et Zucc, *Arnebia euchroma* (Royle) Johnst and *A. guttata* Bunge, and others] [19]. Shikonin (**6**) antagonized CCR1 binding with its chemokines MIP-1a (CCL3) and RANTES (CCL5), but not CCR5 with the same chemokines. It did not inhibit CCR2 with MCP-1 (CCL2), CCR3 with eotaxin (CCL11 or 24), CXCR3 with I-TAC (CXCL11), or CXCR4 with SDF-1 (CXCL5). Moreover, it inhibited RANTES-induced cell migration but not growth factor–induced migration; thus, shikonin (**6**) is a specific inhibitor of CCR1-mediated cell migration. However, the authors stated that this compound is a multiple-target inhibitor since many anti-inflammatory, antitumor, and antiviral effects were reported, all of which are difficult to explain by the antagonistic activity against CCR1. Further studies with shikonin (**6**) and its derivatives are required and now under investigation.

Inhibitors of chemokine receptors are not only potential therapeutics against inflammatory diseases, but also cancer therapeutics. Recently, several studies reported that many cancer cell lines possess a high metastatic index via chemokine-mediated migration [20,21].

9.3.2 Small Molecules That Affect Intracellular Molecular Networks

Rho-Family GTPases and Downstream Signal Transducers Rho-family GTPases Rac and Cdc42, and their respective effector molecules WAVE and WASP, are key regulators of front-membrane protrusion of migrating cells; therefore, several small molecules that specifically regulate them have been designed (Fig. 9-3). Indeed, NSC23766 (**7**) was designed by a structure-based virtual screening of compounds that fit into a surface groove of Rac1 known to be critical for guanine nucleotide exchange factor (GEF) specification. In vitro, NSC23766 (**7**) could effectively inhibit Rac1 binding and activation by Rac-specific GEF, Trio, or Tiam1, in a dose-dependent manner without interfering with closely related Cdc42 or RhoA binding or activation by their respective GEFs. It also did not interfere with Rac1 interaction with BcrGAP or effector PAK1. In cells, it potently blocked serum or PDGF-induced Rac1 activation and lamellipodia formation without affecting the activity of endogenous Cdc42 or RhoA. When applied to human prostate cancer PC-3 cells, as expected, it was able to inhibit proliferation, anchorage-independent

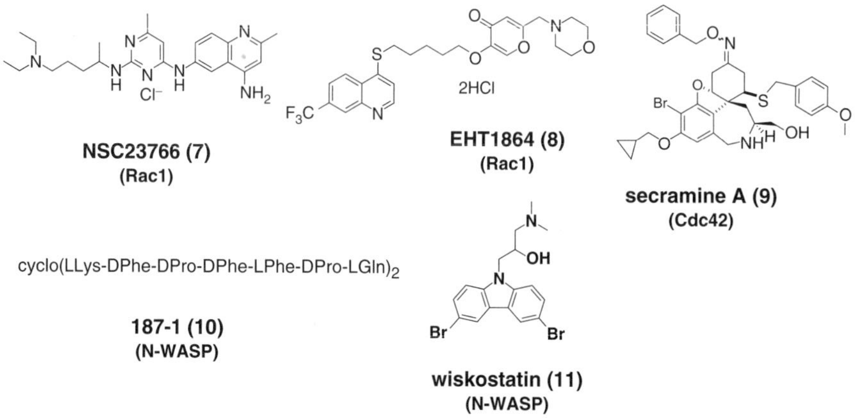

NSC23766 (7)
(Rac1)

EHT1864 (8)
(Rac1)

2HCl

secramine A (9)
(Cdc42)

cyclo(LLys-DPhe-DPro-DPhe-LPhe-DPro-LGln)$_2$

187-1 (10)
(N-WASP)

wiskostatin (11)
(N-WASP)

Figure 9-3 Structures of compounds that affect Rho-family GTPases and downstream signal transductions. Molecules in parentheses are targets of small molecules.

growth, and invasion, phenotypes that require endogenous Rac1 activation [22]. Similarly, the inhibitory effects of a synthesized small molecule, EHT1864 (**8**), on Rac signaling pathways were reported. EHT1864 (**8**) selectively inhibited Rac downstream signaling and transformation by a novel mechanism involving guanine nucleotide displacement. This compound inhibited Rac-dependent amyloid precursor protein processing by γ-secretase and decreased Aβ production in vitro and in vivo [23]. These results reveal that NSC23766 (**7**) and EHT1864 (**8**) could be effective tools to regulate Rac activity and to study Rac1-mediated cellular functions, including cell migration.

Secramine A (**9**), an inhibitor of Cdc42 activation, was discovered from a synthesized chemical library. Secramine A (**9**) inhibits the activation of Cdc42 by a mechanism dependent on the guanine dissociation inhibitor RhoGDI [24]. Although secramine A (**9**) was not discovered as a cell migration inhibitor but a membrane traffic inhibitor, it can be used to study other processes that require Cdc42 activation.

N-WASP is the effector protein of Cdc42, which activates Arp2/3 complex, induces actin polymerization, and mediates filopodium formation at the leading edge. To obtain inhibitors of Cdc42-N-WASP-ARP2/3 complex signal activation, Peterson et al. reconstitute PI(4,5)P$_2$ and Cdc42-induced actin polymerization in cell extracts of *Xenopus* eggs and discovered 187-1 (**10**) and wiskostatin (**11**). N-WASP has a "VCA region" and a GTPase-binding domain (GBD), and VCA activity is negatively regulated by intramolecular binding to the GBD. Binding of activated Cdc42 to the GBD destabilizes autoinhibitory interactions between the GBD and VCA, promoting active conformation and leading to VCA-mediated activation of the Arp2/3 complex. Intriguingly, 187-1 (**10**) is a 14-aa cyclic peptide and wiskostatin (**11**) is a nonpeptide. Despite their structural difference, both compounds allosterically stabilize the autoinhibitory conformation of N-WASP

[9,25]. Although neither 187-1 (**10**) nor wiskostatin (**11**) have been screened as inhibitors of cell migration, they generally can be used as N-WASP inhibitors

As described above, activation of Rho-family GTPases promotes cell migration, and statins affect this step. Statins represent a well-established class of drugs against hypercholesterolemia by inhibiting 3-hydroxy-3-methylglutaryl coenzyme A (HMG-CoA) reductase, which is a rate-limiting enzyme of the cholesterol synthesis pathway. In addition to cholesterol-lowering activity, it is well known that statins inhibit both leukocytes and cancerous cell migration, and have anti-inflammatory activity and antimetastatic activity in vivo, respectively [26]. The most understandable and well-discussed inhibitory mechanism is the blockade of protein prenylation. Rho GTPase, such as Rho, Rac, and Cdc42, play a key role in the migration signal transduction pathway, and post-translational modification with isoprenoids is necessary for their full activation to anchor to the cell membrane. Farnesyl pyrophosphate and geranylgeranyl pyrophosphate are well-known isoprenoids products, and HMG-CoA reductase is also involved in the isoprenoid synthetic pathway.

Integrin-Signaling Transducers Integrins are a large family of cell surface receptors composed of α and β subunits which mediate cell adhesion and cell migration by interacting with the cells and the extracellular matrix (Fig. 9-4). It was revealed that some integrins, including the β_2 integrin subfamily, are required for various chronic inflammatory and autoimmune diseases; therefore,

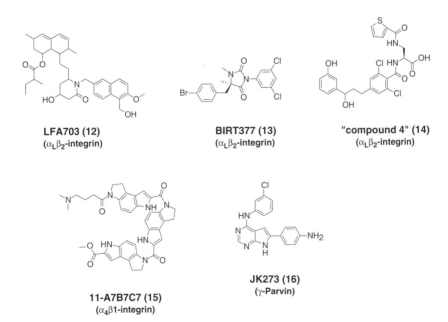

LFA703 (12)
($\alpha_L\beta_2$-integrin)

BIRT377 (13)
($\alpha_L\beta_2$-integrin)

"compound 4" (14)
($\alpha_L\beta_2$-integrin)

11-A7B7C7 (15)
($\alpha_4\beta_1$-integrin)

JK273 (16)
(γ-Parvin)

Figure 9-4 Structures of compounds that affect integrin-signaling transductions. Molecules in parentheses are targets of small molecules.

small-molecule inhibitors that perturb integrin signaling have been under intense investigation as potential anti-inflammatory and anti-autoimmune drugs.

For example, many small-molecule antagonists of integrin $\alpha_L\beta_2$ [e.g., LFA703 (12) and BIRT377 (13)] have been designed. These compounds bind the hydrophobic pocket underneath the α_7 helix of the α_L I domain, block the downward axial movement of the α_7 helix, and inhibit ligand binding of $\alpha_L\beta_2$ allosterically by stabilizing the α_L I domain in low-affinity conformation. These antagonists inhibit lymphocyte transendothelial migration by reducing adhesion [27]. On the other hand, Yang et al. stated that although "compound 4" (14) functioned as an agonist of $\alpha_L\beta_2$ under physiological conditions in Ca^{2+} and Mg^{2+}, compound 4 (14) also inhibited lymphocyte transendothelial migration. It was revealed that agonism by compound 4 (14) resulted in the accumulation of $\alpha_L\beta_2$ in the uropod, extreme uropod elongation, and defective deadhesion [27]. This finding is intriguing because agonists and antagonists inhibit lymphocyte transendothelial migration by distinct mechanisms.

The $\alpha_4\beta_1$ integrin is believed to regulate cellular functions differently from other integrins because the α_4 cytoplasmic tail binds tightly to the signaling adaptor protein paxillin, and α_4/paxillin interaction leads to enhanced rates of cell migration; therefore, Ambroise et al. searched for small-molecule antagonists of this protein–protein interaction, such as 11-A7B7C7 (15), by ELISA, and optimized them structurally. While extracellular inhibitors that inhibited the binding of VCAM-1 or fibronectin to α_4 integrin were under development for the treatment of asthma and sclerosis, the intracellular inhibition of α_4/paxillin binding also disrupted cell migration, offering an alternative target for therapy and a tool for the analysis of α_4 integrin signaling [28].

Integrin signaling is transduced by several molecules. Lee et al. identified JK273 (16) as a small-molecule inhibitor of α_4 integrin-dependent cell migration and revealed that JK273 (16) functioned by interacting with γ-parvin. γ-Parvin is highly expressed in leukocytes, localized to focal adhesions, and is a binding partner with several integrin-signaling related proteins, including paxillin, ILK, α-actinin, and ARHGEF6 [12]. This is a typical example of the identification of molecules that regulate cell migration by a forward chemical genetic approach.

In addition, it was reported that integrin signaling was transduced by several protein kinases, including FAK, Src, and ERK. These protein kinases play a key role in many cellular processes, including cell migration. Indeed, the respective small-molecule inhibitors of FAK, Src, and ERK inhibited cell migration [29–31].

Actin Filaments and Interactions of Actin with Other Molecules   The dynamics of the actin cytoskeleton power cell migration; therefore, actin has also been the primary target of cell migration inhibitors, and many small molecules that target the actin cytoskeleton directly are now available (Fig. 9-5). Such inhibitors can be classified into two broad categories: (1) inhibitors that primarily disrupt actin filament assembly by a variety of mechanisms and effectively destabilize filaments, and (2) inhibitors that stabilize filaments and induce actin polymerization.

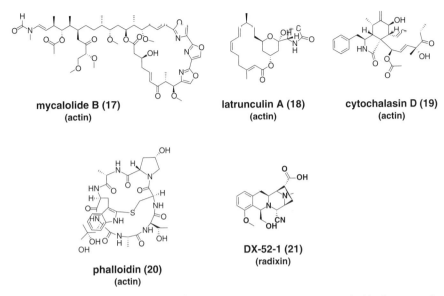

Figure 9-5 Structures of compounds that target actin or an actin-binding protein. Molecules in parentheses are targets of small molecules.

The former category is richer in terms of the number of known compounds: for example, cytochalasins produced by fungi, latrunculins isolated by a Red Sea sponge, and mycalolide B (**17**) isolated from a marine sponge. Cytochalasins, the best known actin-targeted small molecules, bind the barbed end of actin filaments and inhibit the polymerization of actin [32]. On the other hand, latrunculins inhibit actin polymerization, bind G-actin in a 1 : 1 complex, and also inhibit nucleotide exchange in the monomer. Latrunculins, especially latrunculin A (**18**), which is the most potent member of this family, appear to associate only with the actin monomer. They induced changes in cell shape and actin filament organization in cultured mammalian cells in a manner distinct from those caused by cytochalasin D (**19**) [32]. In addition, mycalolide B (**17**) inhibits polymerization and induces rapid depolymerization of F-actin in vitro, apparently by severing F-actin and binding G-actin in a 1 : 1 complex [32]. It is intriguing that some compounds that destabilize F-actin simply bind monomers and prevent polymerization can lead to the apparent disassembly of existing filaments, whereas others destabilize F-actin by severing filaments. Furthermore, the latter category, compounds that stabilize actin filaments and promote actin polymerization, binds and stabilizes actin filaments, shifting the equilibrium between G- and F-actin toward F-actin and lowering the critical concentration for polymerization by an order of magnitude. The best known compound is phalloidin (**20**), from a deadly mushroom. Phalloidin (**20**) has limited use as an inhibitor because of its low cell permeability; however, fluorophore conjugates of phalloidin (**20**) have been used to fluorescently stain and visualize F-actin in fixed and permeabilized cells [32].

Thus, many available compounds stabilize or destabilize actin filaments; however, the mechanisms and the selectivity for actin are different, which is the attraction of studies of small-molecule inhibitors.

In addition, interactions between actin filaments and actin-binding proteins play an important role in cell migration. For example, the interaction of actin with radixin is important, and it is an insight obtained by the elucidation of the molecular mechanism of DX-52-1 (**21**) [33]. Quinocarmycin analog DX-52-1 (**21**) was reported as an inhibitor of epithelial cell migration by Kahsai et al. Because quinocaumycin has been shown to possess DNA-alkylation activity, it was assumed that the effects of its derivative DX-52-1 (**21**) on cells were also mediated by DNA alkylation; however, they showed that it targeted radixin by a pull-down assay with biotinylated-DX-52-1. Radixin is a member of the ezrin/radixin/moesin (ERM) family of membrane–cytoskeleton linkers that are important for organization of the membrane-associated cortical actin cytoskeleton, and thus implicated in cell migration and adhesion. By their efforts it was shown that DX-52-1 (**21**) binds to the C-terminus of radixin and disrupts the interactions of radixin with both actin filaments and the cell adhesion molecule CD44, then inhibits epithelial cell migration.

Actin–Myosin Interaction Regulators Myosin super families are also F-actin-binding proteins that play an important role in cell migration (Fig. 9-6). Interaction of myosin II with actin filaments generates contractile force, which is an important step in cell migration. As described above, the phosphorylation of myosin light-chain (MLC) activates myosin II, which is positively regulated by MLC kinase (MLCK) or Rho kinase (ROCK) and negatively regulated by MLC phosphatase.

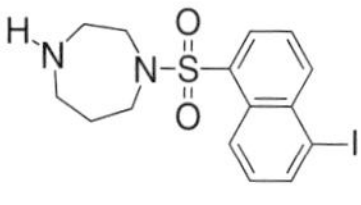

ML-7 (22)
(MLCK)

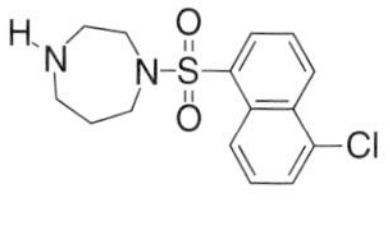

ML-9 (23)
(MLCK)

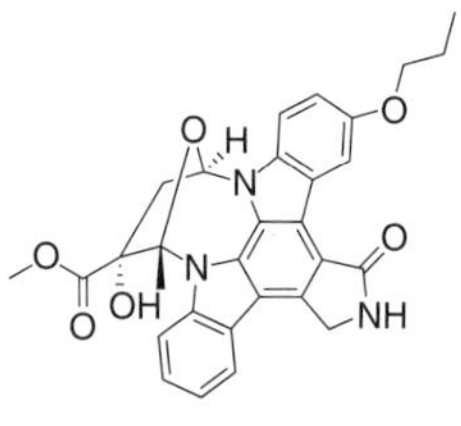

KT5926 (24)
(CaMK)

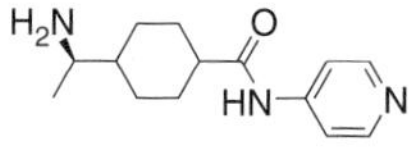

Y27632 (25)
(ROCK)

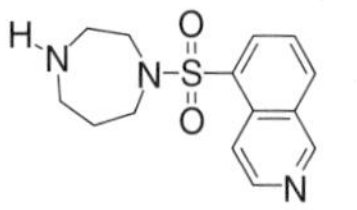

HA1077 (26)
(ROCK)

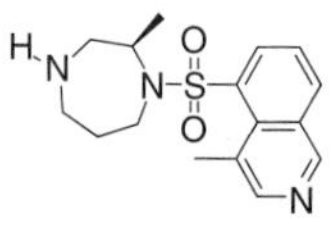

H-1152 (27)
(ROCK)

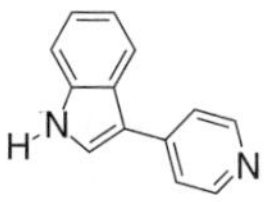

rockout (28)
(ROCK)

Figure 9-6 Structures of compounds that affect actin–myosin interaction. Molecules in parentheses are targets of small molecules.

Indeed, ML-7 (**22**) and ML-9 (**23**), synthetic naphthalene sulfonamides that are available and selective MLCK inhibitors, have been shown to inhibit cell migration in a number of different cell types [34]. On the other hand, KT5926 (**24**) is a synthetic derivative of the broad-spectrum kinase inhibitor K252a, a microbial natural product. KT5926 (**24**) was first reported as a potent ATP-competitive inhibitor of MLCK [35]. A number of other kinases tested were shown to be inhibited with considerably higher K_i, suggesting reasonable selectivity. However, this compound was subsequently found to inhibit Ca^{2+}/calmodulin-dependent protein kinase II with slightly greater potency than MLCK. This demonstrates the potential peril of pharmacological and chemical genetic approaches; discovery of a high-affinity inhibitory interaction does not preclude the possibility that other high-affinity interactions exist, especially with functionally and structurally related proteins. In general, small molecules have several binding proteins, as described below; therefore, the discovery of another target protein of KT5926 (**24**) was not unusual.

Y-27632 (**25**) is a synthetic pyridine derivative that inhibits Rho kinase (ROCK)-I and ROCK-II. As expected, Y27632 (**25**) inhibits the formation of stress fibers in cultured cells and motility in a number of systems [36]. In addition, two isoquinoline sulfonamides, HA1077 (**26**) and H-1152 (**27**), were developed as ROCK inhibitors [37,38].

On the other hand, a ROCK inhibitor, rockout (**28**), was discovered by the wound healing assay, a forward chemical genetic approach. Forward chemical genetics always have problems in identifying the target of the inhibitor; however, T. J. Mitchison's team at ICCB Harvard aimed to resolve these problems by establishing several secondary screening assays to narrow down the potential target of the compounds as follows. First, they collected cell morphology data about inhibitors and categorized them into four groups: (a) decreased migration; (b) aberrant morphology; (c) increased mitotics; (d) disrupted monolayers. They then selected 400 compounds categorized as (a) and (b). Next, they developed several rapid assays to exclude compounds with undesirable effects: (1) actin polymerization in vitro using permeabilized *S. cerevisiae* and *Listeria*, and (2) toxicity by cellular ATP amount evaluation and etidium bromide staining. Furthermore, they established three profiling assays: (1) phagocytosis of *E. coli* by macrophages, (2) cell spreading, and (3) cell blebbing. Many GTPases are involved in cell migration and it was reported that all three biological assays were maintained with several kinds of GTPases. As a result, one compound, named rockout (**28**), 3-(4-pyridyl)indole, was found to be a migratory inhibitor. It was not toxic, was inactive against actin polymerization and phagocytosis, promoted cell spreading, changed the cell shape, and inhibited blebbing. Given all the data about rockout (**28**), it was likely that aberrant contractility and adhesion was induced, and indeed rockout (**28**) was found to decrease stress fibers and focal adhesion complex while membrane ruffling was not affected; therefore, it was strongly suspected that rockout (**28**) acts on the Rho signaling pathway. The authors showed that this compound inhibited the kinase activity of ROCK, both in vivo and in vitro, and concluded that it was a ROCK inhibitor [10].

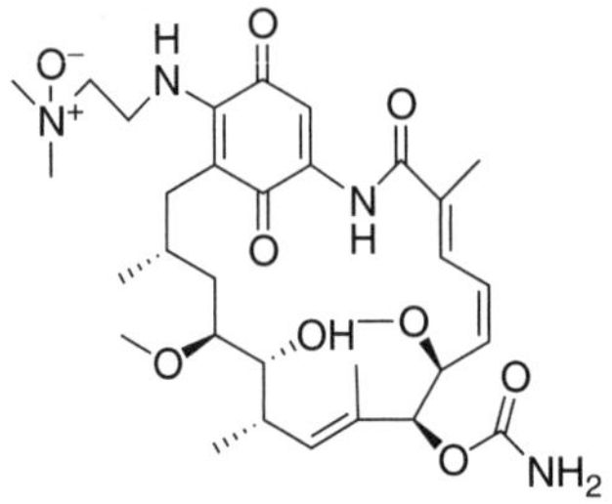

DMAG-N-oxide (29)
(HSP90 at cell surface)

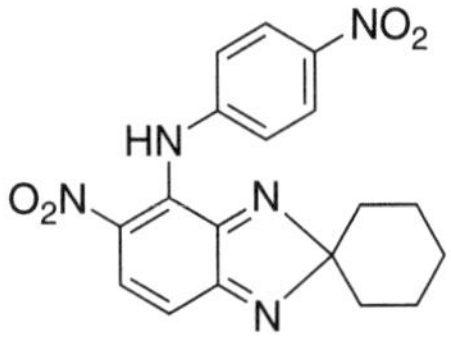

LM11 (30)
(Arf1)

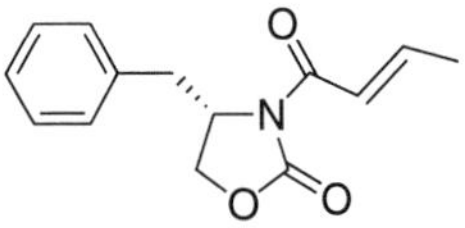

locostatin (31)
(RKIP)

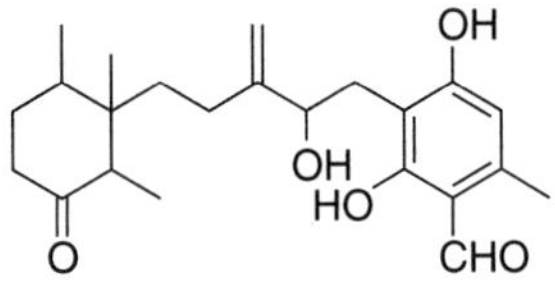

moverastin (32)
(FTase)

Figure 9-7   Structures of compounds that affect cell migration. Molecules in parentheses are targets of small molecules.

Other Molecules   Molecules other than those described above are known to play roles in cell migration (Fig. 9-7); for example, Hsp90, a molecular chaperone, is the target of DMAG-N-oxide (**29**), a cell-impermeant small molecule. Recently, it was revealed that Hsp90 was found not only intracellularly but also on the cell surface. Hsp90 has been important for maintaining the stability and function of numerous client proteins; however, it raised the possibility that cell surface Hsp90 plays a role in cancer cell motility and metastasis. Therefore, DMAG-N-oxide (**29**) was found by screening, and indeed, this compound significantly retarded tumor cell migration and integrin/extracellular matrix-dependent cytoskeletal reorganization while not affecting intracellular Hsp90 function [39]. Thus, DMAG-N-oxide (**29**) would be a useful tool to study the function of cell surface Hsp90, which remains to be elucidated.

Arf1, a small GTPase, is a major regulator of membrane traffic and is activated by the Sec7 catalytic domain of its GEF ARNO. Viaud et al. discovered a noncompetitive inhibitor, LM11 (**30**), by in silico screening of a flexible pocket near the Arf1/ARNO interface. LM11 (**30**) inhibits ARNO-dependent migration of Madin–Darby canine kidey (MDCK) cells. The unique ability of LM11 (**30**) to inhibit Arf1 activation selectively, but not Arf6 activation, by ARNO in vitro addressed the question of the unexpected involvement of Arf1 activation by ARNO in addition to Arf6 in the process of cell migration [40].

There are several cases in which small molecules that affect cell migration have given unexpected cell migration regulators by the forward chemical genetic approach. One example is the identification of Raf kinase inhibitor protein (RKIP) as the target molecule of locostatin (**31**), described by Zhu et al. They screened for the inhibitor by the wound-healing assay and found locostatin (**31**) as an inhibitor against MDCK cell migration. It strongly inhibited membrane protrusion and cell sheet migration, without affecting the gross actin cytoskeletal structure in cells or salt-induced polymerization of pyrene–actin in vitro. It also inhibited the scattering of MDCK epithelial cells, induced by hepatocyte growth factor/scatter factor (HGF/SF), and the migration of B16-BL6 murine melanoma cells, a highly invasive and metastatic cancer cell line. To elucidate the target of locostatin (**31**), they synthesized radiolabeled [^{3}H]locostatin to detect locostatin binding proteins. As a result, four proteins in cytosolic MDCK cell extracts were found and identified as follows: 21 kDa: Raf kinase inhibitor protein (RKIP), 30 kDa: GST omega1-1 (GSTO1-1), 55 kDa: aldehyde dehydrogenase 1A1 (ALDH1A1), 80 kDa: prolyl oligopeptidase (POP). Binding and inhibitory activity of locostatin (**31**) against all these proteins were confirmed by competition analysis and in vitro enzymatic assay; however, only RKIP was implicated in cell migration. Knockdown of RKIP inhibited MDCK cell migration, whereas known inhibitors against all other proteins did not inhibit it. Moreover, locostatin (**31**) disrupted the interaction of RKIP and Raf-1 kinase and blocked the inhibitory activity of RKIP against Raf-1 kinase activity; however, the authors were skeptical that activation of Raf-1 accounted for the inhibition of cell migration with locostatin (**31**), because many reports have shown the malignancy of the Raf-1/MEK/ERK cascade. Thus, they elucidated that the as-yet-unproven effecter of Raf-1 or of RKIP might have an important role in MDCK cell migration. In summary, locostatin (**31**) was shown to inhibit MDCK cell migration by blocking RKIP-mediated migratory signaling [41].

As another example, we studied the identification of farnesyl transferase (FTase) as the target molecule of moverastin (**32**). We screened for cell migration inhibitors of microbial origin with the wound healing assay and obtained moverastin (**32**) from *Aspergillus* sp. F7720, a member of the cylindrol family compounds. Since other cylindrol compounds, cylindrol A and LL-Z1272e, were reported to inhibit FTase, we examined whether moverastin (**32**) inhibited FTase. As a result, moverastin (**32**) inhibited FTase in vitro and also inhibited the membrane localization of H-ras in human esophageal cancer EC-17 cells. In contrast, moverastin (**32**) failed to inhibit GGTase in vitro and N-ras membrane translocation in EC-17 cells. It was reported that H-ras prenylation depended on FTase, whereas N-ras prenylation was regulated by either FTase or GGTase. Thus, our findings indicated that moverastin (**32**) inhibited FTase, but not GGTase, both in vitro and in vivo. The effect of moverastin (**32**) on the downstream transduction of H-ras was also examined. It was reported that H-ras activated the PI3K/Akt pathway preferentially, whereas N-ras activated primarily the Raf/MEK/Erk pathway. Indeed, moverastin (**32**) inhibited Akt phosphorylation but not Erk phosphorylation. Furthermore, PI3K inhibitor LY294002 [42] interrupted EC-17 cell migration, whereas MEK inhibitor U0126 [43] did not. The contribution of the Raf/MEK/Erk pathway and PI3K/Akt pathway to cell migration differs with cell lines and/or stimuli. For example, the Raf/MEK/Erk pathway is necessary in the induction of cell migration by HGF in MDCK epithelial cells. On the other hand, insulin-like growth factor 1–stimulated melanoma cell migration required the PI3K/Akt pathway but not the Raf/MEK/Erk pathway. Our study indicated that the PI3K/Akt signaling pathway dominantly regulated cell migration in EC-17 cells, and moverastin (**32**) inhibited that through the suppression of H-ras farnesylation [44].

As described above, cell migration is regulated complexly by a number of molecules. Generally, many familiar signaling proteins play important roles in a range of cell types and in diverse cellular processes within each cell; however, inhibition of a given signaling protein may sometimes have an effect on a cellular process in one cell and not in another. Similarly, such inhibition may affect the movement of a single cell type following one stimulation but not another; therefore, chemical genomics, identifying inhibitors of each molecule and utilizing them to elucidate the functions of the target molecules in the entire signal transduction network of cells, is a prospective strategy to give new fruitful insights into cellular processes, not only cell migration.

REFERENCES

1. Ridley, A. J., et al. (2003). *Science*, *302*, 1704–1709.
2. Takenawa, T., Suetsugu, S. (2007). *Nat. Rev. Mol. Cell. Biol.*, *8*, 37–48.
3. Sasaki, A. T., Firtel, R. A. (2006). *Eur. J. Cell Biol.*, *85*, 873–895.
4. Lock, J. G., et al. (2008). *Semin. Cancer Biol.*, *18*, 65–76.
5. Yamazaki, D., et al. (2005). *Cancer Sci.*, *96*, 379–386.
6. Savarin-Vuaillat, C., et al. (2007). *Neurotherapeutics*, *4*, 590–601.
7. Owen, C. (2001). *Pulm. Pharmacol. Ther.*, *14*, 193–202.
8. Willis, A. I., et al. (2004). *Vasc. Endovasc. Surg.*, *38*, 11–23.
9. Peterson, J. R., et al. (2001). *Proc. Natl. Acad. Sci. USA*, *98*, 10624–10629.
10. Yarrow, J. C., et al. (2005). *Chem. Biol.*, *12*, 385–395.
11. Sutherland, S. (2003). *Drug Discov. Today*, *8*, 6–7.
12. Lee, J., et al. (2009). *Bioorg. Med. Chem.*, *17*, 977–980.
13. Eggert, U. S., et al. (2006). *Curr. Opin. Chem. Biol.*, *10*, 232–237.
14. White, J. R., et al. (2000). *J. Biol. Chem.*, *275*, 36626–36631.
15. White, J. R., et al. (1998). *J. Biol. Chem.*, *273*, 10095–10098.
16. Schneider, G. P., et al. (2002). *Clin. Cancer Res.*, *8*, 3955–3960.
17. Yang, L., et al. (2007). *J. Med. Chem.*, *50*, 2609–2611.
18. Forbes, I. T., et al. (2000). *Bioorg. Med. Chem. Lett.*, *10*, 1803–1806.
19. Chen, X., et al. (2001). *Int. Immunopharmacol.*, *1*, 229–236.
20. Luker, K. E., et al. (2006). *Cancer Lett.*, *238*, 30–41.
21. Muller, A., et al. (2001). *Nature*, *410*, 50–56.
22. Gao, Y., et al. (2004). *Proc. Natl. Acad. Sci. USA*, *101*, 7618–7623.
23. Shutes, A., et al. (2007). *J. Biol. Chem.*, *282*, 35666–35678.
24. Pelish, H. E., et al. (2006). *Nat. Chem. Biol.*, *2*, 39–46.

25. Peterson, J. R., et al. (2004). *Nat. Struct. Mol. Biol.*, *11*, 747–755.
26. Kusama, T., et al. (2002). *Gastroenterology*, *122*, 308–317.
27. Yang, W., et al. (2006). *J. Biol. Chem.*, *281*, 37904–37912.
28. Ambroise, Y., et al. (2002). *Chem. Biol.*, *9*, 1219–1226.
29. Roberts, W. G., et al. (2008). *Cancer Res.*, *68*, 1935–1944.
30. Kilarski, W.W., et al. (2003). *Exp. Cell Res.*, *291*, 70–82.
31. Bove, P. F., et al. (2008). *J Biol. Chem.*, *283*, 17919–17928.
32. Fenteany, G., Zhu, S. (2003). *Curr. Top. Med. Chem.*, *3*, 593–616.
33. Kahsai, A. W., et al. (2006). *Chem. Biol.*, *13*, 973–983.
34. Levinson, H., et al. (2004). *Wound Repair Regen.*, *12*, 505–511.
35. Hashimoto, Y., et al. (1991). *Biochem. Biophys. Res. Commun.*, *181*, 423–429.
36. Imamura, F., et al. (2000). *Jpn. J. Cancer Res.*, *91*, 811–816.
37. Satoh, S., et al. (1999). *Jpn. J. Pharmacol.*, *80*, 41–48.
38. Ikenoya, M., et al. (2002). *J. Neurochem.*, *81*, 9–16.
39. Tsutsumi, S., et al. (2008). *Oncogene*, *27*, 2478–2487.
40. Viaud, J., et al. (2007). *Proc. Natl. Acad. Sci. USA*, *104*, 10370–10375.
41. Zhu, S., et al. (2005). *Chem. Biol.*, *12*, 981–991.
42. Vlahos, C. J., et al. (1994). *J. Biol. Chem.*, *269*, 5241–5248.
43. Favata, M. F., et al. (1998). *J. Biol. Chem.*, *273*, 18623–18632.
44. Takemoto, Y., et al. (2005). *Chem. Biol.*, *12*, 1337–1347.

10

CHEMICAL BIOLOGY OF CELL SURFACE OLIGOSACCHARIDES

Prabhani U. Atukorale, Sean S. Choi, Udayanath Aich, Christopher T. Campbell, M. Adam Meledeo, and Kevin J. Yarema

Department of Biomedical Engineering, The Johns Hopkins University, Baltimore, Maryland

10.1 INTRODUCTION

10.1.1 Post-translational Modifications, Exemplified by Glycosylation, Increase Protein Diversity

Nature uses 20 common (and one or two specialty) amino acids to construct proteins and achieves a dazzling array of complexity with this relatively modest set of building blocks. Nonetheless, nature (as well as protein engineers!) is not satisfied and seeks to expand protein diversity further through post-translational modifications (PTMs), where the chemical structures of amino acids are modified after assembly of the primary sequence. To date, several dozen PTMs have been discovered and characterized; many involve the addition of small functional chemical groups such as phosphate, sulfate, and acetate (or larger entities that include other peptides, lipids, or sugars) to the peptide backbone of a protein. In other cases the chemical nature of an individual amino acid is changed; for example,

Protein Targeting with Small Molecules: Chemical Biology Techniques and Applications,
Edited by Hiroyuki Osada
Copyright © 2009 John Wiley & Sons, Inc.

a deimination reaction is known that converts arginine to citrulline. Finally, the peptide backbone can be cleaved into fragments, as exemplified by insulin processing, or conversely, covalent linkages are formed between amino acids beyond the primary amino acid sequence, as illustrated by the ubiquitous formation of disulfide bonds between thiols of two appropriately situated cysteine residues.

By far the most diverse category of PTM—based on the number of distinct chemical structures produced—is glycosylation, where carbohydrate moieties ranging in size from a single monosaccharide to highly branched oligosaccharides and finally, to linear polysaccharides hundreds of residues in length are linked covalently to proteins (Fig. 10-1). These sugars, known individually as glycans and collectively as the glycocalyx when displayed on the cell surface, play critical roles in quality control during protein folding [1], modulate the structure and function of mature proteins [2], and endow the cell surface with an immense information-carrying capacity [3]. Somewhat counterintuitively based on their long history of being the most difficult mammalian biopolymer to manipulate both in the laboratory and in living systems, glycans now provide the experimentalist with a facile biomaterial for modification across a wide range of cell types, as well as in living animals, through emerging metabolic oligosaccharide engineering (MOE) technology, as described in detail next.

10.1.2 MOE Allows Glycans to Be Manipulated in Living Cells and Animals

As the dominant roles that glycans play in many biological processes—and their extremely subtle modulation of function in others—have been revealed, modern medical science has sought ways to exploit this knowledge to create new therapeutics. To reach the goal of creating sugar-based medicines, however, significant challenges that are inherent in manipulating glycosylation in living cells and tissues must be overcome [4]. First, from a biological perspective, a particular obstacle is the lack of a template for carbohydrate structures akin to the DNA sequences that specify the primary amino acid compositions of proteins. Consequently, the facile experimental conversions that link DNA, RNA, and proteins, and that have facilitated the genetic engineering revolution of the past three decades, remain elusive for carbohydrates. Quite simply, pending the maturation of computation tools [5–7], there is rarely sufficient knowledge available that connects the transcription of glycan-processing enzymes with the final glycan structures produced by a cell to exploit the tools of modern molecular biology to manipulate carbohydrates.

Moving from a biological to a chemical perspective, the expense of de novo synthesis of complex oligosaccharides coupled with the notoriously poor pharmacological properties of sugars has long hampered conventional drug development for disorders associated with glycans. Recently, however, sugars are gaining newfound appreciation as versatile scaffolds for drug discovery [8] and innovative "chemical biology" strategies based on nonnatural sugar analogs are emerging (Fig. 10-2A). The latter methodology, known as *metabolic oligosaccharide engineering* (MOE) [9,10] (or *metabolic glycoengineering* [11,12]), relies on the biosynthetic incorporation of abiotic building blocks into cellular

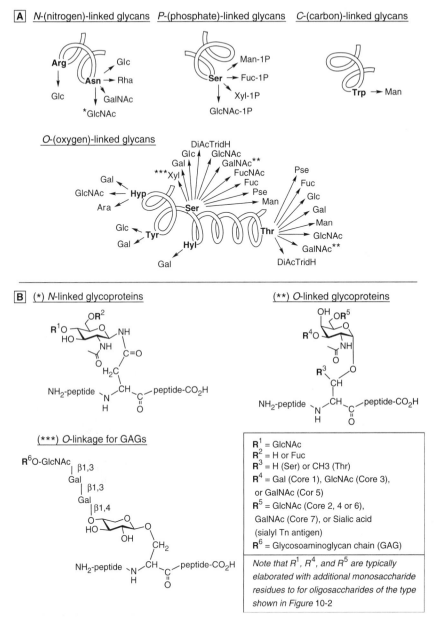

Figure 10-1 Major classes of protein–sugar linkages: (A) carbohydrate PTM of proteins include those formed through N-, P-, C-, and O-linked sugar–peptide bonds; (B) the chemical structures of the major linkages found in mammalian glycoproteins are N-linked GlcNAc-based oligosaccharides (indicated by an asterisk *), O-linked GalNAc based oligosaccharides (**), and xylose-appended polysaccharides found in glycosoaminoglycans (GAGs, ***). Extensive information on protein sugar linkages is provided in recent reviews [155,156].

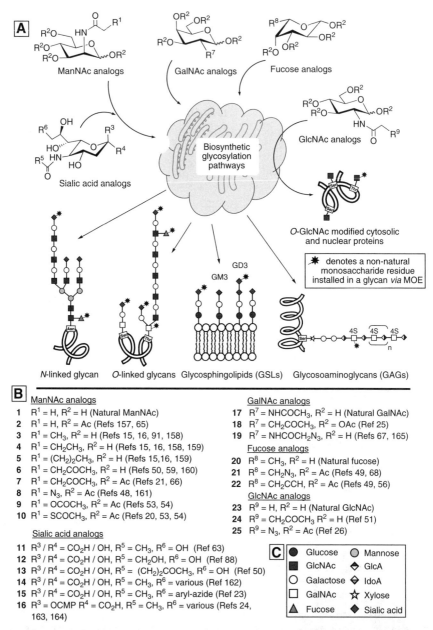

Figure 10-2 Overview of MOE and representative analogs. (A) ManNAc, GalNAc, and fucose analogs intercept mammalian glycosylation pathways and are biosynthetically incorporated into cell surface and secreted N- and O-linked glycoproteins as well as GSLs and GAGs; GlcNAc analogs, by contrast, partition into cytosolic and nuclear O-GlcNAc-modified proteins. (B) A nonexhaustive sampling of analogs falling into each of these classes, along with references to the original reports, is provided. (C) Symbols used to represent monosaccharides in this figure, as well as in Figure 10-3, are shown.

oligosaccharides. Over the past 15 years, tremendous progress in laying a scientific foundation for MOE has been made, which includes the synthesis of dozens of modified monosaccharides (see Section 10.2.1), the characterization of the biological activities of these compounds (Section 10.2.2), and the use of these compounds as chemical probes (Section 10.2.3).

Moving beyond the numerous cell-based studies—and in some cases, rodent testing—that showed the fundamental viability of MOE technology, several promising avenues for the development of human therapies are being pursued (as discussed in Section 10.3). Nonetheless, before MOE-based therapies come to fruition, several practical challenges will need to be overcome to augment the basic scientific technologies. For example, enhanced metabolic flux into glycosylation pathways, the redesign of analogs for greater efficiency, and improved pharmacologic properties are active areas of investigation (as discussed in Section 10.2.4) resulting in second-generation compounds that exhibit greater cellular uptake, enhanced metabolic efficiency, and an expanded range of activities (Section 10.2.5). These ongoing efforts offer opportunities both for tuning sugar-specific responses for therapeutic benefit and for avoiding harmful and unanticipated side effects.

10.2 DEVELOPMENT OF CHEMICAL TOOLS TO STUDY GLYCOSYLATION

10.2.1 Toolkit of Analogs (Pathways Targeted by MOE)

Ten different monosaccharides are incorporated into mammalian glycans: D-glucose (Glc), D-galactose (Gal), D-mannose (Man), L-fucose (Fuc), D-xylose (Xyl), N-acetyl-D-glucosamine (GlcNAc), N-acetyl-D-galactosamine (GalNAc), D-glucuronic acid (GlcA), N-acetyl-D-neuraminic acid (Neu5Ac), and N-glycolyl-D-neuraminic acid (Neu5Gc). Further *natural* diversity is gained through postsynthetic processing steps, which include the epimerization of glucuronic acid to iduronate in glycosoaminoglycans (GAGs) or by the addition of sulfate to certain glycans (e.g., to the sialyl Lewis X tetrasaccharide [13 or proteoglycans [14]); an important goal of MOE is to parallel nature's efforts to provide a broad repertoire of carbohydrate structures by introducing *nonnatural* structural and chemical diversity into glycans. Toward meeting this objective in the laboratory, the pathways responsible for the biosynthesis of four natural sugars have already been demonstrated to permit the incorporation of nonnatural monosaccharide analogs (Fig. 10-2A).

MOE was pioneered with the sialic acid biosynthetic pathway [15], which has proved to be especially amenable at accommodating the flux of nonnatural metabolic intermediates. This path converts N-acetyl-D-mannosamine (ManNAc) into the nine-carbon sugar sialic acid through condensation with phospho-enolpyruvate and then installs the cytidine monophosophate (CMP)–derivatized form of this sugar onto the termini of newly synthesized glycans by the action of one of 20 sialyltransferases (see refs. 16 to 18 for more information on sialic acid metabolism). Almost two decades of experiments have established that sialic acid–based MOE is particularly flexible because this pathway can be intercepted at various stages, allowing experimental strategies to be tailored with fine precision [12].

To illustrate the versatility of sialic acid–based MOE, ManNAc analogs (e.g., compounds 1 to 10, Fig. 10-2B) have been used in the majority of studies, largely from the practical point of view that these compounds can be made from moderately priced starting materials and entail straightforward synthesis (as described in detail elsewhere [19,20]). Conversely, a disadvantage is that ManNAc analogs feed into the sialic acid pathway upstream of the rate-limiting "bottleneck" step of conversion of ManNAc to sialic acid by Neu5Ac synthase [12,21,22]; as a consequence, nonnatural substitutions are restricted to modifications that (1) occur at the N-acyl position and (2) are sterically conservative. By contrast, sialic acid analogs bypass this constriction point, allowing a greater variety of nonnatural N-acyl substituents as well as the positioning of the nonnatural moiety at other sites on the molecule (compounds 11 to 14). These expanded structural features are exemplified by 9-arylazide Neu5Ac (15), an analog capable of photo-activated cross-linking upon glycan incorporation [23]. Going a step further, the nucleotide sugar versions of sialic acid (16), having only to be acted upon by sialyltransferases for incorporation into glycans, show extreme substrate permissivity, allowing enzymatic processing of analogs appended with chemical moieties as large as fluorescein [24]. A downside of CMP–sialic acid analogs that severely limits their use in living cells, however, is their poor membrane permeability. This challenge poses particularly acute problems considering that these nucleotide sugars must not only cross the plasma membrane to enter the cytosol but must then cross the Golgi membrane to enter the lumen of this organelle, where sialic acids are added to nascent glycans.

In the past five years, GalNAc (17 to 19) and fucose (20 to 22) analogs have joined sialic acid pathway intermediates as monosaccharide vehicles for surface display of engineered glycans. In a manner similar to that of sialic acids, these sugars are biosynthetically incorporated into N-linked or O-linked glycoproteins (see Fig. 10-1), and GalNAc analogs are also incorporated into GAGs [25]. By contrast, GlcNAc analogs (23 to 25) show the interesting property of being refractory to surface display but experience ready incorporation into O-GlcNAc-modified nuclear and cytosolic proteins [26]. In the future, it is likely that additional analogs will be found to transit mammalian glycosylation pathways, thereby further increasing the versatility of MOE to modulate glycan biosynthesis, structure, and function. Moreover, microbes that synthesize usual monosaccharides not found in mammals—such as L-rhamnose, a sugar recently chemoselectively bioconjugated to proteins [27]—offer many as-of-yet unexplored opportunities for the expansion of MOE.

10.2.2 Replacement of Natural with Nonnatural Analogs Affects Biological Activity

In the almost two decades since the basic concept of MOE in living systems was demonstrated by the Reutter group's pioneering experiments [28], a wealth of biological responses to nonnatural glycan display have been reported. Most straightforwardly, the replacement of a natural sugar with a nonnatural counterpart directly alters the normal function of the host glycan. For example, interaction of the influenza virus with its preferred cell surface binding epitope, the "human" Neu5Ac form of sialic acid (**11**), is inhibited by the elongated N-acyl side chains of the normally abiotic sialic acids that are newly installed on cells treated with the ManNAc analogs (**3** to **5**) [16] (see further discussion in Section 10.3.2). Similarly, the analog Ac₅ManNGc (**9**) can be used to replace Neu5Ac with the "nonhuman" glycolyl Neu5Gc form of sialic acid (**12**) [29–31] and thereby alter MAG binding and spur neurite outgrowth [32]. Because nature exploits differences between Neu5Ac and Neu5Gc to modulate Siglec-mediated interactions [30] and associated biological activities [33], the ability to control the display of these two sugars on cells is a potentially powerful way to alter cell–cell interactions and signaling by offering a way to recode the vast amount of information stored in cell surface "sugar code" [3].

An outstanding example of the multifaceted and intertwined roles of carbohydrates in modulating protein structure and activity is provided by integrins (Fig. 10-3). The manifold impact of sugars on integrin activity illustrates the layers of complexity potentially introduced into a biological system through MOE. First, the extracellular portions of integrins are heavily glycosylated [34,35]; for example, there are 14 and 12 potential sites of attachment for N-linked oligosaccharides on the α and β subunits, respectively, of the α5β1 integrin [36]. Several

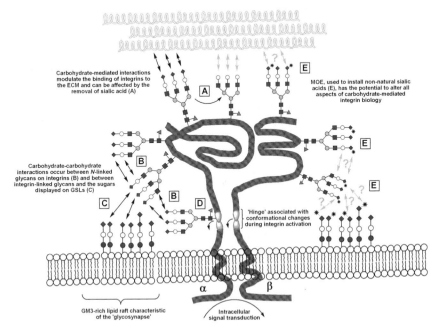

Figure 10-3   Influence of carbohydrates (A to D) and MOE (E) on integrin biology. See the text for a more detailed discussion.

independent experiments have established that modification of these N-linked glycans modulate cell adhesion. In one, a bisecting GlcNAc reduced cell adhesion by down-regulating the binding of activated integrins to extracellular matrix (ECM) components [37]. Ligand binding is particularly sensitive to the presence or absence of sialic acid; sialidase treatment, which removes sialic acids (A) [38], exposes binding domains and increases cell adhesion [39], whereas hypersialylation increases cell motility and is associated with cancer progression [40]. Integrin function is also regulated by *cis* carbohydrate–carbohydrate interactions between *N*-glycans located on the integrin's peptide backbone (B) as well as between the sugars displayed on glycosphingolipids (GSLs) and on the integrin itself (C) [41,42]. The ability of GSLs to modulate protein function is a consequence of an integrin's location within lipid raft assemblages termed *glycosynapses* [43,44]. The glycosynapse is dominated by the ganglioside GM3, a glycosylated trisaccharide epitope (see Fig. 10-2 for symbols used to depict sugar residues) that mediates cell adhesion and signal transduction events and in turn affects cellular phenotypes by modulating the activity of the resident receptor complexes. Finally, the presence of an N-linked glycan in the hinge region of an integrin can stabilize the molecule in the open (activated) position (D) [45].

Unraveling the complex roles that carbohydrates play in integrin biology has been a formidable task, but these experiments yielded sufficient insight to predict that replacement of the natural sugars found on these adhesion molecules with MOE analogs would alter their activity (E). Less certain—and difficult to predict in advance—are the exact nature of these responses as the variables in an already unruly system increase. This complexity was borne out by one of the first experiments to probe the effects of MOE on cell adhesion, wherein Man-NProp (**3**) altered cell adhesion by increasing integrin activity [46]. A vexing aspect of this finding was that the impact of MOE appeared to lie outside the sugar-related activities depicted in Figure 10-3, as compound **3** altered integrin expression at the mRNA level, thereby indirectly affecting activity. In summary, integrins exemplify the many opportunities available for MOE to modulate biological molecules in situ while simultaneously highlighting the many unknowns to be faced.

10.2.3   Metabolically Engineered Glycans Alter the Chemical Properties of the Cell Surface

As just discussed, although MOE has tremendous potential to alter glycan biosynthesis and structure in living systems and thereby modulate corollary biological activities, equally intriguing possibilities lie in the ability of permissive glycosylation pathways to install novel chemical reactivities onto the cell surface. The original intent of these efforts was not to alter the biology of the targeted monosaccharide or cell; indeed, the early experiments regarded the core sugar as a silent delivery vehicle that would transit glycosylation pathways without undue perturbation and ultimately install novel chemical functional groups on the cell surface. In addition to the original molecular targets of MOE [i.e., plasma

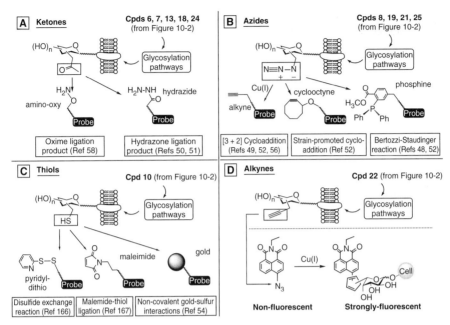

Figure 10-4   Chemoselective ligation reactions used in MOE. Representative coupling reactions are shown along with representative literature references. See the text for a more detailed discussion.

membrane-displayed N-linked and O-linked glycoproteins (see Fig. 10-2A)], this methodology also bestows secreted proteins with new chemical features (Section 10.3.4) and has recently been extended to nuclear and cytosolic O-GlcNAc-modified proteins (Section 10.3.5). As shown in Figure 10-4, monosaccharide analogs (as depicted in Fig. 10-2) now facilitate the incorporation of ketones, azides, thiols, or alkynes—functional groups otherwise absent from the glycocalyx or glycoproteins—into cellular glycans.

Once installed in cell surface–displayed oligosaccharides, novel chemical functional groups can be exploited in an ever-growing range of *chemoselective ligation reactions*, reactions designed for selective joining of two complementary chemical functional groups in the complex milieu of biological systems, including in living cells and animals [47]. Ideally, the two reactive partners are abiotic, undergo nonreversible covalent bond formation under physiological conditions, and recognize only each other, while remaining impervious to potential competing reactions in their cellular surroundings [48]. A sampling of chemoselective coupling reactions that apply to MOE analogs are shown in Figure 10-4 for glycan-displayed ketones (A), azides (B), and thiols (C). The azide–alkyne "click" ligation in part (D) is chemically identical to the Cu(I)-catalyzed [3 + 2] cycloaddition reaction that occurs in part (B) but is shown in detail to emphasize the conversion of a nonfluorescent probe to a highly fluorescent moiety upon ligation within the cellular milieu [49].

The chemoselective ligation strategy has already been employed to deliver a wide range of labeling reagents and molecular probes to glycan-displayed ketones [50,51], azides [48,52], thiols [53], and alkynes [49]. To date, many of these agents have been designed for detection of the nonnatural sugar: for example, via biotinylation followed by subsequent staining with fluorescently derivatized avidin [19,50,53,54] or by ligation of the antigenic FLAG peptide [55]. In other cases, direct chemoselective ligation of fluorescent molecules [49,56] or contrast agents [57] has been reported. Probes have also included therapeutic agents such as toxic proteins [50] and small-molecule drugs [58], additional sugar epitopes [51], quantum dots [54], adenoviruses used for gene delivery [59], as well as polymers [60] and functionalized materials intended to facilitate cell adhesion and serve as a growth substrate [54,61].

10.2.4   Metabolic Flux Considerations for Development of MOE-Based Therapies

The capacity of MOE to modulate biological processes has situated this technology as an attractive platform for drug development. Accordingly, we now discuss briefly metabolic flux considerations that are critical for translation of this methodology from small-scale, pilot cell culture studies to large scale, in vivo testing and, ultimately, to human therapies. At first glance, "back-of-the-envelope" calculations indicate that MOE therapies are feasible from a drug availability perspective. For example, when ManNLev (**6**) was used to install Sia5Lev (**13**) on cancer cells and the newly installed ketone was then to kill the cells [50], efficacy was achieved when there was roughly 1 million ketones on the cell surface (although higher levels, up to 10^7 ketones per cell, could be attained). Considering that there are about 35 to 40 trillion cells in the human body (many fewer if the ca. 30 trillion red blood cells are discounted), a maximum of about 40×10^{18} molecules of Sia5Lev, an amount that corresponds to about 60 µmol of ManNLev, would be needed. In mass terms, the resulting 16 mg is clearly a reasonable dose for a person to consume.

Upon closer examination, however, significant practical problems arise in the development of MOE-based therapies. In particular, sugars are generally not regarded to be "drug-like" (or "druggable" [62]) because they suffer from poor pharmacological properties that include slow uptake by cells and rapid serum clearance. Further complicating MOE effects, cells lack plasma membrane transporters for sugars rarely found in the diet, such as ManNAc, resulting in slow uptake (presumably, by pinocytosis). Even for sugars used in MOE, such as sialic acid analogs (e.g., **11** to **15**), where cellular uptake has been reported [63], it remains unknown what effect nonnatural substitutions have on membrane transport. Consequently, from the perspective of conventional drugs, extremely high levels, ranging into double-digit-millimolar concentrations, are required to achieve robust biological activity in cell culture conditions for a typical ManNAc analog (e.g., **3** to **6**). As a consequence, to extend the ManNLev example given above, where 16 mg was calculated to be a reasonable human therapeutic dose,

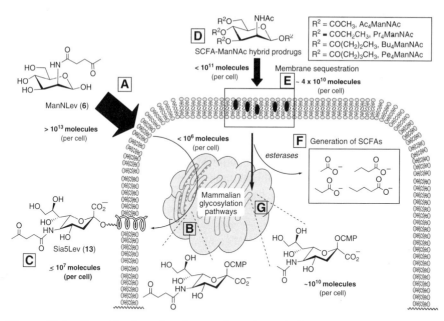

Figure 10-5 Quantification of analog uptake and metabolism. See the text for a more detailed discussion.

extrapolations that take into account the inefficiency of cellular uptake (see Fig. 10-5 and the following discussion) transform the previously attainable 16 mg dose to an absurdly high 80 kg.

Clearly, the low efficiency by which exogenously supplied analogs are converted to their surface-displayed counterparts is a major stumbling block in the development of MOE-based drugs. More positively, significant progress toward overcoming this barrier has already been made by increasing the efficiency of uptake through modification of the "core" sugar with ester-linked short-chain fatty acids (SCFAs; Fig. 10-5). This strategy is based on the ancient design principle—exemplified by aspirin—where prodrugs are rendered more lipophilic and cell permeable by masking their hydrophilic moieties with a SCFA, which is usually acetate [64]. In the context of sugars, peracetylation of ManNAc increases cellular uptake by about 600-fold compared to the perhydroxylated monosaccharide [65], presumably because of the increased lipophilicity and resulting membrane permeability of the prodrug. This premise that lipid solubility assists membrane transit was verified by more lipophilic propanoyl- and butanoyl- protected sugars that experienced even greater increases in metabolic efficiency, about 1800- and 2100-fold, respectively [66].

The numbers and concepts discussed in the preceding paragraph are shown graphically in Figure 10-5. Under routine cell culture conditions (e.g., with ca. 10^6 mammalian cells per milliliter of growth medium), there are about 10^{13} molecules of ManNLev in the effective concentration range 10 to 20 mM (A).

For ManNLev, intracellular pathway intermediates [such as CMP-Sia5Lev, shown in (B)] are below the limit of detection, which is ≤ 2 μM, indicating that there are fewer than 10^6 molecules of this intermediate per cell. Over 3 to 5 days, however, sufficient flux transits the pathway to result in the display of up to 10^7 molecules of Sia5Lev on the cell surface (C) [51]. Derivatization of ManNAc [21,57,65], and other monosaccharides, including GalNAc analogs [25,67], GlcNAc analogs [26], and fucose derivatives [49,68] as well as disaccharides [69] (D), with SCFAs increases the efficiency of cellular uptake by two to three orders of magnitude, reducing the level of analog required for robust metabolic incorporation to $<10^{11}$ molecules per cell. Overall, 50% or more of a SCFA–ManNAc hybrid prodrug can be removed from the culture medium by cells; a large proportion of these molecules, however, appear to become sequestered in cellular membranes (E) [65]. Upon entering the cell, nonspecific esterases liberate active SCFAs from the hybrid molecules (F), allowing the core sugar moiety to enter the glycosylation pathways. In the case of SCFA–ManNAc hybrid molecules, a significant increase in sialic acid intermediates such as CMP–Neu5Ac occurs (**G**) [21].

When applied to the hypothetical example of ManNLev used for human therapy presented above, the ramifications of the three-order-of-magnitude increase in efficiency realized with SCFA–monosaccharide hybrid molecules is that butanoylation—even accounting for the increased mass of the butyrate groups—lowers the quantity of prodrug needed for a human dose from 80 kg to about 100 g. Although still high, this level is not an outlandishly excessive amount that immediately discounts the idea of therapeutic applications of MOE. Encouragingly, a multipronged approach using tactics beyond acylation with SCFA is likely to further decrease the amount of compound required therapeutically; for example, organ and tissue partitioning may negate the need for whole-organism dosing. Another option may be encapsulation of the sugar analogs in biodegradable materials for local controlled release in situ or in nanoparticle-sized entities for selective organ or tissue targeting.

10.2.5 Unanticipated Effects of MOE

Although MOE has a remarkable ability to precisely target individual monosaccharide residues of glycan structures with submolecular microsurgery [16], the biological responses can be anything but highly specific. These off-target effects are exemplified by the already-mentioned ability of ManNProp to modulate mRNA levels of integrins [46], an observation that supports recent observations that simple sugars used as MOE analogs can function as signaling molecules involved in gene regulation [70]. Of note, efforts to increase the metabolic efficiency of analog uptake through SCFA modifications (Fig. 10-5) have the potential to exacerbate perturbations to cellular physiology. To explain briefly, in the past, SCFAs such as acetates have been regarded as innocuous delivery agents; in reality, profound biological activities can be realized when using SCFA–sugar hybrids [12,71–73]. For example, toxicity has long been associated with acetylation of sugars used in MOE, and although it is generally mild and not usually manifest in vivo, the underlying molecular mechanisms proved to be highly intriguing, as various N-acyl modifications of ManNAc were found to "tune" toxicity. In particular, Ac$_4$ManNLev (**7**) was considerably more toxic than structurally similiar compounds without the ketone, such as the fully acetylated form of ManNPent (**5**) or the acetylated form of natural ManNAc (Ac$_4$ManNAc, **2**) [66].

Initial evidence that the toxicity of acetylated sugars was an SCFA effect came from the study by Kim and co-workers that explored propionate (Pr$_4$ManNAc; see Fig. 10-5) and butyrate (Bu$_4$ManNAc) appended to ManNAc [66]. The increase in SCFA chain length was correlated with enhanced metabolic flux compared Ac$_4$ManNAc, indicating that the greater lipophilicity of these compounds enhanced membrane permeability. The longer-chain SCFAs also experienced enhanced toxicity, in line with the higher biological activities of the three- and four-carbon SCFAs compared to acetate. In follow-up studies, our laboratory more thoroughly characterized the SCFA effects of butyrate appended to several monosaccharides, including ManNAc (Bu$_4$ManNAc), mannose (Bu$_5$Man), and GlcNAc (Bu$_4$GlcNAc). We found that each of these hybrid compounds had the canonical activities expected from SCFAs, including histone deacetylase inhibition, up-regulation of p21 gene expression, and transient cycle-cell arrest that resulted in 3- to 5-day growth inhibition of cancer cell lines [73]. Only Bu$_4$ManNAc, however, induced apoptosis in a variety of cancer lines. This result suggested that the combined activities of the butyrate (which was responsible for the SCFA effects) and ManNAc (which supported robust sialic acid overproduction at lethal doses) moieties led to novel cellular responses not achieved by either functionality alone [72].

The connections between sialic acid production and SCFA biology revealed by Bu$_4$ManNAc [72,73] are likely to be extremely difficult to understand in full molecular detail because both SCFA responses and glycosylation are highly complex on their own. When added together, additional intricacies are likely to emerge that link this SCFA metabolism and glycosylation; especially fascinating is long-established evidence of crosstalk between these two systems that has recently undergone renewed investigation. Over three decades ago, Fishman and co-workers reported that butyrate dramatically up-regulated the activity of the sialyltransferase responsible for the production of ganglioside GM3 [74–76], a molecule that readers of this chapter may recall from its effects on integrins (Fig. 10-3). A more detailed examination of the approximately 25 enzymes now known to directly comprise the sialic acid pathway in human cells shows that several of them respond to SCFA mediate-gene expression effects [20]. At the same time, a compound such as Bu$_4$ManNAc dramatically increases the flux of metabolites through the pathway, thereby potentially bringing about a synergistic amplification of biological activity. Clearly, such multifaceted and reinforcing (or counteracting) responses have great potential to confront vexing human diseases, but they also raise the cautionary note that much remains to be learned of the underlying biology before MOE-based therapies reach the clinic.

10.3 IDENTIFICATION OF DRUG TARGETS AND MOE-BASED THERAPIES

In this section, an overview of five promising areas for development of MOE-based drugs and human therapies is provided. These include cancer (Section 10.3.1), infectious disease (Section 10.3.2), tissue engineering and regenerative medicine (Section 10.3.3), recombinant glycoproteins (Section 10.3.4), and glycomic-based diagnostics (Section 10.3.5). In the long term it remains unclear which, if any, of these endeavors will be successful. Other diseases not discussed here are just as likely to emerge as feasible candidates for MOE-based therapies. These particular examples were selected as provocative illustrations of the potentially wide-ranging impact that sugar-based medicines may ultimately have in human health.

10.3.1 Cancer

Abnormal glycan biosynthesis and cell surface display have been associated with cancer for decades [77–79] and have become closely associated with metastasis [80]. In some cases, specific genetic defects have been identified that contribute to altered glycan structures found in cancer [81]; at other times, glycan abnormalities have been reported [82] that have not been explained from a biosynthetic standpoint. Despite the ever-accumulating evidence that "repair" of abnormal glycans has the potential to ameliorate aberrant behaviors associated with cancer, glycosylation-based therapies have been slow to develop, which is not surprising considering the difficulties inherent in manipulating carbohydrates in living systems, as discussed above. However, recent advances, spurred in large part by MOE technologies, are now providing renewed impetus for the development of sugar-based cancer therapies [83,84].

The first attempts to use MOE against cancer were motivated by the overexpression of sialic acid in many types of cancer, especially highly metastatic forms. Consequently, it was reasoned that exogenously supplied ManNAc or sialic acid analogs would be more efficiently metabolized by cancer cells, resulting in overexpression of nonnatural sugar epitopes on their surfaces. Precedent for this premise lay in observations that the Neu5Gc glycolyl form of sialic acid (12) is detectable only in trace amounts in healthy human tissue. As a brief aside, this sugar is virtually absent from normal tissues because human ancestors lost activity of CMP-*N*-acetylneuraminic acid hydroxylase [85], the enzyme responsible for conversion of Neu5Ac to Neu5Gc [86], due to a 93-base mutation several million years ago [31]. Interestingly, Neu5Gc is readily detected on tumors [87] as a consequence of dietary intake (red meat and dairy products are rich in this form of sialic acid), subsequent cellular uptake, and metabolic pathway incorporation of Neu5Gc [88]. In essence, Neu5Gc provides an example of naturally occurring MOE, and the higher display of this sugar on cancerous compared to healthy tissues provides a precedent that abiotic pathway intermediates, such as the analogs shown in Figure 10-2, can be preferentially incorporated into transformed tissues.

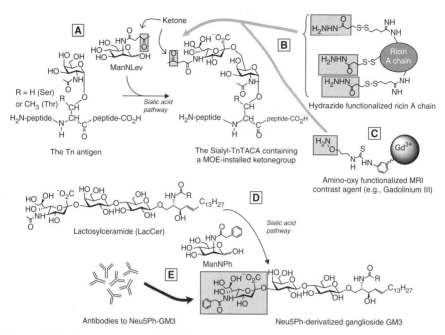

Figure 10-6   Development of MOE-based cancer diagnostic and therapeutic agents. See the text for a more detailed discussion.

As outlined in Figure 10-6, the preferential conversion of ManNAc analogs into nonnatural sialic acids in transformed tissues and their subsequent overexpression on the surfaces of tumor cells provides an extremely versatile platform for cancer treatment and diagnosis. In the first strategies, once a sialic acid is appended with an appropriate chemical tag, such as a ketone group installed via ManNLev (Fig. 10-6a), therapeutics such as toxic proteins (e.g., hydrazide-functionalized ricin as shown in part (B) [89]) or small-molecule drugs such as doxorubicin [58] can be delivered selectively to the target cell. A similar strategy can be used for diagnostics such as MRI contrast agents (e.g., amino-oxy-conjugated gadolinium chelators (C) [57]). When delivering imaging agents or drugs this way, selective internalization or preferential surface binding can be achieved, depending on the precise objective. For example, hydrazide reactions with ketones form hydrazone bonds (Fig. 10-4a) that are pH sensitive, resulting in hydrolysis of a probe in low-pH endosomal vesicles [58]; upon hydrolysis from membrane components, internalization into the cytosol is more likely. By contrast, amino-oxy reactions with ketones result in pH-stable oxime adducts that cycle repeatedly from the cell surface to endosomal vesicles [58]; this type of reaction may be preferred for an imaging agent where cytosolic incorporation and subsequent enzymatic inactivation of the probe are undesired.

A second MOE-based anticancer strategy is depicted in Figure 10-6. In this case, the nonnatural sugar—illustrated by ManNPh, an analog that installs a phenyl group into the N-acyl substituent of sialic acids (D) [90]—constitutes an epitope for immunosurvellience mechanisms that suppress malignant transformation. Importantly, the immunotargeting of nonnatural sialic acids such as Neu5Ph can be enhanced by a passive immunization approach using antibodies generated against this nonnatural sialoside (E) when it is displayed in a cancer-specific glycan such as polysialated NCAM [91] or gangliosides [92]. Finally, in addition to the two modes of anticancer activity realized from surface display of sugar analogs, intracellular responses arising from the combined SCFA–sugar effects discussed above in Sections 10.2.4 and 10.2.5 are anticipated to play important therapeutic functions in the future.

10.3.2 Pathogens and Infectious Agents

Because the molecular landscape of the cell surface is dominated by the glycocalyx, it is not surprising that many pathogens exploit carbohydrate ligands for cell binding and entry. Once again, sialic acid plays a dominant role, exemplified by the high-profile example of "global pandemic" strains of influenza, where the exact linkage and chemical structure of the glycan are important. In particular, various influenza strains can discriminate between $\alpha 2,3$- and $\alpha 2,6$-linked sialic acids as well as between Neu5Ac and Neu5Gc. Many less publicized viral pathogens [93], including adenovirus, the BK virus [94], coronavirus, cytomegalovirus, hepatitis virus, HIV [95], Newcastle disease virus, papovavirus, polyoma virus, rabies virus, reovirus, rhinovirus, and rotavirus [96], also rely on sialic acid in part or wholly as a binding epitope for cell entry. Sialic acids also occupy the interface between the host and commensal or pathogenic microorganisms and participate in molecular mimicry, provide nutrition, and interpret cell signaling [97]. In some cases, polysialic acid capsules also provide pathogenic bacteria with protection against the host immune system [98]. Finally, protozoan parasites, including *Plasmodia, Leishmania, Trypanosoma, Entamoeba*, and *Trichomonas*, also exploit sialic acids as a defense mechanism to survive hostile environments, such as that posed by the mammalian immune system [99].

The ability of MOE to alter sialic acid presentation can potentially modulate pathogen interactions in many ways. Perhaps the easiest to visualize are effects on viral binding, where nonnatural sialic acids usually function in an inhibitory manner. For example, a structural explanation for the inhibition of influenza binding [16] to N-acyl-modified sialic acids lies in the close proximity of the N-acyl group to the side chain of a tryptophan residue [100]. It is reasonable to postulate that the extra steric bulk provided by extended alkyl moieties of Sia5Prop, Sia5But, or Sia5Pent does not allow tight binding of the virus to these modified receptor epitopes. In other cases, the impact of MOE on the viral life cycle may be more convoluted and involve effects on neuraminidases, which have reduced ability to cleave nonnatural sialosides and thereby inhibit liberation of the virus from the originally infected cell. In results that have not been explained mechanistically, nonnatural sialic acids actually increase viral infection, in some cases exemplified by the proliferation of human polyoma virus BK in ManNProp-treated cells [101]. Nonetheless, the idea that appropriate sialic acid precursors used in MOE often influence receptor interactions in a clinically beneficial direction provides an enticing new strategy for the developmental of antiviral agents.

Unlike viruses where nonnatural sialic acids have been tested against several host species, the effects of MOE on pathogenic bacteria that utilize sialic acid as part of the binding interface between themselves and their host during infection have not been investigated intensively. Nonetheless, an intriguing pilot study suggests that sugar analogs may have the capacity to be incorporated in lower organisms and even discriminated between host and pathogen. In particular, nonnatural sialic acids with seven and eight carbons N-acyl side chains were incorporated into the lipooligosaccharides of *Hemophilus ducryei* [102]. By contrast, mammalian cells lack the ability for ready incorporation of alkyl substituents longer than six carbons [21,66], potentially offering a means to differentiate between host and pathogen sugars. In a second example of using exogenous precursors to influence bacterial carbohydrate biosynthesis, the Nishimura laboratory synthesized UDP–MurNAc pentapeptide derivatives that were metabolically incorporated into cell-wall components of *Lactobacilli plantarum* [103], thereby changing the adhesive properties of these bacteria [104]. Both strategies offer a way to identify the pathogen to the immune system for eradication, or to adhesive constructs for removal from the body [60,61].

The preceding two examples, which used MOE to confront viral and bacterial pathogens, diverged on the display of the nonnatural sugar. First, viral inhibition occurred when the abiotic sugar was displayed on the host cells not the pathogen—whereas the nonnatural sugar was situated on the pathogen in the case of the bacteria. A more complex scenario arises for a third type of pathogen, represented by protozoan parasites of the *Trypanosoma* genus. The species *T. cruzi* has been particularly well studied and known to have a surface *trans*-sialidase that removes sialic acids from the glycocalyx of host cells and transfers these sugars onto its own surface mucins (a mucin is a glycoprotein rich in O-linked glycans). The scavenged sialic acids on *T. cruzi* participate in host–parasite interactions that are critical in the initial stages of the invasion of host cells [105]. Although completely speculative at present, MOE may engender a number of novel strategies to prevent infection by this trypanosome. For example, if the *trans*-sialidase has reduced activity against nonnatural N-acyl-derivatized sialic acids, the host may be able to consume ManNAc analogs and replace his or her natural sialic acids with nonnatural sialosides refractory to parasite *trans*-sialidase processing, thereby hindering the invasion of host cells. By contrast, if nonnatural sialic acids are efficiently transferred from the host to parasite, this would still be a desirable outcome if the modified glycan structures do not support the exact ligand–receptor interactions required for cellular entry of the pathogen. Alternatively, the surface display of nonnatural sialosides may render the protozoan more detectable by the immune system. Finally, a third strategy is based on a new class of MOE analogs comprised of *para*-nitrophenyl glycosides of sialic acids with modified polyol chains (the C7 to C9 positions). The Thiem group recently showed that these compounds were readily transferred by the *trans*-sialidase of *T. cruzi* [106], and therefore systemic administration of such compounds in vivo should result in metabolic labeling of the parasites. Such glycosides, however, are probably refractory to the mammalian sialic acid machinery—if for no other reason than poor uptake by cells—promising highly specific display on the parasite. Clearly, many questions must be answered before it can be determined if MOE is a viable therapy, but once again many intriguing possibilities lie ahead.

10.3.3   Tissue Engineering and Regenerative Medicine

Tissue engineering, the effort to regenerate or produce replacement tissues and organs in the laboratory for diseased and worn-out body parts, has advanced significantly in recent years, in part through an ever-increasing focus on stem cells. Stem cells are characterized by their ability to self-renew and by their potential to differentiate into multiple lineages and, as a result, are especially promising in situations where alternate cell sources are scarce [107–109]. Tissue engineering and stem cell biology stand to gain much from exploiting MOE because of the issues discussed in the next two paragraphs.

First, glycans, which have long been expressed on specific stem cell markers (e.g., SSEA-3 and SSEA-4 [110–112], TRA-1-60 and TRA-1-81 [110,111,113,114], and AC133 [110,115]), are now known to play determinative roles in cell fates. Even without any metabolic perturbation, the glycosylated regions of membrane proteins and lipids intricately couple cell adhesion with signaling, forming functional assemblies referred to as *glycosynapses* [43], exemplified here by integrins (Fig. 10-3). The carbohydrate–carbohydrate interactions involved in glycosynapse-mediated processes serve to enhance formation of the dynamic plasma membrane complexes necessary for signal transduction [116]. The ability of MOE to alter glycosylation and thereby provide an additional layer of control in cellular differentiation was first shown in the Reutter laboratory in neonatal rat studies, where the metabolism of ManNProp increased the proliferation of astrocytes and microglia and enhanced the expression of the oligodendroglial progenitor marker A2B5 [117]. The influence of MOE over neural cell fates was reinforced by experiments reported by the Schnaar laboratory, which demonstrated the use of Ac$_5$ManNGc (**9**) to convert Neu5Ac (**11**) to Neu5Gc (**12**), thereby inhibit the binding of the sialic acid–dependent lectin MAG [32]. Inhibition of MAG, in turn, induces neurite outgrowth in neuroblastoma–glioma hybrid cells, a finding that ultimately could be exploited for spinal cord regeneration.

A second avenue for the use of MOE in tissue engineering is based on the position of surface sugars at the outer periphery of the cell surface. Under natural conditions, these sugars—if anything—are antiadhesive because they comprise the glycocalyx, which is a dense layer of carbohydrates that extends up to 500 nm from the plasma membrane and sterically shields the underlying integrins and cadherins [118], preventing attachment to RGD-derivatized tissue engineering scaffolds [119–121]. Despite their normally antiadhesive nature, surface sugars are ideally positioned to form a binding interface between a cell and a biomaterial that functions as a scaffold for tissue formation. Consequently, an obvious application of MOE technology in tissue engineering was to endow cell surface glycans with adhesive properties and thereby shift their behavior from inhibiting cell attachment to facilitating adhesion to a complementary surface.

The use of nonnatural sugars to modulate the chemical composition of the outermost region of the glycocalyx to achieve cell adhesion to chemically compatible materials has been demonstrated in several studies. First, as mentioned earlier, Villavicencio-Lorini and colleagues demonstrated that the treatment of HL-60 neutrophils with ManNProp enhanced the activation of β_1-integrins and subsequently increased cell adhesion to the extracellular protein fibronectin [46]. A direct way to utilize MOE for cell attachment became available after the Bertozzi laboratory introduced chemical functional groups onto the cell surface [48,50]. De Bank and colleagues were among the first to exploit these "chemical handles" situated in sugars at the cell's periphery for cell adhesion by demonstrating that rat myoblasts formed into clusters after treatment of the cells with ManNLev, tagging the resulting cell surface Sia5Lev-displayed ketones with biotin hydrazide, and allowing aggregation to occur in the presence of avidin [61].

In addition to cell–cell interactions, the use of MOE to incorporate functional groups into the outer periphery of the glycocalyx can be complemented by the simultaneous engineering of the growth substrate for cell adhesion. Iwasaki and colleagues implemented this strategy by treating HL-60 neutrophils with ManNLev and culturing these ketone-expressing cells on polymer surfaces functionalized with hydrazides. The normally nonadhesive cells selectively attached to the modified polymer surfaces [60]. These studies, although directed at the separation of cancer cells, lay a scientific foundation for the use of MOE analogs for tissue engineering where cell attachment to biopolymeric materials used as scaffolds is a critical first step in the formation of artificial tissues.

The use of MOE in tissue engineering became even more attractive after our laboratory recently demonstrated that this method could be used for more than the simple attachment of a cell to a growth substrate. Instead, a further layer of control over cell fate can be achieved by modulating the microscale interactions between a cell and a substrate to achieve both cell–substrate adhesion and stem cell differentiation. In a pilot study demonstrating this concept, human embryonic cells (the hEBD LVEC line [122]) were incubated with Ac5ManNTGc to install thiols into the glycocalyx (Fig. 10-7) while being grown on a gold surface to provide a high-affinity substrate for the glycocalyx-displayed thiols. These conditions led to neuronlike morphology, and the accumulation of β-catenin [54], a key player in the Wnt signaling pathway involved in neuronal differentiation, was observed in these cells. A stronger differentiation response was observed in Ac5ManNTGc-treated embryonic cells cultured on gold substrates relative to their glass or tissue culture plastic counterparts. These results highlighted the importance of selective cell adhesion (e.g., high-affinity thiol-gold linkages in comparison to weaker or nonexistent thiol–glass interactions) for cell fate

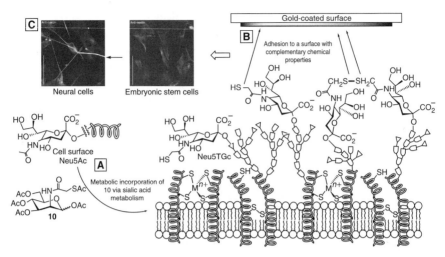

Figure 10-7 Cell adhesion and control of stem cell fate through MOE: (A) metabolic incorporation of Ac5ManNTGc (**10**) results in the replacement of the natural cell surface sialic acid Neu5Ac with the thiol-containing counterpart Neu5TGc; (B) glycocalyx-displayed thiols are ideally situated to attach to exogenous surfaces with complementary chemical properties such as gold; (C) Novel carbohydrate-mediated thiol–gold cell adhesion leads to the differentiation of human embryonic cells to neurons as described in detail in ref. 32.

determination and establish Ac5ManNTGc as a representative of a new small-molecule class of glycosylation-based tools for tissue engineering based on its capacity to control stem cell adhesion and differentiation [54].

In summary, MOE techniques provide important methods to modulate glycosylation, which, by itself, plays vital roles in tissue formation. By providing a way to remodel cell surface carbohydrates, MOE allows the mediation of cell fate determination and adhesion, both of which are essential tools in tissue engineering. Finally, an attractive aspect of using MOE for tissue engineering is that sugar analogs can be used to initiate the desired outcomes (e.g., cell attachment and differentiation) and then removed from the system. Upon removal, natural sugar metabolism and surface display resume [54], thereby avoiding the potential immune complications and other unwanted side effects as discussed above in Sections 10.2.4 and 10.2.5.

10.3.4 Protein Therapeutics

Post-translation modifications are critical for the activity of recombinant proteins intended to be used as therapeutic agents, and glycosylation is the most frequent, most complex, and least understood PTM [123]. Inappropriate glycosylation can result in reduced biological activity, limited half-life in circulation, and unwanted immunogenicity. Consequently, the inability of high-productivity, low-cost microbial, fungal, or insect hosts commonly used for recombinant protein expression to produce humanlike glycans poses a major challenge for the biotechnology industry [124]. However, although bacteria fail to produce glycosylated proteins, they do append oligo- and polysaccharide structures to lipids and antibiotics [125]), and although yeast and insect cells produce only truncated versions of N-linked glycans [126], they do have the basic glycosylation machinery, which, at least in theory, could be tweaked to produce human glycan patterns. Hence, efforts to build on the basic sugar-processing enzymes in order to humanize glycoprotein production in hosts such as bacteria, yeast, and insect cells have been intense in recent years [124,127,128].

The glycoengineering of organisms for the production of humanized glycans typically depends on the expression of glycosyltransferases that assemble human sugar linkages. To accompany the enzymes, the production of the nucleotide sugars used as substrates by these enzymes is also a necessity [129,130]. In some cases, especially in bacterial systems where polysaccharides or sucrose can be used to produce nucleotide sugar donors, elegant coupling schemes for cofactor regeneration have been devised to facilitate recombinant glycoprotein production [131]. Comparable efforts to manufacture nucleotide sugars are more difficult in eukaryotes, leading to the application of MOE technology, where exogenously supplied sugars are fed to cells to supply metabolic flux into the native or engineered glycosylation pathways. As an example of the latter, both the free monosaccharide and the acetylated forms of ManNAc successfully supplied a nascent sialic acid biosynthetic pathway that had been genetically engineered into insect cells with the necessary metabolic intermediates for sialoside production [22].

In addition to using MOE to improve the presentation of the normal sugars on glycoproteins, such as the human Neu5Ac form of sialic acid derived from ManNAc feeding, this methodology can be extended to control the type of sialic acid added to recombinant therapeutics. For example, Viswanathan and co-workers showed that mannose-6-phosphate, used in place of ManNAc, allowed their glycoengineered insect cells to produce the 2-keto-3-deoxy-D-*glycero*-D-*galacto*-nononic acid (KDN) variant of sialic acid [22]. Although KDN is one of the natural forms of sialic acid, this experiment set the precedent for feeding abiotic precursors into host cells used for recombinant glycoprotein production to install nonnatural sialosides into their glycans. Luchansky and co-workers report feeding ManNAz to cells producing recombinant interferon-β and GlyCAM-Ig, resulting in the replacement of between 4 and 41% of the natural sialic acids on these glycoproteins with N-azidoacetylsialic acid (Sia5Az) [132]. This study showed that metabolic functionalization of recombinant glycoproteins via MOE was a general strategy that applied to both N-linked and O-linked glycans and constituted a starting point to augment protein structures with properties not found in nature to improve their biological activities. More specifically, the azides installed in the sialosides provide bioorthogonal chemical reactivity (see Fig. 10-4) that can be used to further modify the protein structure and properties, and bulky groups could be appended to nonnatural sialic acids to render them refractory to removal by sialidases, thereby extending serum half-life of the host molecule by blocking clearance via asialoglycan-mediated mechanisms.

10.3.5 Glycomics: Glycan Profiling and In Situ Function Aided by MOE

Based on the growing interest in using high-throughput "omics" techniques for diagnosis and treatment of disease, exemplified by mRNA profiling, which shows promise in cancer diagnosis [133,134], burgeoning glycomics methods [135–138] may achieve similar diagnostic power for cancer as well as other disease [139]. To date, progress in glycomics has relied on advanced technological developments involving separations [140,141], mass spectrometry [142,143] and microarrays [144–146]. Now, MOE technology allows glycans to be labeled and visualized in living cells and animals [147–149], and provides a facile means for the enrichment of specific types of glycoconjugates for proteomic analysis [55]. In the past, the side effects emanating from the peracylated SCFA–monosaccharide hybrid molecules used in MOE (see Section 10.2.5) have had the potential to introduce unwanted artifacts into the labeling procedures. This pitfall has recently been ameliorated by the discovery of structure–activity relationships where triacylated analogs lacking a SCFA at the C6-OH position maintain a high flux into glycosylation pathways with negligible toxicity [12] and reduced perturbation of gene expression [71]. All in all, the pieces are falling into place to position MOE as an attractive technique to augment current glycomics methods.

Although doubtless valuable for addressing any of the numerous diseases in which carbohydrates are implicated [95], or unraveling the myriad roles of these complex molecules in healthy cells, the use of MOE in glycomics appears particularly promising for the characterization of O-GlcNAc protein modification aberrations associated with metabolic disorder and diabetes [150,151]. In particular, as outlined in Figure 10-8, the azide-derivatized GlcNAc analog Ac$_4$GlcNAc can be incorporated into O-GlcNAc-modified proteins via salvage into the hexosamine pathway (see Fig. 10-8a). The "tagged" proteins can then be further derivatized with the FLAG epitope (B) or biotin (C) by using phosphine-based reagents that enact the modified Staudinger reaction [48]; alternatively, similar chemoselective ligation reactions can be performed using "click chemistry" (see Fig. 10-4). Regardless of the route of installation, the antigenic epitope or biotin can be visualized after separating the labeled glycoproteins by polyacrylamide gel electrophoresis by using anti-FLAG-horseradish peroxidase [HRP, (D)] or streptavidin–HRP (E), respectively. The tagged separated proteins can then be isolated from the gel, or isolated directly by using anti-FLAG (F)- or streptavidin (G)-coated beads (e.g., magnetic particles) and identified by mass spectrometry. This procedure provides a means to metabolically tag glycosylated moleclues in situ and identify specific proteins, out of the hundreds that potentially undergo this modification, that are associated with a particular cellular process or disease state.

The power of this approach was demonstrated in a pilot study in which 199 putative O-GlcNAc modified proteins were isolated from GlcNAz-treated HeLa cells, and among the first set of 23 proteins identified in this experiment, 10 were

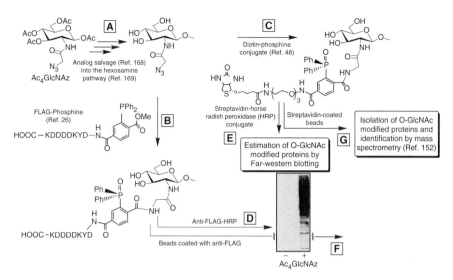

Figure 10-8 MOE facilitates glycomic technologies. As discussed in detail in the text (and the references cited), the MOE analog Ac₄GlcNAz can be used to incorporate biologically orthogonal azide functional groups into cellular proteins that can then be exploited to separate, visualize, and identify the host glycoprotein.

previously reported and 13 were newly identified as bearing *O*-GlcNAc [152]. In a variation of this approach where enzymatic tagging of *O*-GlcNAc was accomplished with keto-sugars, several previously unknown *O*-GlcNAc-modified proteins were identified in brain [153]. Considering that obesity-related disease conditions are rapidly increasing and now rival the deleterious impact of cancer and vascular disease on human health, together with many other emerging biological processes influenced by *O*-GlcNAc [154], molecular tools to understand this PTM are sorely needed and MOE technology constitutes a welcome new approach.

10.4 CONCLUSIONS

As outlined in this chapter, MOE has the capacity to alter the properties of carbohydrates and their associated proteins in living cells and animals in ways that provide enticing developmental leads for novel therapeutics for vexing human illnesses ranging from cancer to diabetes to infectious disease. At the same time, obstacles to the translation of this sugar-based technology from the laboratory to the clinic remain formidable. Encouragingly, significant challenges, which include the potential toxicity and the poor pharmacological properties of MOE analogs are being actively investigated and are now beginning to be understood mechanistically, offering hope that solutions can be developed in the near term. In conclusion, we confidently predict that even if a fraction of therapeutic applications now under consideration ultimately reach fruition, MOE has a bright future.

Acknowledgments

The authors would like to thank the National Cancer Institute (5R01CA1123 14-03), the National Institute for Biomedical Imaging and Bioengineering (1R01EB005692-01A1), and the Susan G. Komen Breast Cancer Foundation for financial support.

REFERENCES

1. Helenius, A., Aebi, M. (2001). Intracellular functions of N-linked glycans. *Science*, *291*, 2364–2369.

2. Wormald, M. R., Petrescu, A. J., Pao, Y.-L., Glithero, A., Elliot, T., Dwek, R. A. (2002). Conformational studies of oligosaccharides and glycopeptides: complementarity of NMR, x-ray crystallography, and molecular modelling. *Chem. Rev.*, *102*, 371–386.

3. Gabius, H.-J., Siebert, H.-C., André, S., Jiménez-Barbero, J., Rüdiger, H. (2004). Chemical biology of the sugar code. *ChemBioChem*, *5*, 740–764.

4. Dove, A. (2001). The bittersweet promise of glycobiology. *Nat. Biotechnol.*, *19*, 913–917.

5. Murrell, M. P., Yarema, K. J., Levchenko, A. (2004). The systems biology of glycosylation. *ChemBioChem*, *5*, 1334–1447.

6. Krambeck, F. J., Betenbaugh, M. J. (2005). A mathematical model of N-linked glycosylation. *Biotechnol. Bioeng.*, *92*, 711–728.

7. Lau, K. S., Partridge, E. A., Grigorian, A., et al. (2007). Complex N-glycan number and degree of branching cooperate to regulate cell proliferation and differentiation. *Cell*, *129*, 123–134.

8. Meutermans, W., Le, G. T., Becker, B. (2006). Carbohydrates as scaffolds in drug discovery. *ChemMedChem*, *1*, 1164–1194.

9. Campbell, C. T., Sampathkumar, S.-G., Weier, C., Yarema, K. J. (2007). Metabolic oligosaccharide engineering: perspectives, applications, and future directions. *Mol. Biosyst.*, *3*, 187–194.

10. Dube, D. H., Bertozzi, C. R. (2003). Metabolic oligosaccharide engineering as a tool for glycobiology. *Curr. Opin. Chem. Biol.*, *7*, 616–625.

11. Wang, Q., Zhang, J., Guo, Z. (2007). Efficient glycoengineering of GM3 on melanoma cell and monoclonal antibody-mediated selective killing of the glycoengineered cancer cell. *Bioorg. Med. Chem.*, *15*, 7561–7567.

12. Aich, U., Campbell, C. T., Elmouelhi, N., et al. (2008). Regioisomeric SCFA attachment to hexosamines separates metabolic flux from cytotoxcity and MUC1 suppression. *ACS Chem. Biol.*, *3*, 230–240.

13. Bistrup, A., Bhakta, S., Lee, J. K., et al. (1999). Sulfotransferases of two specificities function in the reconstitution of high endothelial cell ligands for L-selectin. *J. Cell Biol.*, *145*, 899–910.

14. Bishop, J. R., Schuksz, M., Esko, J. D. (2007). Heparan sulphate proteoglycans fine-tune mammalian physiology. *Nature*, *446*, 1030–1037.

15. Kayser, H., Zeitler, R., Kannicht, C., Grunow, D., Nuck, R., Reutter, W. (1992). Biosynthesis of a nonphysiological sialic acid in different rat organs, using N-propanoyl-D-hexosamines as precursors. *J. Biol. Chem.*, *267*, 16934–16938.

16. Keppler, O. T., Horstkorte, R., Pawlita, M., Schmidt, C., Reutter, W. (2001). Biochemical engineering of the N-acyl side chain of sialic acid: biological implications. *Glycobiology*, *11*, 11R–18R.

17. Tanner, M. E. (2005). The enzymes of sialic acid biosynthesis. *Bioorg. Chem.*, *33*, 216–228.

18. Hinderlich, S., Oetke, C., Pawlita, M. (2005). Biochemical engineering of sialic acids. In *Handbook of Carbohydrate Engineering*, K. J. Yarema, ed. Francis & Taylor/CRC Press, Boca Raton, FL, pp. 387–405.

19. Jacobs, C. L., Yarema, K. J., Mahal, L. K., Nauman, D. A., Charters, N., Bertozzi, C. R. (2000). Metabolic labeling of glycoproteins with chemical tags through unnatural sialic acid biosynthesis. *Methods Enzymol.*, *327*, 260–275.

20. Sampathkumar, S.-G., Li, A. V., Yarema, K. J. (2006). Synthesis of non-natural ManNAc analogs for the expression of thiols on cell surface sialic acids. *Nat. Protocols*, *1*, 2377–2385.

21. Jacobs, C. L., Goon, S., Yarema, K. J., et al. (2001). Substrate specificity of the sialic acid biosynthetic pathway. *Biochemistry*, *40*, 12864–12874.

22. Viswanathan, K., Lawrence, S., Hinderlich, S., Yarema, K. J., Lee, Y. C., Betenbaugh, M. (2003). Engineering sialic acid synthetic ability into insect cells: identifying metabolic bottlenecks and devising strategies to overcome them. *Biochemistry*, *42*, 15215–15225.

23. Han, S., Collins, B. E., Bengtson, P., Paulson, J. C. (2005). Homo-multimeric complexes of CD22 revealed by in situ photoaffinity protein–glycan crosslinking. *Nat. Chem. Biol.*, *1*, 93–97.

24. Gross, H. J., Brossmer, R. (1988). Enzymatic introduction of a fluorescent sialic acid into oligosaccharide chains of glycoproteins. *Eur. J. Biochem.*, *177*, 583–589.

25. Hang, H. C., Bertozzi, C. R. (2001). Ketone isoteres of 2-*N*-acetamidosugars as substrates for metabolic cell surface engineering. *J. Am. Chem. Soc.*, *123*, 1242–1243.

26. Vocadlo, D. J., Hang, H. C., Kim, E.-J., Hanover, J. A., Bertozzi, C. R. (2003). A chemical approach for identifying *O*-GlcNAc-modified proteins in cells. *Proc. Natl. Acad. Sci. USA*, *100*, 9116–9121.

27. Pozsgay, V., Vieira, N. E., Yergey, A. (2002). A method for bioconjugation of carbohydrates using Diels–Alder cycloaddition. *Org. Lett.*, *4*, 3191–3194.

28. Kayser, H., Geilen, C. C., Paul, C., Zeitler, R., Reutter, W. (1993). New amino sugar analogues are incorporated at different rates into glycoproteins of mouse organs. *Experientia*, *49*, 885–887.

29. Chou, H. H., Takematsu, H., Diaz, S., et al. (1998). A mutation in human CMP–sialic acid hydroxylase occurred after the Homo–Pan divergence. *Proc. Natl. Acad. Sci. USA*, *95*, 11751–11756.

30. Brinkman-Van der Linden, E. C., Sjoberg, E. R., Juneja, L. R., Crocker, P. R., Varki, N., Varki, A. (2000). Loss of *N*-glycolylneuraminic acid in human evolution: implications for sialic acid recognition by siglecs. *J. Biol. Chem.*, *275*, 8633–8640.

31. Varki, A. (2002). Loss of *N*-glycolylneuraminic acid in humans: mechanisms, consequences, and implications for hominid evolution. *Am. J. Phys. Anthropol.*, *116*, 54–69.

32. Collins, B. E., Fralich, T. J., Itonori, S., Ichikawa, Y., Schnaar, R. L. (2000). Conversion of cellular sialic acid expression from *N*-acetyl- to *N*-glycolylneuraminic acid using a synthetic precursor, *N*-glycolylmannosamine pentaacetate: inhibition of myelin-associated glycoprotein binding to neural cells. *Glycobiology*, *10*, 11–20.

33. Crocker, P. R. (2002). Siglecs: sialic acid-binding immunogloulin-like lectins in cell–cell interactions and signalling. *Curr. Opin. Struct. Biol.*, *12*, 609–615.

34. Nakagawa, H., Zheng, M., Hakomori, S., Tsukamoto, Y., Kawamura, Y., Takahashi, N. (1996). Detailed oligosaccharide structures of human integrin α5β1 analyzed by a three-dimensional mapping technique. *Eur. J. Biochem.*, *237*, 76–85.

35. Sujan, A. H., Manilla, G. A., Skinner, M. (2003). Gel electrophoresis and mass spectrometric analysis of β1 integrin to identify occupied N-linked glycosylation sites. *Glycobiology*, *13*, 861.

36. Gu, J., Taniguchi, N. (2004). Regulation of integrin functions by *N*-glycans. *Glycoconjug. J.*, *21*, 9–15.

37. Isaji, T., Gu, J., Nishiuchi, R., et al. (2004). Introduction of bisecting GlcNAc into integrin α5β1 reduces ligand binding and down-regulates cell adhesion and cell migration. *J. Biol. Chem.*, *279*, 19747–19754.

38. Seales, E. C., Jurado, G. A., Brunson, B. A., Bellis, S. L. (2003). α2,6 Sialylation regulates β1 integrin function. *Glycobiology*, *13*, 860–861.

39. Pretzlaff, R. K., Xue, V. W., Rowin, M. E. (2000). Sialidase treatment exposes the β1−integrin active ligand binding site on HL60 cells and increases binding to fibronectin. *Cell Adhes. Commun.*, *7*, 491–500.

40. Seales, E. C., Jurado, G. A., Brunson, B. A., Wakefield, J. K., Frost, A. R., Bellis, S. L. (2005). Hypersialylation of β1 integrins, observed in colon adenocarcinoma, may contribute to cancer progression by up-regulating cell motility. *Cancer Res.*, *65*, 4645–4652.

41. Wang, X., Sun, P., Al-Qamari, A., Tai, T., Kawashima, I., Paller, A. S. (2001). Carbohydrate-carbohydrate binding of ganglioside to integrin α5 modulates α5β1 function. *J. Biol. Chem.*, *276*, 8436–8444.

42. Kato, K., Shiga, K., Yamaguchi, K., et al. (2006). Plasma membrane–associated sialidase (NEU3) differentially regulates integrin-mediated cell proliferation through laminin- and fibronectin-derived signaling. *Biochem. J.*, *394*, 647–656.

43. Hakomori, S.-I. (2002). The glycosynapse. *Proc. Natl. Acad. Sci. USA*, *99*, 225–232.

44. Toledo, M. S., Suzuki, E., Handa, K., Hakomori, S. (2005). Effect of ganglioside and tetraspanins in microdomains on interaction of integrins with fibroblast growth factor receptor. *J. Biol. Chem.*, *280*, 16227–16234.

45. Luo, B.-H., Springer, T. A., Takagi, J. (2003). Stabilizing the open conformation of the integrin headpiece with a glycan wedge increases affinity for ligand. *Proc. Natl. Acad. Sci. USA*, *100*, 2403–2408.

46. Villavicencio-Lorini, P., Laabs, S., Danker, K., Reutter, W., Horstkorte, R. (2002). Biochemical engineering of the acyl side chain of sialic acids stimulates integrin-dependent adhesion of HL60 cells to fibronectin. *J. Mol. Med.*, *80*, 671–677.

47. Lemieux, G. A., Bertozzi, C. R. (1998). Chemoselective ligation reactions with proteins, oligosaccharides and cells. *Trends Biotechnol.*, *16*, 506–513.

48. Saxon, E., Bertozzi, C. R. (2000). Cell surface engineering by a modified Staudinger reaction. *Science*, *287*, 2007–2010.

49. Sawa, M., Hsu, T.-L., Itoh, T., et al. (2006). Glycoproteomic probes for fluorescent imaging of fucosylated glycans in vivo. *Proc. Natl. Acad. Sci. USA*, *103*, 12371–12376.

50. Mahal, L. K., Yarema, K. J., Bertozzi, C. R. (1997). Engineering chemical reactivity on cell surfaces through oligosaccharide biosynthesis. *Science*, *276*, 1125–1128.

51. Yarema, K. J., Mahal, L. K., Bruehl, R. E., Rodriguez, E. C., Bertozzi, C. R. (1998). Metabolic delivery of ketone groups to sialic acid residues: application to cell surface glycoform engineering. *J. Biol. Chem.*, *273*, 31168–31179.

52. Agard, N. J., Baskin, J. M., Prescher, J. A., Lo, A., Bertozzi, C. R. (2006). A comparative study of bioorthogonal reactions with azides. *ACS Chem. Biol.*, *1*, 644–648.

53. Sampathkumar, S.-G., Jones, M. B., Yarema, K. J. (2006). Metabolic expression of thiol-derivatized sialic acids on the cell surface and their quantitative estimation by flow cytometry. *Nat. Protocols*, *1*, 1840–1851.

54. Sampathkumar, S.-G., Li, A. V., Jones, M. B., Sun, Z., Yarema, K. J. (2006). Metabolic installation of thiols into sialic acid modulates adhesion and stem cell biology. *Nat. Chem. Biol.*, *2*, 149–152.

55. Laughlin, S. T., Agard, N. J., Baskin, J. M., et al. (2006). Metabolic labeling of glycans with azido sugars for visualization and glycoproteomics. *Methods Enzymol.*, *415*, 230–250.

56. Hsu, T.-L., Hanson, S. R., Kishikawa, K., Wang, S.-K., Sawa, M., Wong, C.-H. (2007). Alkynyl sugar analogs for the labeling and visualization of glycoconjugates in cells. *Proc. Natl. Acad. Sci. USA*, *104*, 2614–2619.

57. Lemieux, G. A., Yarema, K. J., Jacobs, C. L., Bertozzi, C. R. (1999). Exploiting differences in sialoside expression for selective targeting of MRI contrast reagents. *J. Am. Chem. Soc.*, *121*, 4278–4279.

58. Nauman, D. A., Bertozzi, C. R. (2001). Kinetic parameters for small-molecule drug delivery by covalent cell surface targeting. *Biochim. Biophys. Acta*, *1568*, 147–154.

59. Lee, J. H., Baker, T. J., Mahal, L. K., et al. (1999). Engineering novel cell surface receptors for virus-mediated gene transfer. *J. Biol. Chem.*, *274*, 21878–21884.

60. Iwasaki, Y., Tabata, E., Kurita, K., Akiyoshi, K. (2005). Selective cell attachment to a biomimetic polymer surface through the recognition of cell-surface tags. *Bioconjug. Chem.*, *16*, 567–575.

61. De Bank, P. A., Kellam, B., Kendall, D. A., Shakesheff, K. M. (2003). Surface engineering of living myoblasts via selective periodate oxidation. *Biotechnol. Bioeng.*, *81*, 800–808.

62. Keller, T. H., Pichota, A., Yin, Z. (2006). A practical view of "druggability." *Curr. Opin. Chem. Biol.*, *10*, 357–361.

63. Oetke, C., Hinderlich, S., Brossmer, R., Reutter, W., Pawlita, M., Keppler, O. T. (2001). Evidence for efficient uptake and incorporation of sialic acid by eukaryotic cells. *Eur. J. Biochem.*, *268*, 4553–4561.

64. Lavis, L. D. (2008). Ester bonds in prodrugs. *ACS Chem. Biol.*, *3*, 203–206.

65. Jones, M. B., Teng, H., Rhee, J. K., Baskaran, G., Lahar, N., Yarema, K. J. (2004). Characterization of the cellular uptake and metabolic conversion of acetylated

N-acetylmannosamine (ManNAc) analogues to sialic acids. *Biotechnol. Bioeng.*, *85*, 394–405.

66. Kim, E. J., Sampathkumar, S.-G., Jones, M. B., Rhee, J. K., Baskaran, G., Yarema, K. J. (2004). Characterization of the metabolic flux and apoptotic effects of O-hydroxyl- and N-acetylmannosamine (ManNAc) analogs in Jurkat (human T-lymphoma-derived) cells. *J. Biol. Chem.*, *279*, 18342–18352.

67. Dube, D. H., Prescher, J. A., Quang, C. N., Bertozzi, C. R. (2006). Probing mucin-type O-linked glycosylation in living animals. *Proc. Natl. Acad. Sci. USA*, *103*, 4819–4824.

68. Rabuka, D., Hubbard, S. C., Laughlin, S. T., Argade, S. P., Bertozzi, C. R. (2006). A chemical reporter strategy to probe glycoprotein fucosylation. *J. Am. Chem. Soc.*, *128*, 12078–12079.

69. Sarkar, A. K., Brown, J. R., Esko, J. D. (2000). Synthesis and glycan priming activity of acetylated disaccharides. *Carbohydr. Res.*, *329*, 287–300.

70. Kontou, M., Bauer, C., Reutter, W., Horstkorte, R. (2008). Sialic acid metabolism is involved in the regulation of gene expression during neuronal differentiation of PC12 cells. *Glycoconjug. J.*, *25*, 237–244.

71. Campbell, C. T., Aich, U., Weier, C. A., et al. (2008). Targeting pro-invasive oncogenes with short chain fatty acid-hexosamine analogs inhibits the mobility of metastatic MDA-MB-2331 breast cancer cell. *J. Med. Chem.*, *51*, 8135–8147.

72. Sampathkumar, S.-G., Campbell, C. T., Weier, C., Yarema, K. J. (2006). Short-chain fatty acid-hexosamine cancer prodrugs: The sugar matters! *Drugs Future*, *31*, 1099–1116.

73. Sampathkumar, S.-G., Jones, M. B., Meledeo, M. A., et al. (2006). Targeting glycosylation pathways and the cell cycle: sugar-dependent activity of butyrate–carbohydrate cancer prodrugs. *Chem. Biol.*, *13*, 1265–1275.

74. Fishman, P. H., Simmons, J. L., Brady, R. O., Freese, E. (1974). Induction of glycolipid biosynthesis by sodium butyrate in HeLa cells. *Biochem. Biophys. Res. Commun.*, *59*, 292–299.

75. Fishman, P. H., Brady, R. O. (1976). Biosynthesis and function of gangliosides. *Science*, *194*, 906–915.

76. Simmons, J. L., Fishman, P. H., Freese, E., Brady, R. O. (1975). Morphological alterations and ganglioside sialyltransferase activity induced by small fatty acids in HeLa cells. *J. Cell Biol.*, *66*, 414–424.

77. Ripka, J., Shin, S., Stanley, P. (1986). Decreased tumorigenicity correlates with expression of altered cell surface carbohydrates in Lec9 CHO cells. *Mol. Cell. Biol.*, *6*, 1268–1275.

78. Sell, S. (1990). Cancer-associated carbohydrates identified by monoclonal antibodies. *Hum. Pathol.*, *21*, 1003–1019.

79. Varki, A., Cummings, R., Esko, J., Freeze, H., Hart, G., Marth, J. (1999). Glycosylation changes in cancer. In *Essentials of Glycobiology*. Cold Spring Harbor Laboratory, Cold Spring Harbor, NY, p. 537.

80. Demetriou, M., Nabi, I. R., Coppolino, M., Dedhar, S., Dennis, J. W. (1995). Reduced contact-inhibition and substratum adhesion in epithelial cells expressing GlcNAc-transferase V. *J. Cell Biol.*, *130*, 383–392.

81. Dennis, J. W., Laferte, S., Waghorne, C., Breitman, M. L., Kerbel, R. S. (1987). β1-6 Branching of Asn-linked oligosaccharides is directly associated with metastasis. *Science*, *236*, 236–239.

82. Naka, R., Kamoda, S., Ishizuka, A., Kinoshita, M., Kakehi, K. (2006). Analysis of total *N*-glycans in cell membrane fractions of cancer cells using a combination of serotonin affinity chromatography and normal phase chromatography. *J. Proteome Res.*, *5*, 88–97.

83. Dube, D. H., Bertozzi, C. R. (2005). Glycans in cancer and inflammation: potential for therapeutics and diagnostics. *Nat. Rev. Drug Discov.*, *4*, 477–488.

84. Fuster, M. M., Esko, J. D. (2005). The sweet and sour of cancer: glycans as novel therapeutic targets. *Nat. Rev. Cancer*, *5*, 526–542.

85. Kawano, T., Koyama, S., Hiromu, T., et al. (1995). Molecular cloning of cytidine monophospho-*N*-acetylneuraminic acid hydroxylase: regulation of species- and tissue-specific expression of *N*-glycolylneuraminic acid. *J. Biol. Chem.*, *270*, 16458–16463.

86. Irie, A., Koyama, S., Kozutsumi, Y., Kawasaki, T., Suzuki, A. (1998). The molecular basis for the absence of *N*-glycolylneuraminic acid in humans. *J. Biol. Chem.*, *273*, 15866–15871.

87. Malykh, Y. N., Schauer, R., Shaw, L. (2001). *N*-Glycolylneuraminic acid in human tumours. *Biochimie*, *83*, 623–634.

88. Bardor, M., Nguyen, D. H., Diaz, S., Varki, A. (2005). Mechanism of uptake and incorporation of the non-human sialic acid *N*-glycolylneuraminic acid into human cells. *J. Biol. Chem.*, *280*, 4228–4237.

89. Mahal, L. K., Yarema, K. J., Lemieux, G. A., Bertozzi, C. R. (1999). Chemical approaches to glycobiology: engineering cell surface sialic acids for tumor targeting. In *Sialobiology and Other Novel Forms of Glycosylation*, Y. Inoue, Y. C. Lee, and F. A. Troy II, eds. Gakushin Publishing Company, Osaka, Japan, pp. 273–280.

90. Chefalo, P., Pan, Y.-B., Nagy, N., Harding, C., Guo, Z.-W. (2004). Preparation and immunological studies of protein conjugates of *N*-acylneuraminic acids. *Glycoconjug. J.*, *20*, 407–414.

91. Liu, T., Guo, Z., Yang, Q., Sad, S., Jennings, H. J. (2000). Biochemical engineering of surface α 2,8 polysialic acid for immunotargeting tumor cells. *J. Biol. Chem.*, *275*, 32832–32836.

92. Pan, Y., Chefalo, P., Nagy, N., Harding, C., Guo, Z. (2005). Synthesis and immunological properties of N-modified GM3 antigens as therapeutic cancer vaccines. *J. Med. Chem.*, *48*, 875–883.

93. Lehmann, F., Tiralongo, E., Tiralongo, J. (2006). Sialic acid–specific lectins: occurrence, specificity and function. *Cell. Mol. Life Sci.*, *63*, 1331–1354.

94. Dugan, A. S., Eash, S., Atwood, W. J. (2006). Update on BK virus entry and intracellular trafficking. *Transplant Infect. Dis.*, *8*, 62–67.

95. Tong, L., Baskaran, G., Jones, M. B., Rhee, J. K., Yarema, K. J. (2003). Glycosylation changes as markers for the diagnosis and treatment of human disease. In *Biochemical and Genetic Engineering Reviews*, S. Harding, ed. Intercept Ltd., Andover, Hampshire, UK, pp. 199–244.

96. López, S., Arias, C. F. (2004). Multistep entry of rotavirus into cells: a Versaillesque dance. *Trends Microbiol.*, *12*, 271–278.

97. Vimr, E. R., Kalivoda, K. A., Deszo, E. L., Steenbergen, S. M. (2004). Diversity of microbial sialic acid metabolism. *Microbiol. Mol. Biol. Rev.*, *68*, 132–153.

98. Mühlenhoff, M., Eckhardt, M., Gerardy-Schahn, R. (1998). Polysialic acid: three-dimensional structure, biosynthesis, and function. *Curr. Opin. Struct. Biol.*, *8*, 558–564.

99. Chava, A. K., Bandyopadhyay, S., Chatterjee, M., Mandal, C. (2004). Sialoglycans in protozoal diseases: their detection, modes of acquisition and emerging biological roles. *Glycoconjug. J.*, *20*, 199–206.

100. Cross, K. J., Burleigh, L. M., Steinhauer, D. A. (2001). Mechanisms of cell entry by influenza virus. *Expert Rev. Mol. Med.*, *6*, 1–18.

101. Keppler, O. T., Stehling, P., Herrmann, M., et al. (1995). Biosynthetic modulation of sialic acid–dependent virus–receptor interactions of two primate polyoma viruses. *J. Biol. Chem.*, *270*, 1308–1314.

102. Goon, S., Schilling, B., Tullius, M. V., Gibson, B. W., Bertozzi, C. R. (2003). Metabolic incorporation of unnatural sialic acids into *Haemophilus ducreyi* lipooligosaccharides. *Proc. Natl. Acad. Sci. USA*, *18*, 3089–3094.

103. Sadamoto, R., Niikura, K., Sears, P. S., et al. (2002). Cell-wall engineering of living bacteria. *J. Am. Chem. Soc.*, *124*, 9018–9019.

104. Sadamoto, R., Niikura, K., Ueda, T., Monde, K., Fukuhara, N., Nishimura, S.-I. (2004). Control of bacteria adhesion by cell-wall engineering. *J. Am. Chem. Soc.*, *126*, 3755–3761.

105. Schenkman, S., Eichinger, D., Pereira, M. E. A., Nussenzweig, V. (1994). Structural and functional properties of *Trypanosoma trans*-sialidase. *Annu. Rev. Microbiol.*, *48*, 499–523.

106. Neubacher, B., Scheid, S., Kelm, S., Frasch, A. C., Meyer, B., Thiem, J. (2006). Synthesis of Neu5Ac oligosaccharides and analogues by transglycosylation and their binding properties as ligands to MAG. *ChemBioChem*, *7*, 896–899.

107. Shieh, S. J., Vacanti, J. P. (2005). State-of-the-art tissue engineering: from tissue engineering to organ building. *Surgery*, *137*, 1–7.

108. Lalan, S., Pomerantseva, I., Vacanti, J. P. (2001). Tissue engineering and its potential impact on surgery. *World J. Surg.*, *25*, 1458–1466.

109. Bianco, P., Robey, P. G. (2001). Stem cells in tissue engineering. *Nature*, *414*, 118–121.

110. Carpenter, M. K., Rosler, E., Rao, M. S. (2003). Characterization and differentiation of human embryonic stem cells. *Cloning Stem Cells*, *5*, 79–88.

111. Reubinoff, B. E., Pera, M. F., Fong, C.-Y., Trounson, A., Bongso, A. (2000). Embryonic stem cell lines from human blastocysts: somatic differentiation in vitro. *Nat. Biotechnol.*, *18*, 399–404.

112. Thomson, J. A., Itskovitz-Eldor, J., Shapiro, S. S., et al. (1998). Embryonic stem cell lines derived from human blastocysts. *Science*, *282*, 1145–1147.

113. Andrews, P. W., Banting, G., Damjanov, I., Arnaud, D., Avner, P. (1984). Three monoclonal antibodies defining distinct differentiation antigens associated with different high molecular weight polypeptides on the surface of human embryonal carcinoma cells. *Hybridoma*, *3*, 347–361.

114. Badcock, G., Pigott, C., Goepel, J., Andrews, P. W. (1999). The human embryonal carcinoma marker antigen TRA-1-60 is a sialylated keratan sulfate proteoglycan. *Cancer Res.*, *59*, 4715–4719.

115. Kaufman, D. S., Hanson, E., Lewis, R. L., Auerbach, R., Thomson, J. A. (2001). Hematopoietic colony-forming cells derived from human embryonic stem cells. *Proc. Nat. Acad. Sci. USA*, *98*, 10716–10721.

116. Allende, M. L., Proia, R. L. (2002). Lubricating cell signaling pathways with gangliosides. *Curr. Opin. Struct. Biol.*, *12*, 587–592.

117. Schmidt, C., Stehling, P., Schnitzer, J., Reutter, W., Horstkorte, R. (1998). Biochemical engineering of neural cell surfaces by the synthetic N-propanoyl-substituted neuraminic acid precursor. *J. Biol. Chem.*, *273*, 19146–19152.

118. Weinbaum, S., Zhang, X., Han, Y., Vink, H., Cowin, S. C. (2003). Mechanotransduction and flow across the endothelial glycocalyx. *Proc. Natl. Acad. Sci. USA*, *100*, 7988–7995.

119. Yang, F., Williams, C. G., Wang, D. A., Lee, H., Manson, P. N., Elisseeff, J. (2005). The effect of incorporating RGD adhesive peptide in polyethylene glycol diacrylate hydrogel on osteogenesis of bone marrow stromal cells. *Biomaterials*, *26*, 5991–5998.

120. Hersel, U., Dahmen, C., Kessler, H. (2003). RGD modified polymers: biomaterials for stimulated cell adhesion and beyond. *Biomaterials*, *24*, 4385–4415.

121. Shu, X. Z., Ghosh, K., Liu, Y., et al. (2004). Attachment and spreading of fibroblasts on an RGD peptide-modified injectable hyaluronon hydrogel. *J. Biomed. Mater. Res. A*, *68*, 365–375.

122. Shamblott, M. J., Axelman, J., Littlefield, J. W., et al. (2001). Human embryonic germ cell derivatives express a broad range of developmentally distinct markers and proliferate extensively in vitro. *Proc. Natl. Acad. Sci. USA*, *98*, 113–118.

123. Walsh, G., Jefferis, R. (2006). Post-translational modifications in the context of therapeutic proteins. *Nat. Biotechnol.*, *24*, 1241–1252.

124. Brooks, S. A. (2006). Protein glycosylation in diverse cell systems: implications for modification and analysis of recombinant proteins. *Expert Rev. Proteom.*, *3*, 345–359.

125. Luzhetskyy, A., Vente, A., Bechthold, A. (2005). Glycosyltransferases involved in the biosynthesis of biologically active natural products that contain oligosaccharides. *Mol. Biosyst.*, *1*, 117–126.

126. Tomiya, N., Narang, S., Lee, Y. C., Betenbaugh, M. J. (2004). Comparing N-glycan processing in mammalian cell lines to native and engineered lepidopteran insect cell lines. *Glycoconjug. J.*, *21*, 343–360.

127. Wildt, S., Gerngross, T. U. (2005). The humanization of N-glycosylation pathways in yeast. *Nat. Rev. Microbiol.*, *3*, 119–128.

128. Jarvis, D. L. (2003). Developing baculovirus–insect cell expression systems for humanized recombinant glycoprotein production. *Virology*, *310*, 1–7.

129. Lawrence, S. M., Huddleston, K. A., Tomiya, N., et al. (2001). Cloning and expression of human sialic acid pathway genes to generate CMP–sialic acids in insect cells. *Glycoconjug. J.*, *18*, 205–213.

130. Shiraishi, N., Natsume, A., Togayachi, A., et al. (2001). Identification and characterization of three novel β1,3 − N-acetylglucosaminyltransferases structurally related to the β1,3-galactosyltransferase family. *J. Biol. Chem.*, *276*, 3498–3507.

131. Koizumi, S. (2005). Large-scale production of oligosaccharides using engineered bacteria. In *Handbook of Carbohydrate Engineering*, K. J. Yarema, ed. CRC Press/Taylor & Francis, Boca Raton, FL, pp. 325–338.

132. Luchansky, S. J., Argade, S., Hayes, B. K., Bertozzi, C. R. (2004). Metabolic functionalization of recombinant glycoproteins. *Biochemistry*, *43*, 12358–12366.

133. Quackenbush, J. (2006). Microarray analysis and tumor classification. *N. Engl. J. Med.*, *354*, 2463–2472.

134. Mazumder, A., Wang, Y. (2006). Gene-expression signatures in oncology diagnostics. *Pharmacogenomics*, *7*, 1167–1173.

135. Mrksich, M. (2004). An early taste of functional glycomics. *Chem. Biol.*, *11*, 739–747.

136. Ratner, D. M., Adams, E. W., Disney, M. D., Seeberger, P. H. (2004). Tools for glycomics: mapping interactions of carbohydrates in biological systems. *ChemBioChem*, *5*, 1375–1383.

137. Campbell, C. T., Yarema, K. J. (2005). Large-scale approaches for glycobiology. *Genome Biol.*, *6*, article 236, doi: 10.1186/gb-2005-6-11-236.

138. Raman, R., Raguram, S., Venkataraman, G., Paulson, J. C., Sasisekharan, R. (2005). Glycomics: an integrated systems approach to structure–function relationships of glycans. *Nat. Methods*, *2*, 817–824.

139. Miyamoto, S. (2006). Clinical applications of glycomic approaches for the detection of cancer and other diseases. *Curr. Opin. Mol. Ther.*, *8*, 507–513.

140. Hirabayashi, J., Kasai, K.-I. (2002). Separation technologies for glycomics. *J. Chromatogr. B*, *771*, 67–87.

141. Kuster, B., Krogh, T. N., Mortz, E., Harvey, D. J. (2001). Glycosylation analysis of gel-separated proteins. *Proteomics*, *1*, 350–361.

142. Nishimura, S.-I., Niikura, K., Kurogochi, M., et al. (2004). High-throughput protein glycomics: combined use of chemoselective glycoblotting and MALDI-TOF/TOF mass spectrometry. *Angew. Chem. Int. Ed.*, *44*, 91–96.

143. Cooper, C. A., Gasteiger, E., Packer, N. H. (2001). GlycoMod: a software tool for determining glycosylation compositions from mass spectrometric data. *Proteomics*, *1*, 340–349.

144. Feizi, T., Fazio, F., Chai, W., Wong, C.-H. (2004). Carbohydrate microarrays: a new set of technologies at the frontiers of glycomics. *Curr. Opin. Struct. Biol.*, *13*, 637–645.

145. Comelli, E. M., Head, S. R., Gilmartin, T., et al. (2006). A focused microarray approach to functional glycomics: transcriptional regulation of the glycome. *Glycobiology*, *16*, 117–131.

146. Hsu, K.-L., Mahal, L. K. (2006). A lectin microarray approach for the rapid analysis of bacterial glycans. *Nat. Protocols*, *1*, 543–549.

147. Prescher, J. A., Dube, D. H., Bertozzi, C. R. (2003). Probing azido sugar metabolism in vivo using the Staudinger ligation. *Glycobiology*, *13*, 894.

148. Prescher, J. A., Dube, D. H., Bertozzi, C. R. (2004). Chemical remodelling of cell surfaces in living animals. *Nature*, *430*, 873–877.

149. Laughlin, S. T., Baskin, J. M., Amacher, S. L., Bertozzi, C. R. (2008). In vivo imaging of membrane-associated glycans in developing zebrafish. *Science*, *320*, 664–667.

150. Buse, M. G. (2006). Hexosamines, insulin resistance, and the complications of diabetes: current status. *Am. J. Physiol. Endocrinol. Metab.*, *290*, E1–E8.

151. Akimoto, Y., Hart, G. W., Hirano, H., Kawakami, H. (2006). *O*-GlcNAc modification of nucleocytoplasmic proteins and diabetes. *Med. Mol. Morphol.*, *38*, 84–91.

152. Nandi, A., Sprung, R., Barma, D. K., et al. (2006). Global identification of *O*-GlcNAc-modified proteins. *Anal. Chem.*, *78*, 452–458.

153. Khidekel, N., Ficarro, S. B., Peters, E. C., Hsieh-Wilson, L. C. (2004). Exploring the *O*-GlcNAc proteome: direct identification of *O*-GlcNAc-modified proteins from the brain. *Proc. Natl. Acad. Sci. USA*, *101*, 13132–13137.

154. Zachara, N. E., Hart, G. W. (2006). Cell signaling, the essential role of *O*-GlcNAc! *Biochim. Biophys. Acta*, *1761*, 599–617.

155. Spiro, R. G. (2002). Protein glycosylation: nature, distribution, enzymatic formation, and disease implications of glycopeptide bonds. *Glycobiology*, *12*, 43R–56R.

156. Chen, H., Wang, Z., Sun, Z., Kim, E. J., Yarema, K. J. (2005). Mammalian glycosylation: an overview of carbohydrate biosynthesis. In *Handbook of Carbohydrate Engineering*, K. J. Yarema, ed. Francis & Taylor/CRC Press, Boca Raton, FL, pp. 1–48.

157. Kim, E. J., Jones, M. B., Rhee, J. K., Sampathkumar, S.-G., Yarema, K. J. (2004). Establishment of *N*-acetylmannosamine (ManNAc) analogue-resistant cell lines as improved hosts for sialic acid engineering applications. *Biotechnol. Prog.*, *20*, 1674–1682.

158. Faure, C., Chalazonitis, A., Rhéaume, C., et al. (2007). Gangliogenesis in the enteric nervous system: roles of the polysialylation of the neural cell adhesion molecule and its regulation by Bone Morphogenetic Protein-4. *Dev. Dynam.*, *236*, 44–59.

159. Mahal, L. K., Charter, N. W., Angata, K., Fukuda, M., Koshland, D. E., Jr., Bertozzi, C. R. (2001). A small-molecule modulator of poly-α2,8-sialic acid expression on cultured neurons and tumor cells. *Science*, *294*, 380–382.

160. Yarema, K. J., Goon, S., Bertozzi, C. R. (2001). Metabolic selection of glycosylation defects in human cells. *Nat. Biotechnol.*, *19*, 553–558.

161. Saxon, E., Luchansky, S. J., Hang, H. C., Yu, C., Lee, S. C., Bertozzi, C. R. (2002). Investigating cellular metabolism of synthetic azidosugars with the Staudinger ligation. *J. Am. Chem. Soc.*, *124*, 14893–14902.

162. Oetke, C., Brossmer, R., Mantey, L. R., et al. (2002). Versatile biosynthetic engineering of sialic acid in living cells using synthetic sialic acid analogues. *J. Biol. Chem.*, *277*, 6688–6695.

163. Brossmer, R., Gross, H. J. (1994). Fluorescent and photoactivatable sialic acids. *Methods Enzymol.*, *247*, 177–193.

164. Brossmer, R., Gross, H. J. (1994). Sialic acid analogs and application for preparation of neoglycoconjugates. *Methods Enzymol.*, *247*, 153–176.

165. Hang, H. C., Yu, C., Kato, D. L., Bertozzi, C. R. (2003). A metabolic labeling approach toward proteomic analysis of mucin-type O-linked glycosylation. *Proc. Natl. Acad. Sci. USA*, *100*, 14846–14851.

166. Carlsson, J., Drevin, H., Axén, R. (1978). Protein thiolation and reversible protein–protein conjugation: *N*-succinimidyl 3-(2-pyridyldithio)propionate, a new heterobifunctional reagent. *Biochem. J.*, *173*, 723–737.

167. Ni, J., Singh, S., Wang, L. X. (2003). Synthesis of maleimide-activated carbohy-drates as chemoselective tags for site-specific glycosylation of peptides and proteins. *Bioconjug. Chem.*, *14*, 232–238.

168. Weihofen, W. A., Berger, M., Chen, H., Saenger, W., Hinderlich, S. (2006). Structures of human *N*-acetylglucosamine kinase in two complexes with *N*-acetylglucosamine and with ADP/glucose: insights into substrate specificity and regulation. *J. Mol. Biol.*, *364*, 388–399.

169. Hanover, J. A. (2001). Glycan-dependent signaling: O-linked *N*-acetylglucosamine. *FASEB J.*, *15*, 1865–1876.

11

CHEMICAL GENOMICS BASED ON YEAST GENETICS

SHINICHI NISHIMURA, YOKO YASHIRODA, AND MINORU YOSHIDA

Chemical Genomics Research Group, RIKEN Advanced Science Institute, Wako, Saitama, Japan

11.1 INTRODUCTION

Secondary metabolites, which are generally nonessential for the producing organism, are a tremendous source of active compounds that induce drastic effects on other organisms. Such compounds have been explored in a wide range of organisms for the development of research tools and medicinal agents. Identification of a cellular target of bioactive secondary metabolites can elucidate the complicated mechanisms underlying the unique biological phenomena and provide fundamental information for the design of new therapeutic agents with improved safety and efficacy.

However, identification of small-molecule targets remains one of the most difficult and challenging tasks for researchers. Biochemical approaches for target identification rely mainly on in vitro binding assays that are both time consuming and labor intensive. In this procedure, analogs of a test compound harboring a radioactive isotope or a linker to a plastic bead should be synthesized. Target proteins are purified either by the bound radiolabeled compound or by the bead-immobilized compound. The limitation of this approach is the necessity to

Protein Targeting with Small Molecules: Chemical Biology Techniques and Applications,
Edited by Hiroyuki Osada
Copyright © 2009 John Wiley & Sons, Inc.

modify the structure of the test compound to introduce these requisite elements without interfering with biological activity. In many cases, this proves to be difficult or impossible. Even if chemical modification was successful, the affinity between the test compound and the target protein is sometimes insufficient for successful isolation of the target. Therefore, novel strategies alternative to these conventional methods are currently required.

11.1.1 Chemical Genetics in the Postgenomic Era

In a recent decade, complete genome sequences for many organisms, including six eukaryotes—two yeast species [1,2], *Caenorhabditis elegans* [3], *Drosophila melanogaster* [4], *Arabidopsis thaliana* [5], and *Homo sapiens* [6,7]—have been determined. This enabled one to do comprehensive analyses of the entire genome, called *genomics*. Classically, genetics is an approach in which responsible genes are identified based on the phenotype observed, induced by perturbation of gene expression such as gene mutation (*forward genetics*). Chemical genetics is an alternative genetic approach using small molecules to cause perturbation in place of gene manipulation. In the postgenomic era, the reverse direction of genetics (*reverse genetics*), in which phenotype is determined by manipulating a defined gene, has been developed and widely used. Once we cloned all the cDNAs or ORFs, or prepared deletion mutants for all genes in an organism, reverse genetics could be carried out comprehensively.

Systematic matching of chemical genetic phenotypes with genetic phenotypes also provides a valuable starting point for many investigations [8]. For example, Hughes et al. identified a novel target of the commonly used drug dyclonine by preparing a compendium of transcription-level profiles corresponding to 300 diverse genetic and chemical treatments in *Saccharomyces cerevisiae* [9]. In this study, 276 deletion mutants, 11 tetracycline-regulatable alleles of essential genes, and 13 well-characterized compounds were profiled. Observation of effects of defined gene deletion and determination of phenotypes by compounds were combined to investigate functionally unknown genes and compounds. When such a comprehensive study is carried out by combining forward and reverse genetics, it will be a powerful methodology to dissect roles of bioactive compounds and gene function. For this purpose, yeast is an attractive model organism because of its ease of genetic manipulation.

11.1.2 Yeast as a Model Organism for Chemical Genomics

Yeast is a unicellular eukaryote, one of the most representative model organisms to decipher complex pathways conserved from fungi to human. Yeast propagates either by budding (as in *S. cerevisiae*) or by fission (as in *Schizosaccharomyces pombe*). Cellular processes of yeast share many features with those of higher eukaryotes. Importantly, there are several advantages, such as simple growth requirements, rapid cell division, and an ease of genetic manipulation. Although classical yeast genetics remains an essential element, the completion of whole genome sequences has greatly accelerated the pace of biological discovery with a wealth of experimental tools for genome-wide analysis.

Budding yeast *S. cerevisiae* has been used successfully for many years as a model organism for mammalian diseases and pathways. For example, the cellular target of rapamycin, an immunosuppressant used as an antirejection drug in tissue transplants, was first identified in *S. cerevisiae* and subsequently verified in humans [10,11]. In 1996, the first complete sequence of a eukaryotic organism, the budding yeast *S. cerevisiae*, was reported [1]. This led to an effort to disrupt each gene in the genome systematically by gene replacement [12,13]. The resulting deletion strain collections provide new avenues for genetic screens for defects in drug sensitivity [14], cell size and morphology [13,15], bud site selection [16], cell surface function [17], and vacuolar protein sorting [18].

The fission yeast *S. pombe* has been used as an excellent model organism for the study of cell-cycle control, mitosis and meiosis, DNA repair and recombination, and the checkpoint controls important for genome stability. From gene sequence comparisons and phylogenetic analyses, it has been suggested that fission yeast diverged from budding yeast around 330 to 420 million years ago, and from metazoans and plants around 1000 to 1100 million years ago [19]. Actually, some gene sequences are equally diverged between the two yeasts as they are from their human homologs. Completion of the annotated genome sequence of the two yeast species clearly demonstrated their difference [2]. For example, compared with *S. cerevisiae, S. pombe* has significantly more intron sequences (roughly 4700 compared with 275), very few transposable-mobile-genetic elements, and large centromeres. It is also noteworthy that *S. pombe* has a smaller gene set than *S. cerevisiae* (roughly 5000 compared with 6000), as a consequence of fewer duplication events. Because duplication is frequently accompanied by divergence, this means that *S. pombe* gene sequences are often more similar than the corresponding *S. cerevisiae* sequences to their higher eukaryotic orthologs. Furthermore, it should not be overlooked that a substantial number of fission yeast genes are conserved in higher eukaryotes but absent in *S. cerevisiae* [20]. These differences could make *S. pombe* a better model than *S. cerevisiae* for understanding some eukaryotic processes. However, at present, *S. pombe* suffers from a relative paucity of tools and resources for genome-wide analyses compared to *S. cerevisiae*.

Both yeast species are suitable for genetic manipulation to change quality or quantity of gene products (see Fig. 11-1). With this advantage, chemical genomics approaches are explored using reverse genomics or proteomics tools. Here we review progress in chemical genomics using budding or fission yeast.

11.2   CHEMICAL GENOMICS USING BUDDING YEAST

Yeast genetics has been used successfully for target identification of bioactive compounds. More than 20 years ago, the use of gene-dosage effects to identify target proteins of small molecules was documented [21]. In that study, target

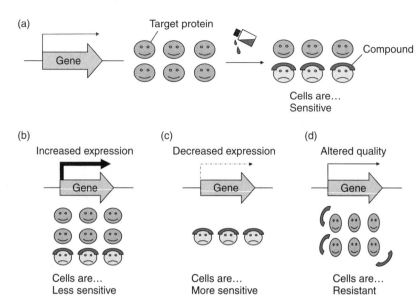

Figure 11-1 Relationships between gene expression and drug sensitivity. Quantity or quality of the target protein (shown as a face) is critical for the corresponding inhibitory compound (shown as an arch). When the expression level of a gene encoding the target protein is increased (b) or decreased (c) compared to that in the control cell (a), the cell becomes less sensitive or more sensitive against the compound, respectively. A mutation in the binding site that causes loss of binding activity renders the cell resistance to the compound (d).

proteins for tunicamycin and compactin were identified by screening a genomic library for genes that confer drug resistance when overexpressed. Yeast clones harboring multiple copies of a target gene were less sensitive to corresponding molecules than those expressing nontarget genes. On the other hand, in the case of an immunosuppressant rapamycin, deletion of the target protein FK506-binding protein (FKBP; Fpr1p in *S. cerevisiae*) was fully resistant to rapamycin [10]. Thus, changing the dosage of genes has been proved to be an effective strategy for drug target discovery in yeast (Fig. 11-1).

After completion of the budding yeast genome sequence in 1996, more systematic ways have been explored: chemical genetic study using genome-wide resources, termed *chemical genomics*. For budding yeast, a complete set of deletion mutants has been constructed by polymerase chain reaction (PCR)–based homologous recombination [12,13]. This project was carried out by an international consortium of laboratories, which identified about 1000 genes (ca. 20% of the entire genome) essential for viability in haploids under standard laboratory conditions. Now, entire sets of deletion mutants—about 6000 heterozygous diploid strains for almost all genes and about 5000 haploid and homozygous diploid strains for nonessential genes—have been made available to the public, enabling a systematic and comprehensive approach to phenotypic analysis.

Two phenomena, *haploinsufficiency* and *synthetic lethality*, have been applied to comprehensive chemical genetic studies.

During construction of the yeast deletion alleles, strain-specific DNA barcodes, consisting of two unique 20-nucleotide oligomers of DNA sequence flanked by common PCR primer sites, were engineered on either side of the drug-resistant marker that replaced the deleted gene [12]. PCR amplification of the bar codes allows the detection of all mutants with a DNA microarray-based detection system; hence, it is possible to quickly assess the presence or absence of each deletion mutant in a mixed population simply by examining the barcode pattern of a population sample [13]. Mutants that show hypersensitivity to the drug are cleared from the drug-treated population, and thus genes encoding candidate drug target proteins or proteins involved in the target pathways can be identified. Pools of haploid or diploid deletion mutants can be examined in relatively small culture volumes, thereby providing a high-throughput system for linking compounds to their intracellular targets.

11.2.1 Haploinsuffiency for Target Discovery

Lowering the dosage of a single gene from two copies to one copy in a diploid cell results in a heterozygote that is more sensitive to a compound that inhibits the gene product (Fig. 11-1c). Thus, by examining drug sensitivity tests using a set of yeast strains with defined gene deletions, we might expect that the most sensitive strain would be a heterozygote that harbors a deletion in the gene encoding the target protein. In a landmark study, Giaever et al. demonstrated that simply reducing the copy number of a target gene from two to one can significantly sensitize a diploid cell to the corresponding inhibitor [22]. This phenomenon, known as *drug-induced haploinsufficiency*, was used to identify the known targets of five well-characterized compounds. Genome sequence information was used to generate molecularly tagged heterozygous yeast strains that were pooled, grown competitively with a compound, and analyzed for sensitivity using high-density oligonucleotide arrays. In that proof-of-principle study, parallel analysis identified the known target and two hypersensitive loci in a mixed culture of 233 strains in the presence of tunicamycin. The use of heterozygotes was proved to be particularly powerful for drug target identification in vivo. This finding has been extended to larger strain pools and diverse compound libraries to identify novel drug targets [23–25].

In a study by Lum et al., a pool of about 3500 heterozygotes was tested against 78 diverse chemical treatments [24]. Statistical analysis of the resulting microarray data revealed a small number of highly specific compound-sensitive strains for about 50 of the compounds tested. Targets were identified correctly for most of the compounds with known protein targets. A novel candidate target of the therapeutic compound, the antianginal compound molsidomine [26], was uncovered and confirmed with conventional biochemical methods. Yeast heterozygous for deletion of *ERG7* showed hypersensitivity to molsidomine. The yeast *ERG7* gene encodes lanosterol synthase, a highly conserved enzyme involved in sterol biosynthesis. Purified lanosterol synthase was inhibited in vitro by SIN-1, the first metabolic derivative of molsidomine. This finding may explain the observation that molsidomine treatment lowers cholesterol levels in both rats and humans [27,28]. In another study, by Giaever et al., a comprehensive set of 5916 heterozygous strains were examined for hypersensitivity to 10 diverse compounds [25]. Interestingly, a similar set of heterozygous diploid strains was hypersensitive to three seemingly disparate drugs: alverine citrate, a muscle relaxant; fenpropimorph, an antifungal; and dyclonine, an anesthetic. This observation prompted a closer examination of these drugs and revealed that they shared a common core chemical structure. Thus, haploinsufficiency profiling clustered three therapeutically distinct compounds, suggesting that analysis of a compendium of profiles may reveal many new structure–activity relationships. Both groups studied 5-FU, which is an antiproliferative drug used in cancer treatment and is thought to inhibit thymidylate synthase [29], but neither group identified a deletion mutant of thymidylate synthase in their haploinsufficiency profile. Although the simple interpretation is that thymidylate synthase levels are not affected by haploinsufficiency, both groups identified a large set of genes involved in ribosomal RNA processing, consistent with evidence suggesting that the antiproliferative effects of 5-FU may result from inhibition of RNA metabolism [30]. The former group showed that 5-FU may inhibit cell growth through perturbation of rRNA processing by the exosome complex, whereas the latter group suggested that 5-FU may be misincorporated into RNA, leading to impairment of essential RNA processing function. In any case, the yeast experiments ought to provoke additional experiments into the mechanism of action of 5-FU, with an eye to new therapeutic options in cancer treatment.

11.2.2 Synthetic Lethality for Target Prediction

In a traditional yeast synthetic lethal screen, one searches for combinations of gene deletions or mutations that together cause cell death but that individually have little or no effect on the organism. In this case, the term *synthetic lethal* is used to denote that the lethal event derives from the synthesis of two different deletions rather than from a single deletion. Such synthetic lethal screens have been useful for identifying genetic interactions and mapping networks and pathways regulating cellular processes [31]. To find such synthetic lethal relationships comprehensively, Tong et al. developed a genome-wide screening system termed *synthetic genetic array* (SGA) *analysis* [32]. SGA analysis offered an efficient approach to the systematic construction of double mutants and enables a global analysis of synthetic lethal genetic interactions. In this screen, a query mutation is crossed with an ordered array of approximately 5000 viable gene deletion mutants such that meiotic progeny harboring both mutations can be scored for growth defects. If the activity of a nonessential pathway is required for cellular fitness when a particular query gene is compromised functionally, all the components of the pathway should be identified in a comprehensive synthetic lethal screen. Thus, application of the SGA system identifies a set of synthetic genetic interactions that are enriched for the components of pathways and complexes. For example, *BIM1* encodes a protein that associates with the plus end of microtubules and participates in nuclear positioning and spindle orientation [33]. An SGA screen with the *BIM1* gene deletion as the query identified genetic interactions with kinetochore components, spindle checkpoints, and the dynein–dynactin spindle orientation pathway [32]. These interactions suggest that Bim1p participates in the attachment of microtubules to the kinetochore, possibly supported by two-hybrid interaction with the kinetochore component Duo1p [34]. As the genetic network expands, complexes and pathways are expected to show a unique pattern of genetic interactions, whereas the molecular function of previously uncharacterized genes can be thus inferred from the connectivity and the position within the network.

Parsons et al. applied this concept by synthesizing gene deletion and chemical treatment to identify genes involved in target pathways of the test compound [35]. Ten chemical treatments were done with the same set of mutant strains to generate drug-sensitivity profiles in parallel on the solid media. By comparing the compound–gene and gene–gene interaction profiles generated by an SGA system, they found that benomyl treatment and the *TUB2* mutant (tubulin 2) have similar synthetic lethal properties, as do cyclosporine A, FK506, and the *CNB1* (calcineurin regulatory B subunit) deletion, or fluconazole and the *ERG11* (lanosterol-14α-demethylase) mutant. It should be noted that SGA profile for *TUB2* and *ERG11* were generated using their temperature-sensitive mutant strains possessing *TUB2-403* and *ERG11-DHFR*, respectively [36], since they are essential genes. Chemical genetic profiles shared a similar, but not identical, synthetic lethal profile with a mutation or a deletion in the gene encoding the protein targeted by the compound, perhaps because of incomplete inactivation of the target protein by the compounds or inherent differences in genetic versus chemical mechanisms of target inhibition. When expanded to the full-genome level, synthetic lethal profiles of every gene in the genome could theoretically be used on all query compounds with unknown mechanisms of action. This can be facilitated by computational analysis (e.g., pattern matching). Compounds with similar chemical genetic profiles can target the same pathway and therefore have a similar mode of action, whereas deletion mutants that show a similar signature of chemical sensitivities often have similar cellular functions (Fig. 11-2).

Parsons et al. explored this possibility by crossing 82 compounds and crude natural product extracts with about 4800 haploid deletion mutants [37]. In this study, a huge number of data points obtained by taking advantage of the DNA microarray technology were analyzed by two-dimensional hierarchical clustering analysis and probabilistic sparse matrix factorization analysis to classify compounds. Both analyses identified similarities among chemical genetic profiles that reflect a common biological target of mode of action. From the compendium generated, the robustness of the barcode-based chemical genetic profiling in yeast was demonstrated. This approach also identifies genes whose products buffer the cells from defects in the target pathway of the compound. Analysis of the genes identified by their function sometimes gives clues for modes of action

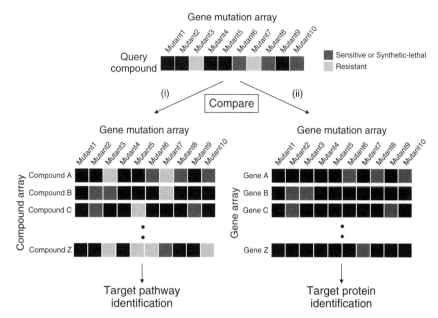

Figure 11-2 From phenotype observation to target pathway and target protein identification. Profiling of chemical genetic interactions is of help for identification of target pathway and target protein. Once phenotypes of gene mutants for a query compound are defined, modes of action of the compound can be expected from statistical analysis using phenotype compendia or functional analysis using public databases. For example, the mutants 6, 8, and 10 are sensitive (shown in red) to the query compound and the mutants 3 and 7 are resistant (shown in green) to the query compound. (i) Comparison with the compendia of chemical genetic interaction profiles reveals that the query compound has a similar target pathway or modes of action with those of the compound A. (ii) Comparison with the synthetic lethal profiles shows that the gene A product is predicted to be a target protein of the query compound. This strategy is particularly useful for reverse genomics approaches using yeast. (*See insert for color representation of figure.*)

of the test compound. For example, papuamide B, a cycliclipopeptide with an unknown target, exhibited more than 300 chemical genetic interactions, which was enriched for genes with certain Gene Ontology (GO) Consortium annotations [38], including vesicle-mediated transport, cell wall organization and biogenesis, and protein modification, suggesting that this compound may affect intracellular transport or perturb some target on the cell surface. After conventional studies, they discovered that papuamide B impaired the yeast cell membrane through direct interaction with phosphatidylserine.

In contrast to the haploinsufficiency screening, which generally identifies a relatively small number of potential target genes, chemical genetic profiling of haploid deletion mutants often identifies tens to hundreds of deletion-mutant hypersensitive strains. Thus, although the haploinsufficiency screening has a great potential to identify the target genes, there is less functional information associated with the mutant hypersensitivity signature than in the haploid deletions. Consequently, the two approaches are complementary.

11.3   CHEMICAL GENETICS AND GENOMICS USING FISSION YEAST

Fission yeast has also contributed to chemical genetics. One of the examples to be noted is target identification of an antibiotic leptomycin B (LMB). LMB originally isolated from a *Streptomyces* sp. [39,40] causes characteristic inhibition of the fission yeast cell proliferation being accompanied by abnormal nuclear structure [41] and the arrest of the cell cycle at both the G1 and G2 phases in mammalian cells [42]. For the target identification, a genomic library from the fission yeast LMB-resistant mutant was screened for LMB resistance genes, and the open reading frame (ORF) identified was found to be a mutant form of the *crm1* gene [43]. Either a single copy of the mutated gene or overexpression of the wild-type *crm1* conferred LMB resistance [43]. Crm1 was originally identified from a genetic screen for an altered higher chromosome structure in cold-sensitive mutants of fission yeast [44]. Morphological and biochemical phenotypes of the cold-sensitive strain at the nonpermissive temperature were identical to those of LMB-treated wild-type cells. Taken together, *S. pombe* Crm1 was proposed to be a cellular target of LMB. However, the function of the Crm1 protein remained to be unveiled, and hence the mode of action of LMB was not understood at that moment.

Amazingly, a breakthrough occurred after a report that LMB was found as an inhibitor of nuclear export of the human immunodeficiency virus type 1 regulatory protein Rev and Rev-dependent mRNA [45]. Rev is required for unspliced and viral mRNAs to appear in the cytoplasm and thus viral replication. This finding implied that Crm1 functions in the process of protein export from nucleus. Crm1 was subsequently identified as a nuclear export receptor that recognizes the nuclear export signal (NES) sequence in the cargo protein, and LMB as an inhibitor of Crm1–NES binding [46–49] (Fig. 11-3).

LMB inactivates Crm1 by covalent binding to a cysteine residue, which is conserved in higher enkaryotes but not in *S. cerevisiae* [50] (Fig. 11-3). The *S. pombe crm1* mutant in which the cysteine residue is replaced with a serine residue is completely insensitive to LMB. The sulfhydryl functionality attacks the α, β-unsaturated δ-lactone of LMB, which is followed by the Michael-type addition. It should be noted that *S. cerevisiae* lacks the corresponding cysteine residue, which accounts for its intrinsic resistance to LMB. Since the molecular recognition between LMB and Crm1 was proved to be high selectivity, this reagent is an indispensable tool to dissect the nucleocytoplasmic transport system.

After completion of the fully annotated genome of *S. pombe* in 2002, cloning of ORFeome, an entire set of an organism's protein coding ORFs, and localizome (whole proteome localization determination) were reported. In this study, each ORF was PCR-amplified, and the entire library covered about 99% of the genes

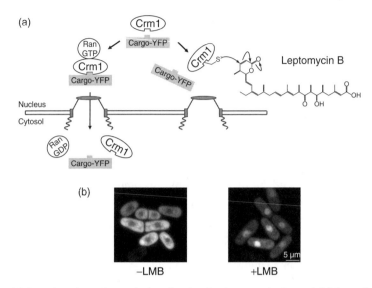

Figure 11-3 Alteration of protein localization by leptomycin B, an inhibitor of nuclear protein export. (a) Crm1-mediated nuclear export. Crm1 exports cargo proteins with the aid of a small GTPase, Ran, to the cytosol (left). Leptomycin B inactivates Crm1 irreversibly and causes nuclear accumulation of the cargo proteins (right). (b) Example images of a localization-altered protein by leptomycin B (LMB) treatment. A YFP-fused cargo protein is localized to the cytosol before (−) LMB treatment and the protein is nuclear accumulated after (+) LMB treatment.

predicted. The library of the cloned *S. pombe* ORFs was swapping into a variety of expression plasmids. Each cloned ORF was fused at the 3′ end to the yellow fluorescent protein (YFP) and a relatively small tag (FLAGx2-His$_6$, termed FFH) by using the *S. pombe* expression vectors pDUAL-YFH1c and pDUAL-FFH1c, respectively [51]. 3′-Tagged ORFs were integrated individually into the *leu1* locus of the *leu1-32* strain, keeping the wild-type copy of the gene intact (Fig. 11-4). The cloned ORFs were expressed under the control of the thiamine-regulatable *nmt1* promoter [52].

Using the strain collection expressing each YFP-tagged protein (ca. 5000 strains), the intracellular localization for nearly 90% of the fission yeast proteome was determined by fluorescence microscopy, including >70% of the proteome which was not characterized previously, making this data set of protein localization the most comprehensive among eukaryotic organisms [53,54]. Crm1 is an evolutionarily conserved protein that mediates nuclear export of proteins containing the NES sequence, with a loosely conserved pattern of three or four hydrophobic residues [55]. Using the *S. pombe* localizome data set and LMB, an inhibitor of Crm1, a chemical genomic study on the protein nuclear export was carried out. The localizome data set prepared in the presence of LMB was compared with the "native" localizome data set, which revealed about 300 putative Crm1 cargo proteins. The number of the proteins that shuttle between the

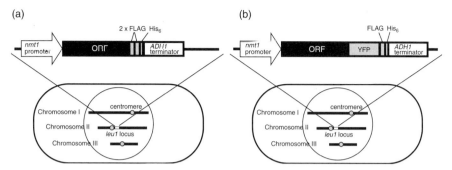

Figure 11-4 Schematic drawing of overexpression strains of *S. pombe*. In fission yeast, two types of strain collections for genome-wide studies have been constructed. Each ORF is 3′-terminally tagged: with two FLAG epitopes and one hexahistidine (His$_6$) tag as depicted in (a) or with YFP, one FLAG epitope, and one His$_6$ tag in (b). The tagged ORF is integrated into the *leu1* locus and expressed under the control of the *nmt1* promoter.

nucleus and the cytoplasm is larger than the existing literature would suggest [56–60]. Analysis of these proteins furnished several novel issues.

First, only about 45% of the proteins identified contained the canonical NES sequence, although it is unclear whether this sequence functions as an NES. Contrary to widely held expectations, half of the proteins identified in this study do not have a consensus NES sequence, implying that these proteins are transported by an adapter that binds Crm1 and/or that there should be alternative NES sequences. In any case, further examination of this class of proteins will lead to a better understanding of the mechanisms of nuclear protein export.

Second, these proteins include not only the cytosolic proteins expected but also proteins localized to other cellular structures, including cell tip, septum, and microtubules. Actually, based on grouping analysis of LMB-sensitive proteins using the GO biological process [38], Crm1 appears to be involved in various cellular processes. These observations suggest that transport into and out of the nucleus of these proteins or their regulators may be important for their proper function at their correct locations.

Finally, 80 proteins localized to an abnormal intranuclear microtubule structure upon LMB treatment, while others simply occupied nucleus or nucleolus. Crm1-dependent dynamic interchange of proteins between the nucleus and the microtubules may regulate microtubule organization during the cell cycle in fission yeast. Further analysis of individual proteins will reveal the involvement of Crm1 in the process for microtubule assembly and disassembly.

Thus, the *S. pombe* ORFeome library in which each ORF is expressed under the thiamine-regulatable strong promoter in the yeast cell is a powerful and versatile tool for chemical genomics. Since dosage levels of a gene product in a yeast cell affect sensitivity to small molecules, profiling the sensitivities to small molecules of each overexpression strain will lead to identification of drug targets and understanding the modes of action of drugs.

11.4 PERSPECTIVES

Neither target identification of a bioactive compound nor discovering a novel compound that specifically inactivates a protein is the goal of chemical biology, but they are the starting point. Small molecules with defined modes of action are essential tools to dissect biological phenomena and generate potential therapeutics. However, we do not have so many useful compounds to challenge the chaos of biology and there is no single robust approach to connect bioactive compounds with their targets, thereby limiting their experimental and therapeutic use.

The systematic general tools for functional genomics can help unveil connections between compounds and disparate cellular events. All of the chemical genomic studies using yeast reviewed in this chapter provide powerful and unique strategies, whereas they are only part of the chemical genomic studies. The use of deletion mutant sets has proven to be a particularly powerful way to observe the gene–gene and compound–gene interactions comprehensively as reported using *S. cerevisiae*. For *S. pombe*, deletion mutants for all genes are currently being prepared [61]. Overexpression is also validated as a useful comprehensive methodology for target protein and target pathway discovery [62,63], while integration of the deletion and overexpression approaches could furnish more informative results [64]. Collectively, these yeast technologies would enable us to identify target proteins or pathways of compounds without a priori knowledge of the underlying mechanisms of action. Finally, dissecting biological phenomena in yeast will provide a template to understand the more complex systems in metazoans, since gene functions are often highly conserved.

REFERENCES

1. Anon. (1997). The yeast genome directory. *Nature*, *387*, 5.
2. Wood, V., Gwilliam, R., Rajandream, M. A., et al. (2002). The genome sequence of *Schizosaccharomyces pombe*. *Nature*, *415*, 871–880.
3. *C. elegans* Sequencing Consortium (1998). Genome sequence of the nematode *C. elegans:* a platform for investigating biology. *Science*, *282*, 2012–2018.
4. Adams, M. D., Celniker, S. E., Holt, R. A., et al. (2000). The genome sequence of *Drosophila melanogaster*. *Science*, *287*, 2185–2195.
5. *Arabidopsis* Genome Initiative (2000). Analysis of the genome sequence of the flowering plant *Arabidopsis thaliana*. *Nature*, *408*, 796–815.
6. Lander, E. S., Linton, L. M., Birren, B., et al. (2001). Initial sequencing and analysis of the human genome. *Nature*, *409*, 860–921.
7. Venter, J. C., Adams, M. D., Myers, E. W., et al. (2001). The sequence of the human genome. *Science*, *291*, 1304–1351.
8. Butcher, R. A., Schreiber, S. L. (2005). Using genome-wide transcriptional profiling to elucidate small-molecule mechanism. *Curr. Opin. Chem. Biol.*, *9*, 25–30.
9. Hughes, T. R., Marton, M. J., Jones, A. R., et al. (2000). Functional discovery via a compendium of expression profiles. *Cell*, *102*, 109–126.

10. Heitman, J., Movva, N. R., Hall, M. N. (1991). Targets for cell cycle arrest by the immunosuppressant rapamycin in yeast. *Science*, *253*, 905–909.

11. Crespo, J. L., Hall, M. N. (2002). Elucidating TOR signaling and rapamycin action: lessons from *Saccharomyces cerevisiae*. *Microbiol. Mol. Biol. Rev*., *66*, 579–591.

12. Winzeler, E. A., Shoemaker, D. D., Astromoff, A., et al. (1999). Functional characterization of the *S. cerevisiae* genome by gene deletion and parallel analysis. *Science*, *285*, 901–906.

13. Giaever, G., Chu, A. M., Ni, L., et al. (2002). Functional profiling of the *Saccharomyces cerevisiae* genome. *Nature*, *418*, 387–391.

14. Armour, C. D., Lum, P. Y. (2005). From drug to protein: using yeast genetics for high-throughput target discovery. *Curr. Opin. Chem. Biol*., *9*, 20–24.

15. Jorgensen, P., Nishikawa, J. L., Breitkreutz, B. J., Tyers, M. (2002). Systematic identification of pathways that couple cell growth and division in yeast. *Science*, *297*, 395–400.

16. Ni, L., Snyder, M. (2001). A genomic study of the bipolar bud site selection pattern in *Saccharomyces cerevisiae*. *Mol. Biol. Cell*, *12*, 2147–2170.

17. Page, N., Gerard-Vincent, M., Menard, P., et al. (2003). A *Saccharomyces cerevisiae* genome-wide mutant screen for altered sensitivity to K1 killer toxin. *Genetics*, *163*, 875–894.

18. Bonangelino, C. J., Chavez, E. M., Bonifacino, J. S. (2002). Genomic screen for vacuolar protein sorting genes in *Saccharomyces cerevisiae*. *Mol. Biol. Cell*, *13*, 2486–2501.

19. Sipiczki, M. (2000). Where does fission yeast sit on the tree of life? *Genome Biol*., *1*, Reviews 1011.

20. Lespinet, O., Wolf, Y. I., Koonin, E. V., Aravind, L. (2002). The role of lineage-specific gene family expansion in the evolution of eukaryotes. *Genome Res*., *12*, 1048–1059.

21. Rine, J., Hansen, W., Hardeman, E., Davis, R. W. (1983). Targeted selection of recombinant clones through gene dosage effects. *Proc. Natl. Acad. Sci. USA*, *80*, 6750–6754.

22. Giaever, G., Shoemaker, D. D., Jones, T. W., et al. (1999). Genomic profiling of drug sensitivities via induced haploinsufficiency. *Nat. Genet*., *21*, 278–283.

23. Baetz, K., McHardy, L., Gable, K., et al. (2004). Yeast genome-wide drug-induced haploinsufficiency screen to determine drug mode of action. *Proc. Natl. Acad. Sci. USA*, *101*, 4525–4530.

24. Lum, P. Y., Armour, C. D., Stepaniants, S. B., et al. (2004). Discovering modes of action for therapeutic compounds using a genome-wide screen of yeast heterozygotes. *Cell*, *116*, 121–137.

25. Giaever, G., Flaherty, P., Kumm, J., et al. (2004). Chemogenomic profiling: identifying the functional interactions of small molecules in yeast. *Proc. Natl. Acad. Sci. USA*, *101*, 793–798.

26. Reden, J. (1990). Molsidomine. *Blood Vessels*, *27*, 282–294.

27. Chassoux, G. (1989). Molsidomine and lipid metabolism. *J. Cardiovasc. Pharmacol*., *14* (Suppl. 11), S137–S138.

28. Granzer, E., Ostrowski, J. (1984). [Lipoprotein regulatory effect of molsodomine in rats.] *Arzneimittelforschung*, *34*, 191–193.

29. Parker, W. B., Cheng, Y. C. (1990). Metabolism and mechanism of action of 5-fluorouracil. *Pharmacol. Ther.*, *48*, 381–395.

30. Scherf, U., Ross, D. T., Waltham, M., et al. (2000). A gene expression database for the molecular pharmacology of cancer. *Nat. Genet.*, *24*, 236–244.

31. Bader, G. D., Heilbut, A., Andrews, B., Tyers, M., Hughes, T., Boone, C. (2003). Functional genomics and proteomics: charting a multidimensional map of the yeast cell. *Trends Cell Biol.*, *13*, 344–356.

32. Tong, A. H., Evangelista, M., Parsons, A. B., et al. (2001). Systematic genetic analysis with ordered arrays of yeast deletion mutants. *Science*, *294*, 2364–2368.

33. Bloom, K. (2000). It's a kar9ochore to capture microtubules. *Nat. Cell Biol.*, *2*, E96–E98.

34. Uetz, P., Giot, L., Cagney, G., et al. (2000). A comprehensive analysis of protein–protein interactions in *Saccharomyces cerevisiae*. *Nature*, *403*, 623–627.

35. Parsons, A. B., Brost, R. L., Ding, H., et al. (2004). Integration of chemical-genetic and genetic interaction data links bioactive compounds to cellular target pathways. *Nat. Biotechnol.*, *22*, 62–69.

36. Dohmen, R. J., Wu, P., Varshavsky, A. (1994). Heat-inducible degron: a method for constructing temperature-sensitive mutants. *Science*, *263*, 1273–1276.

37. Parsons, A. B., Lopez, A., Givoni, I. E., et al. (2006). Exploring the mode-of-action of bioactive compounds by chemical-genetic profiling in yeast. *Cell*, *126*, 611–625.

38. Ashburner, M., Ball, C. A., Blake, J. A., et al. (2000). Gene ontology: tool for the unification of biology. The Gene Ontology Consortium. *Nat. Genet.*, *25*, 25–29.

39. Hamamoto, T., Gunji, S., Tsuji, H., Beppu, T. (1983). Leptomycins A and B, new antifungal antibiotics: I. Taxonomy of the producing strain and their fermentation, purification and characterization. *J. Antibiot. (Tokyo)*, *36*, 639–645.

40. Hamamoto, T., Seto, H., Beppu, T. (1983). Leptomycins A and B, new antifungal antibiotics: II. Structure elucidation. *J. Antibiot. (Tokyo)*, *36*, 646–650.

41. Hamamoto, T., Uozumi, T., Beppu, T. (1985). Leptomycins A and B, new antifungal antibiotics: III. Mode of action of leptomycin B on *Schizosaccharomyces pombe*. *J. Antibiot. (Tokyo)*, *38*, 1573–1580.

42. Yoshida, M., Nishikawa, M., Nishi, K., Abe, K., Horinouchi, S., Beppu, T. (1990). Effects of leptomycin B on the cell cycle of fibroblasts and fission yeast cells. *Exp. Cell Res.*, *187*, 150–156.

43. Nishi, K., Yoshida, M., Fujiwara, D., Nishikawa, M., Horinouchi, S., Beppu, T. (1994). Leptomycin B targets a regulatory cascade of crm1, a fission yeast nuclear protein, involved in control of higher order chromosome structure and gene expression. *J. Biol. Chem.*, *269*, 6320–6324.

44. Adachi, Y., Yanagida, M. (1989). Higher order chromosome structure is affected by cold-sensitive mutations in a *Schizosaccharomyces pombe* gene *crm1*$^{+}$, which encodes a 115-kD protein preferentially localized in the nucleus and its periphery. *J. Cell Biol.*, *108*, 1195–1207.

45. Wolff, B., Sanglier, J. J., Wang, Y. (1997). Leptomycin B is an inhibitor of nuclear export: inhibition of nucleo-cytoplasmic translocation of the human immunodeficiency virus type 1 (HIV-1) Rev protein and Rev-dependent mRNA. *Chem. Biol.*, *4*, 139–147.

46. Fukuda, M., Asano, S., Nakamura, T., et al. (1997). CRM1 is responsible for intracellular transport mediated by the nuclear export signal. *Nature*, *390*, 308–311.

47. Ossareh-Nazari, B., Bachelerie, F., Dargemont, C. (1997). Evidence for a role of CRM1 in signal-mediated nuclear protein export. *Science*, *278*, 141–144.

48. Fornerod, M., Ohno, M., Yoshida, M., Mattaj, I. W. (1997). CRM1 is an export receptor for leucine-rich nuclear export signals. *Cell*, *90*, 1051–1060.

49. Stade, K., Ford, C. S., Guthrie, C., Weis, K. (1997). Exportin 1 (Crm1p) is an essential nuclear export factor. *Cell*, *90*, 1041–1050.

50. Kudo, N., Matsumori, N., Taoka, H., et al. (1999). Leptomycin B inactivates CRM1/exportin 1 by covalent modification at a cysteine residue in the central conserved region. *Proc. Natl. Acad. Sci. USA*, *96*, 9112–9117.

51. Matsuyama, A., Shirai, A., Yashiroda, Y., Kamata, A., Horinouchi, S., Yoshida, M. (2004). pDUAL, a multipurpose, multicopy vector capable of chromosomal integration in fission yeast. *Yeast*, *21*, 1289–1305.

52. Maundrell, K. (1990). *nmt1* of fission yeast: a highly transcribed gene completely repressed by thiamine. *J. Biol. Chem.*, *265*, 10857–10864.

53. Huh, W. K., Falvo, J. V., Gerke, L. C., et al. (2003). Global analysis of protein localization in budding yeast. *Nature*, *425*, 686–691.

54. Ding, D. Q., Tomita, Y., Yamamoto, A., Chikashige, Y., Haraguchi, T., Hiraoka, Y. (2000). Large-scale screening of intracellular protein localization in living fission yeast cells by the use of a GFP-fusion genomic DNA library. *Genes Cells*, *5*, 169–190.

55. Kutay, U., Guttinger, S. (2005). Leucine-rich nuclear-export signals: born to be weak. *Trends Cell Biol.*, *15*, 121–124.

56. Kudo, N., Taoka, H., Toda, T., Yoshida, M., Horinouchi, S. (1999). A novel nuclear export signal sensitive to oxidative stress in the fission yeast transcription factor Pap1. *J. Biol. Chem.*, *274*, 15151–15158.

57. Nguyen, A. N., Ikner, A. D., Shiozaki, M., Warren, S. M., Shiozaki, K. (2002). Cytoplasmic localization of Wis1 MAPKK by nuclear export signal is important for nuclear targeting of Spc1/Sty1 MAPK in fission yeast. *Mol. Biol. Cell*, *13*, 2651–2663.

58. Qin, J., Kang, W., Leung, B., McLeod, M. (2003). Ste11p, a high-mobility-group box DNA-binding protein, undergoes pheromone- and nutrient-regulated nuclear-cytoplasmic shuttling. *Mol. Cell. Biol.*, *23*, 3253–3264.

59. Sutani, T., Yuasa, T., Tomonaga, T., Dohmae, N., Takio, K., Yanagida, M. (1999). Fission yeast condensin complex: essential roles of non-SMC subunits for condensation and Cdc2 phosphorylation of Cut3/SMC4. *Genes Dev.*, *13*, 2271–2283.

60. Paoletti, A., Chang, F. (2000). Analysis of mid1p, a protein required for placement of the cell division site, reveals a link between the nucleus and the cell surface in fission yeast. *Mol. Biol. Cell*, *11*, 2757–2773.

61. Wixon, J., Wood, V. (2006). Tools and resources for *Sz. pombe*: a report from the 2006 European Fission Yeast Meeting. *Yeast*, *23*, 901–903.

62. Luesch, H., Wu, T. Y., Ren, P., Gray, N. S., Schultz, P. G., Supek, F. (2005). A genome-wide overexpression screen in yeast for small-molecule target identification. *Chem. Biol.*, *12*, 55–63.

63. Butcher, R. A., Bhullar, B. S., Perlstein, E. O., Marsischky, G., LaBaer, J., Schreiber, S. L. (2006). Microarray-based method for monitoring yeast overexpression strains reveals small-molecule targets in TOR pathway. *Nat. Chem. Biol.*, *2*, 103–109.

64. Hoon, S., Smith, A. M., Wallace, I. M., et al. (2008). An integrated platform of genomic assays reveals small-molecule bioactivities. *Nat. Chem. Biol.*, *4*, 498–506.

12

DATA ON SMALL MOLECULES AND THEIR TARGET PROTEINS

TAKEO USUI

Graduate School of Life and Environmental Sciences, University of Tsukuba, Tsukuba, Ibaraki, Japan

AKIKO SAITO AND HIROYUKI OSADA

Chemical Biology Department, RIKEN Advanced Science Institute, Wako, Saitama, Japan

12.1 CHROMECEPTIN

Structure See Figure 12-1.

Discovery and Biochemical Studies Chromeceptin (**1**) was identified as an inhibitor of insulin-induced adipogenesis of 3T3-L1 cells [1]. Chromeceptin exhibited cytotoxicity to IGF2-producing hepatocarcinoma cells with selectivity similar to that of a neutralizing antibody against IGF2. Reporter gene transcription assays showed that chromeceptin inhibits the promoter of IGF2 in the hepatocarcinoma cells, suggesting that the compound blocks the autocrine loop of IGF2. DNA microarray analysis revealed that chromeceptin activates the expression of IGFBP-1, a secreted IGF-binding polypeptide that inhibits the metabolic and mitogenic functions of IGF2, and STAT6 was identified as a chromeceptin-responsive element binding transcription factor in IGFBP-1

Protein Targeting with Small Molecules: Chemical Biology Techniques and Applications,
Edited by Hiroyuki Osada

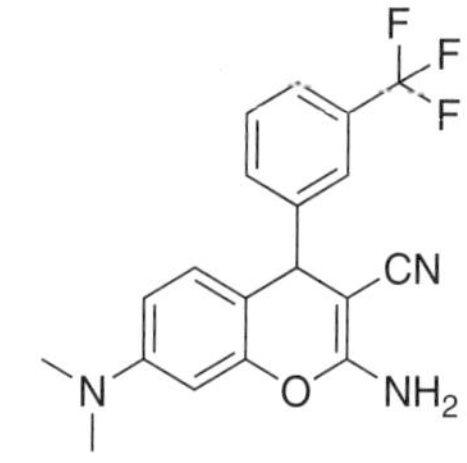

Figure 12-1   Chromeceptin.

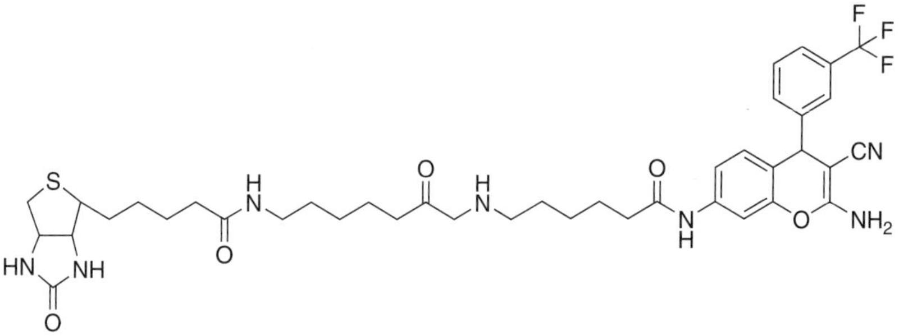

[2]. Furthermore, chromoceptin induced the expression of SOCS-3 (suppressor of cytokine signaling-3) via STAT-6 activation. These results indicate that the expression of SOCS-3 and IGFBP-1, and perhaps that of the other genes that are controlled by STAT6, is up-regulated by chromeceptin and may work together to suppress the function of IGF2.

Identification of Molecular Target and Binding Site   Based on the structure–activity relationships of 65 chromeceptin analogs which showed that the dimethylamino group has limited roles in selective biological activity, biotinylated chromeceptin (**2**) was synthesized. An affinity purified protein that binds specifically to the biotinylated chromeceptin was identified as MFP-2 (multifunctional protein-2) by peptide microsequencing. Studies employing siRNA demonstrate that MFP-2 mediates the chromeceptin-induced expression of IGFBP-1 via STAT6 activation [2]. These results strongly suggest the importance of MFP-2 in STAT6 activation.

REFERENCES

1. Choi, Y., Kawazoe, Y., Murakami, K., Misawa, H., Uesugi, M. (2003). Identification of bioactive molecules by adipogenesis profiling of organic compounds. *J. Biol. Chem.*, *278*, 7320–7324.

2. Choi, Y., Shimogawa, H., Murakami, K., et al. (2006). Chemical genetic identification of the IGF-linked pathway that is mediated by STAT6 and MFP2. *Chem. Biol.*, *13*, 241–249.

12.2 DIMINUTOL

Structure   See Figure 12-2.

Discovery   In the course of screening mitotic spindle assembly inhibitors in *Xenopus* egg extract from the combinatorial library of 2,6,9-trisubstituted purines (**1**), diminutol (**2**) was identified [1].

Biochemical Studies   Diminutol not only inhibited the microtubule polymerization in both interphase and mitosis extract, but also caused microtubules to shorten in interphase and prevented proper spindle formation in mitosis in *Xenopus* XL177 cells at a concentration of 50 μM. The effects on cells were reversible, and, after diminutol was washed out, normal microtubule arrays in

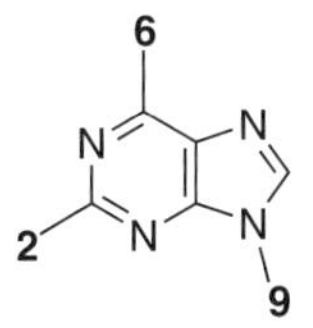

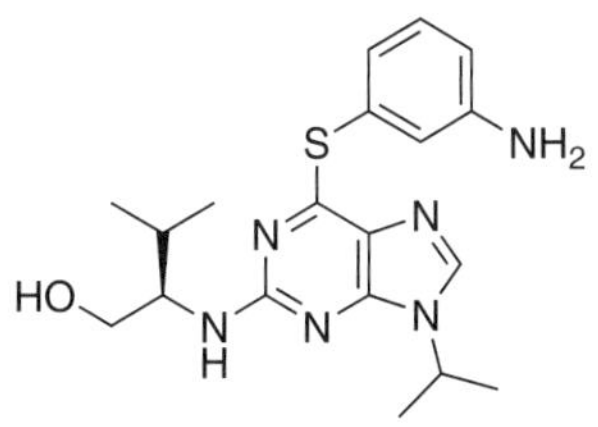

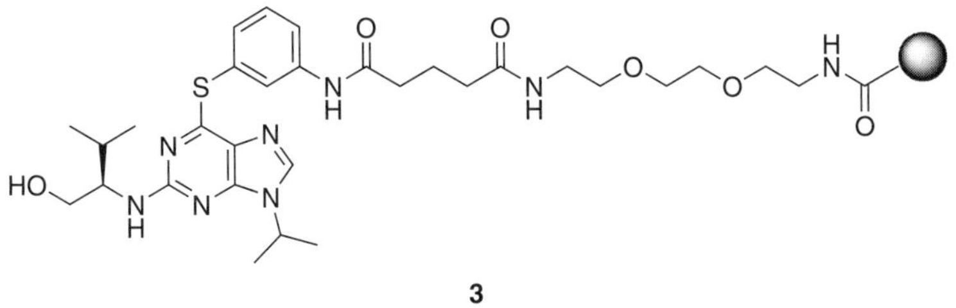

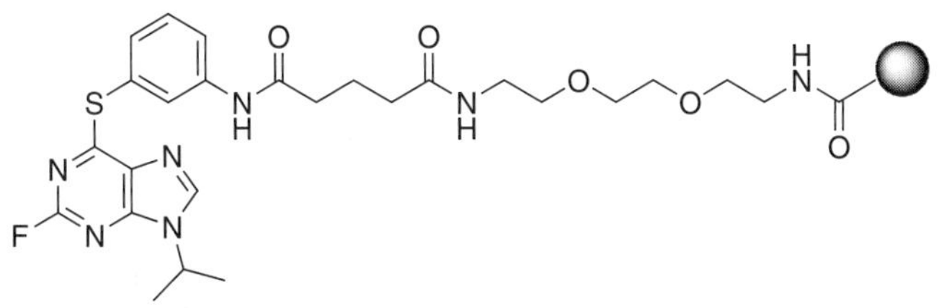

Figure 12-2   Diminutol.

interphase and bipolar spindles in mitosis were re-formed. Because diminutol did not affect CDK1 kinase activity, DNA synthesis, actin polymerization, and chromosomes/nuclear morphology, the compound seems to have a more specific effect on the microtubule cytoskeleton. Interestingly, diminutol did not inhibit tubulin polymerization in vitro; it was thought that there was an unknown functional target molecule.

Identification of Molecular Target   Structure–activity relationship analyses using a focused library consisting of diverse substituents at either the C6 or C2 position of the diminutol scaffold revealed that a C6-thioether is important for a short microtubule phenotype. Based on this SAR results, affinity matrices were synthesized by coupling agarose beads with diminutol (**3**) and one of the inactive derivatives (**4**) through the C6-thiophenyl substituent. After checking the linker-modified version of diminutol-retained activity in *Xenopus* egg extract, two diminutol-binding proteins on matrix were determined as NADP-dependent quinone oxidoreductase (NQO1) and leukotriene B4-12 hydroxydehydrogenase by MALDI-TOF MS. Because depletion or addition of leukotriene B4-12 hydroxydehydrogenase did not affect microtubule dynamics, NQO1 seemed to be the responsible protein. However, there are no detectable interactions between NQO1 and microtubules, diminutol inhibited NQO1 competitively with respect to NADH, and depletion of NQO1 resulted in microtubule depolymerization. Therefore, further investigation will be required to reveal the connection between NQO1 function and microtubule regulation.

REFERENCE

1. Wignall, S. M., Gray, N. S., Chang, Y.-T., et al. (2004). Identification of a novel protein regulating microtubule stability through a chemical approach. *Chem. Biol.*, *11*, 135–146.

12.3   ENT-15-OXOKAURENOIC ACID

Structure   See Figure 12-3.

Discovery and Biochemical Studies   *ent*-15-oxokaurenoic acid (EKA, **1**) was identified as a chemical that causes prolonged mitotic arrest at a stage resembling prometaphase [1]. EKA inhibits the association of the mitotic motor protein centromeric protein E with kinetochores, and inhibits chromosome movement. Unlike most antimitotic agents, EKA does not inhibit the polymerization or depolymerization of tubulin.

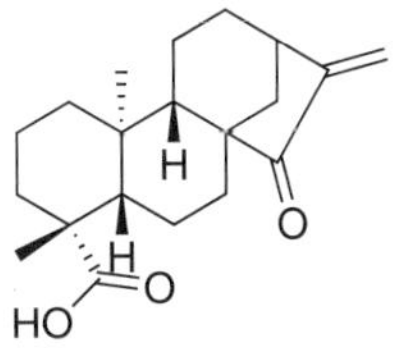

Figure 12-3 *Ent*-kaurene.

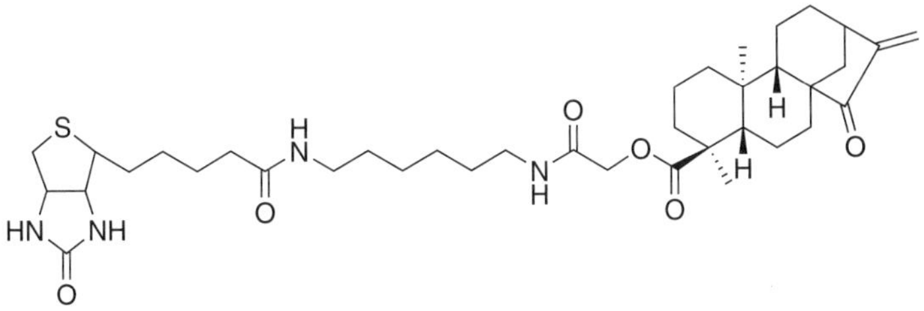

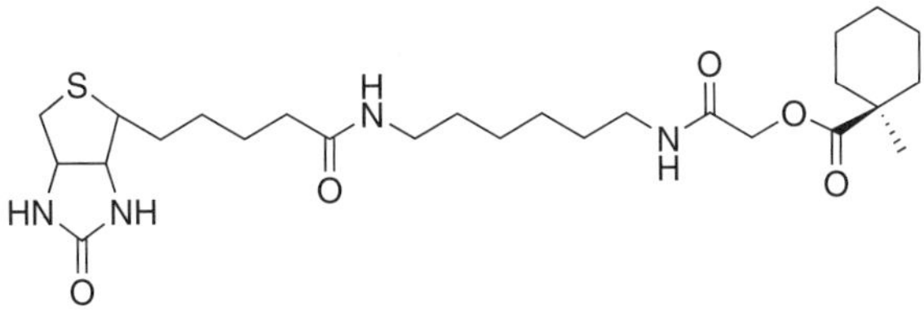

Identification of Molecular Target and Binding Site To identify EKA-interacting proteins, a cell-permeable biotinylated EKA (B-EKA, **2**) that retains biological activity was synthesized to isolate binding proteins from living cells. Six proteins were covalently modified by B-EKA, but not by control compound **3**, and determined as FRAP1, phosphate carrier precursor isoform 1a, Parc, RanBP2 (Ran-binding protein 2), and two unnamed proteins by mass spectrometric analysis. One of these proteins, RanBP2, is a kinetochore protein, and most of the RanBP2 was co-localized with B-EKA at the base of the mitotic spindle and kinetochores. Furthermore, depletion of RanBP2 by siRNA causes a similar mitotic arrest phenotype. These results suggest that EKA causes mitotic arrest at least in part by inhibiting RanBP2 function.

REFERENCE

1. Rundle, N. T., Nelson, J., Flory, M. R., Joseph, J., et al. (2006). An *ent*-kaurene that inhibits mitotic chromosome movement and binds the kinetochore protein Ran-binding protein 2. *ACS Chem. Biol.*, *1*, 443–450.

12.4 EPOLACTAENE

Structure   See Figure 12-4.

Discovery   Epolactaene (**1**) was isolated from the fungal strain *Penicillium* sp. BM 1689-P as a neuritogenic compound that induced neurite outgrowth in the human neuroblastoma cell line SH-SY5Y, which lacks significant TRK family mRNAs [1].

Biochemical Studies   Epolactaene promoted neurite outgrowth and arrested cell-cycle progression at the G0/G1 phase in the neuroblastoma cell line [1]. Different from lactacystin, a compound that promotes neurite outgrowth and arrests cell cycle in both the G0/G1 and G2/M phases in mouse neuroblastoma Neuro2A cells, epolactaene did not inhibit the proteasome peptidase activities [2] and the other proteases [3]. Mizushina et al. focused on the effects on DNA metabolic enzymes in vitro and found that epolactaene was not only a potent inhibitor of DNA polymerases α and β, but also of DNA topoisomerase II [3]. Epolactaene also induced apoptosis in a human leukemia B-cell line, BALL-1 cells, in a dose- and time-dependent manner [4]. Further SAR investigation showed compound **2** to be a strongest inhibitor of DNA polymerase α and β [5,6].

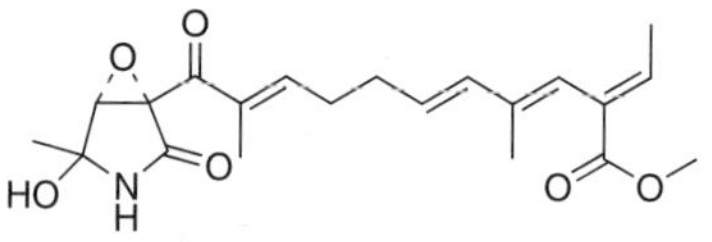

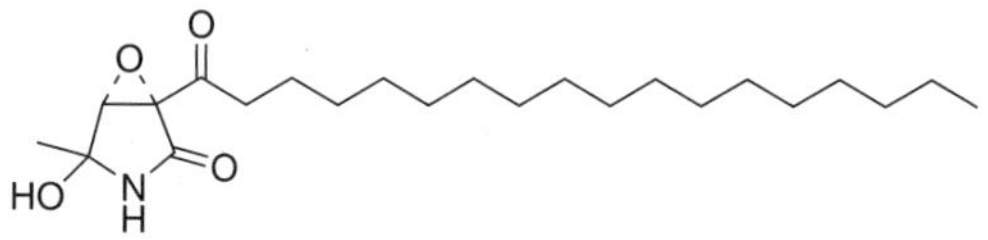

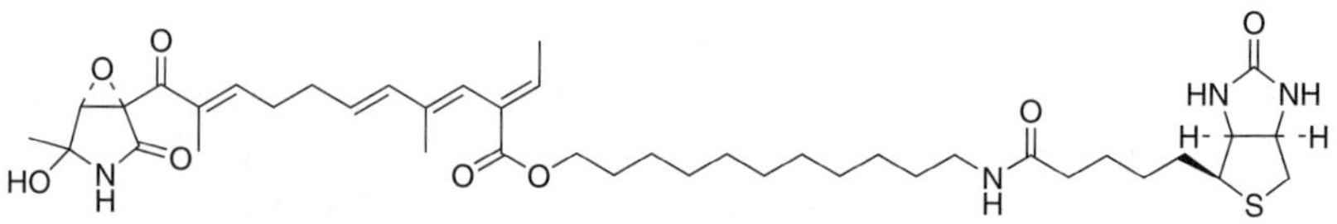

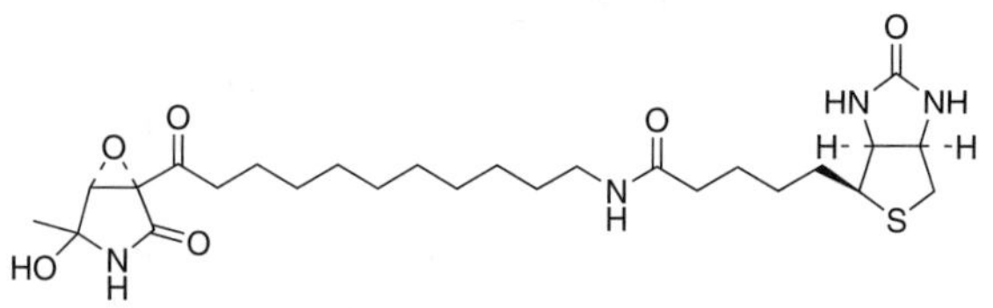

Figure 12-4   Epolactaene.

Identification of Molecular Target Nagumo et al. found that the methyl ester moiety could be substituted by a bulky group such as tertiary butyl ester by structure–activity relationship analyses, and synthesized compound **3** by conjugating a biotin linker at the ester position [2]. Using compound **3**, human HSP60 was identified as a binding protein of epolactaene [7]. Epolactaene inhibits chaperone activity of HSP60 by covalent binding on the Cys442. The binding is reversible; therefore, it was speculated that epolactaene reacts with the α, β-unsaturated ketone via Michael addition.

Recently, Kuramochi et al. reported the other molecular target proteins and binding mechanism. It has been reported that the epolactaene derivatives lacking α, β-unsaturated ketone showed cytotoxicity [2,4,6]. Therefore, they prepared biotinylated derivative Bio-Epo-C12 (**4**), which lacks α, β-unsaturated ketone but possesses α, β-epoxy-γ-lactam moiety, and identified several binding proteins (e.g., fatty acid synthase, ATP citrate lyase, HSP60, adenine nucleotide translocator 2) [8]. Because cerulenin, a structurally related antibitotic, binds covalently to the active-site cysteine of fatty acid synthase at the α-position of α, β-epoxy-γ-lactam moiety, they assumed that epolactaene also reacts with target molecules at the α-position of α, β-epoxy-γ-lactam moiety. They further investigated the binding mechanism and proposed that epolactaene induces intramolecular or intermolecular disulfide formation between protein cysteines via retro-Claisen reaction.

REFERENCES

1. Kakeya, H., Takahashi, I., Okada, G., Isono, K., Osada, H. (2005). Epolactaene, a novel neuritegenic compound in human neuroblastoma cells, produced by a marine fungus. *J. Antibiot.*, *48*, 733–735.

2. Nagumo, Y., Kakeya, H., Yamaguchi, J., et al. (2004). Structure–activity relationships of epolactaene derivatives: structural requirements for inhibition of Hsp60 chaperone activity. *Bioorg. Med. Chem. Lett.*, *14*, 4425–4429.

3. Mizushina, Y., Kobayashi, S., Kuramochi, K., Nagata, S., Sugawara, F., Sakaguchi, K. (2000). Epolactaene, a novel neuritogenic compound in human neuroblastoma cells, selectively inhibits the activities of mammalian DNA polymerases and human DNA topoisomerase II. *Biochem. Biophys. Res. Commun.*, *273*, 784–788.

4. Nakai, J., Kawada, K., Nagata, S., et al. (2002). A novel lipid compound, epolactaene, induces apoptosis: Its action is modulated by its side chain structure. *Biochim. Biophys. Acta*, *1581*, 1–10.

5. Kuramochi, K., Mizushina, Y., Nagata, S., Sugawara, F., Sakaguchi, K., Kobayashi, S. (2004). Structure–activity relationships of epolactaene analogs as DNA polymerase inhibitors. *Bioorg. Med. Chem.*, *12*, 1983–1989.

6. Kuramochi, K., Matsui, R., Matsubara, Y., et al. (2006). Apoptosis-inducing effect of epolactaene derivatives on BALL-1 cells. *Bioorg. Med. Chem.*, *14*, 2151–2161.

7. Nagumo, Y., Kakeya, H., Shoji, M., Hayashi, Y., Dohmae, N., Osada, H. (2005). Epolactaene binds human Hsp60 Cys442, resulting in the inhibition of chaperone activity. *Biochem. J.*, *387*, 835–840.

8. Kuramochi, K., Yukizawa, S., Ikeda, S., et al. (2008). Syntheses and applications of fluorescent and biotinylated epolactaene derivatives: Epolactaene and its derivative induce disulfide formation. *Bioorg. Med. Chem.*, *16*, 5039–5049.

12.5 FK506 (TACROLIMUS)

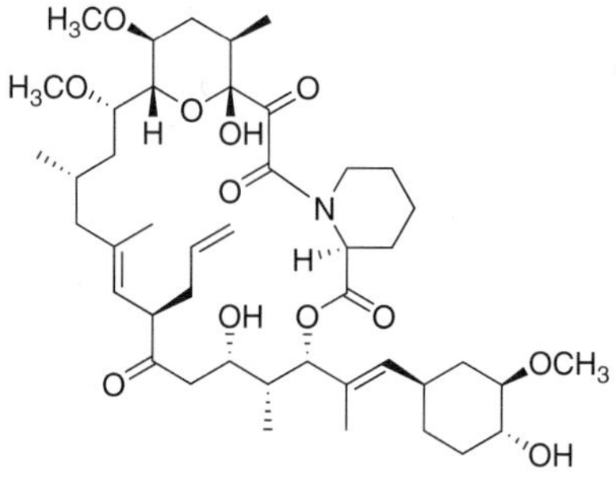

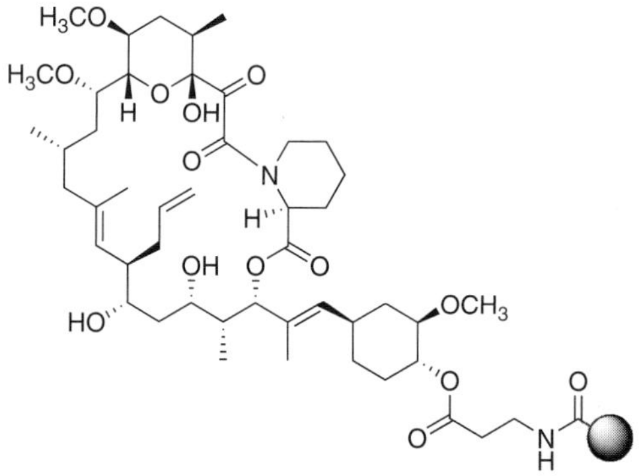

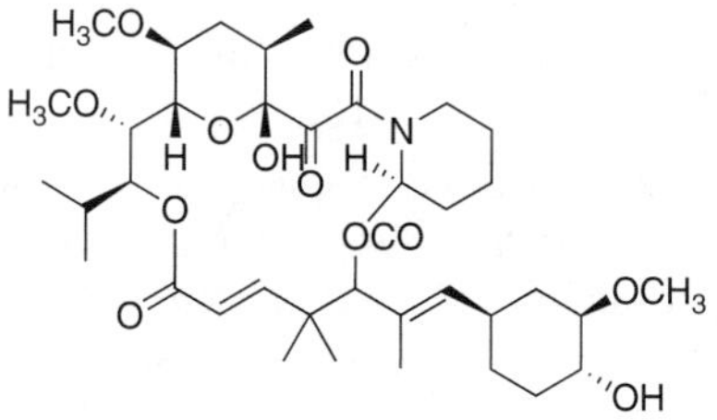

Figure 12-5 FK506.

Structure   See Figure 12-5.

Discovery and Biochemical Studies   FK506 was isolated in the course of the search for immunosuppressant from *Streptomyces tsukubaensis* [1,2]. The structure was determined as shown in **1** [3]. FK506 inhibited mixed lymphocyte reaction, cytotoxic T-cell generation, and expression of early T-cell activation genes [2,4].

Identification of Molecular Target   Harding et al. prepared an FK506 affinity matrix (**2**) and purified a binding protein for FK506 from bovine thymus and human spleen [5]. The FK506-binding protein (FKBP) catalyzes the *cis-trans* isomerization of the proline amide in a tetrapeptide substrate, and FK506 inhibits the action of this new isomerase [5,6]. Therefore, it was speculated that FKBP regulates T-cell activation and other metabolic processes, perhaps by the recognition of proline-containing epitopes in target proteins. However, Bierer et al. found that the inhibition of peptidyl–prolyl isomerase activity is insufficient for immunosuppressive effects using unnatural immunophilin ligand, 506BD (**3**), which contains only the common structural element of FK506 and rapamycin [7]. Liu et al. finally identified calcineurin as the target protein of the FKBP–FK506 complex [8]. They showed that the FKBP–FK506 complex, but not FKBP–506BD, binds competitively to, and inhibits, the phosphatase activity of calcineurin. The cyclophilin–cyclosporin A complex, the other immunosuppressant complex, also binds to and inhibit calcineurin. These results strongly suggested that calcineurin is involved in a common step associated with T-cell and IgE receptor signaling pathways. Inhibition of calcineurin phosphatase activity is strongly correlated with the inhibition of transcriptional activation by NF-AT, a T-cell-specific transcription factor that regulates IL-2 gene synthesis in human T cells [9].

REFERENCES

1. Kino, T., Hatanaka, H., Hashimoto, M., et al. (1987). FK-506, a novel immunosuppressant isolated from a *Streptomyces*: I. Fermentation, isolation, and physico-chemical and biological characteristics. *J. Antibiot.*, *40*, 1249–1255.

2. Kino, T., Hatanaka, H., Miyata, S., et al. (1987). FK-506, a novel immunosuppressant isolated from a *Streptomyces*: II. Immunosuppressive effect of FK-506 in vitro. *J. Antibiot.*, *40*, 1256–1265.

3. Tanaka, H., Kuroda, A., Marusawa, H., et al. (1987). Structure of FK506: a novel immunosuppressant isolated from *Streptomyces*. *J. Am. Chem. Soc.*, *109*, 5031–5033.

4. Tocci, M. J., Matkovich, D. A., Collier, K. A., et al. (1989). The immunosuppressant FK506 selectively inhibits expression of early T cell activation genes. *J. Immunol.*, *143*, 718–726.

5. Harding, M. W., Galat, A., Uehling, D. E., Schreiber, S. L. (1989). A receptor for the immunosuppressant FK506 is a *cis-trans* peptidyl–prolyl isomerase. *Nature*, *341*, 758–760.

6. Rosen, M. K., Standaert, R. F., Galat, A., Nakatsuka, M., Schreiber, S. L. (1990). Inhibition of FKBP rotamase activity by immunosuppressant FK506: twisted amide surrogate. *Science*, *248*, 863–866.

7. Bierer, B. E., Somers, P. K., Wandless, T. J., Burakoff, S. J., Schreiber, S. L. (1990). Probing immunosuppressant action with a nonnatural immunophilin ligand. *Science*, *250*, 556–559.

8. Liu, J., Farmer, J. D. Jr., Lane, W. S., Friedman, J., Weissman, I., Schreiber, S. L. (1991). Calcineurin is a common target of cyclophilin–cyclosporin A and FKBP–FK506 complexes. *Cell*, *66*, 807–815.

9. Liu, J., Albers, M. W., Wandless, T. J., et al. (1992). Inhibition of T cell signaling by immunophilin–ligand complexes correlates with loss of calcineurin phosphatase activity. *Biochemistry*, *31*, 3896–3901.

12.6 FUMAGILLIN

Structure   See Figure 12-6.

Discovery   Fumagillin (**1**) was isolated as active concentrates that were capable of inhibiting *Staphylococcus aureus* 209 bacteriophage [1] and was also found as an extremely potent amebicidal antibiotic [2].

Biochemical Studies   Ingber et al. re-isolated fumagillin from *Aspergillus fumigatus* as an active compound which has potent endothelial cell-rounding activity

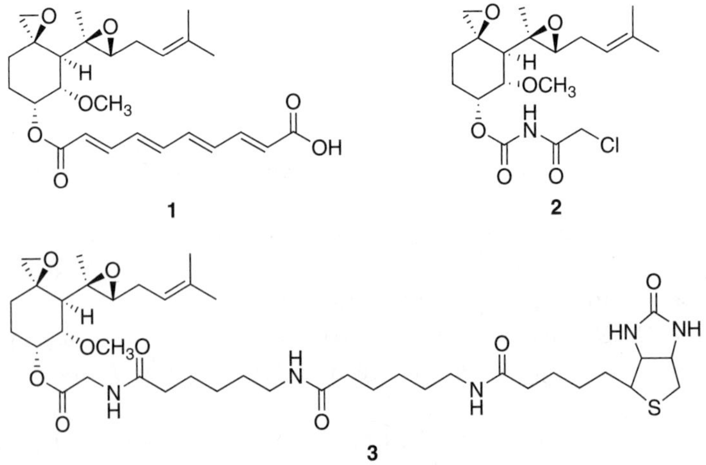

Figure 12-6 Fumagillin.

and inhibits angiogenesis in the growing chick chorioallantoic membrane [3]. The compound completely inhibited endothelial cell proliferation and suppressed tumor-induced neovascularization in the mouse dorsal air sac. However, the severe side effect limited the in vivo effectiveness. They synthesized several fumagillin derivatives and identified AGM-1470 (**2**), one of the most potent compounds.

Identification of Molecular Target Since epoxides of fumagillin are thought to play an important role in antiangiogenic activity [4], it was thought that the compound binds its target protein covalently. Sin et al. [5] and Griffith et al. [6] synthesized biotinylated fumagillin (**3** for [5]) and showed that fumagillin binds covalently and inhibits type II methionine aminopeptidase (MetAP2). Fumagillin covalently modified the active-site histidine of MetAP2 [7,8], and irreversibly inhibited the aminopeptidase activity of MetAP2 [8]. X-ray crystallography revealed that a covalent bond formation between a reactive epoxide of fumagillin and histidine-231 in the active site of MetAP2, and that extensive hydrophobic and water-mediated polar interactions with other parts of fumagillin provide additional affinity [9].

REFERENCES

1. Hanson, F. R., Eble, E. (1949). An antiphage agent isolated from *Aspergillus* sp. *J. Bacteriol.*, *58*, 527–529.
2. McCowen, M. C., Callender, M. E., Lawlis, J. F., Jr. (1951). Fumagillin (H-3), a new antibiotic with amebicidal properties. *Science*, *113*, 202–203.
3. Ingber, D., Fujita, T., Kishimoto, S., et al. (1990). Synthetic analogues of fumagillin that inhibit angiogenesis and suppress tumour growth. *Nature*, *348*, 555–557.
4. Marui, S., Kishimoto, S. (1992). Chemical modification of fumagillin: II. 6-Amino-6-deoxyfumagillol and its derivatives. *Chem. Pharm. Bull.*, *40*, 575–579.
5. Sin, N., Meng, L., Wang, M. Q. W., Wen, J. J., Bornmann, W. G., Crews, C. M. (1997). The anti-angiogenic agent fumagillin covalently binds and inhibits the methionine aminopeptidase, MetAP-2. *Proc. Nat. Acad. Sci. USA*, *94*, 6099–6103.
6. Griffith, E. C., Su, Z., Turk, B. E., et al. (1997). Methionine aminopeptidase (type2) is the common target for angiogenesis inhibitors AGM-1470 and ovalicin. *Chem. Biol.*, *4*, 461–471.
7. Loweher, W. T., McMillen, D. A., Orville, A. M., Matthews, B. W. (1998). The anti-angiogenic agent fumagillin covalently modifies a conserved active-site histidine in the *Escherichia coli* methionine aminopeptidase. *Proc. Nat. Acad. Sci. USA*, *95*, 12153–12157.

8. Griffith, E. C., Su, Z., Niwayama, S., Ramsay, C. A., Chang, Y.-H., Liu, J. O. (1998). Molecular recognition of angiogenesis inhibitors fumagillin and ovalicin by methionine aminopeptidase 2. *Proc. Nat. Acad. Sci. USA, 95*, 15183–15188.

9. Liu, S., Widom, J., Kemp, C. W., Crews, C. M., Clardy, J. (1998). Structure of human methionine aminopeptidase-2 complexed with fumagillin. *Science, 282*, 1324–1327.

12.7   GELDANAMYCIN

Structure   See Figure 12-7.

Discovery   Geldanamycin was isolated from a culture broth of *Streptomyces hygroscopicus* as a compound showing moderate inhibitory activity against protozoans, bacteria, fungi, and tumor cells [1]. The structure was determined as **1** [2].

Biochemical Studies   Geldanamycin shows antitumor activity in vitro and in situ [1,3]. Uehara et al. found that geldanamycin and other benzoquinonoid ansamycins reversibly decrease the autophosphorylation of p60^{src} and reverse the cell morphology from the transformed to the normal phenotype [4].

Identification of Molecular Target   Because geldanamycin reverses transformed morphology of v-*src*-transformed fibroblasts without inhibiting in vitro p60^{src} kinase activity, it was thought that the molecular target of the compound was not p60^{src}. To identify the target protein, Whitesell et al. prepared the

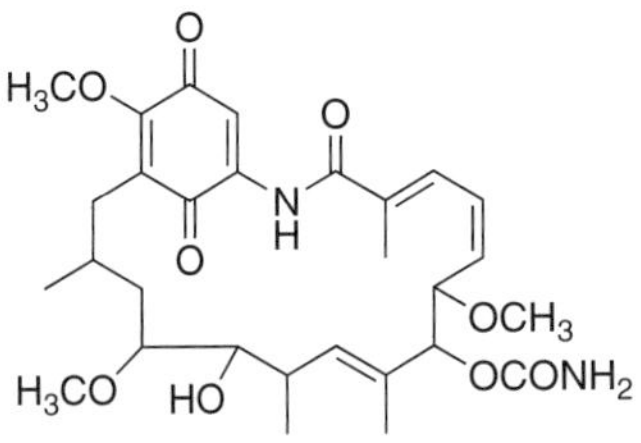

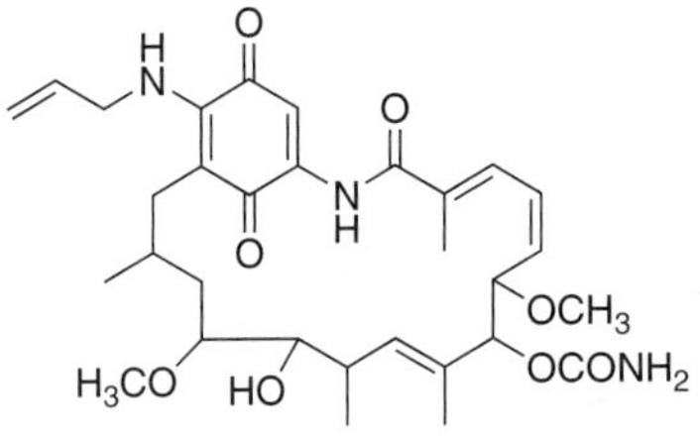

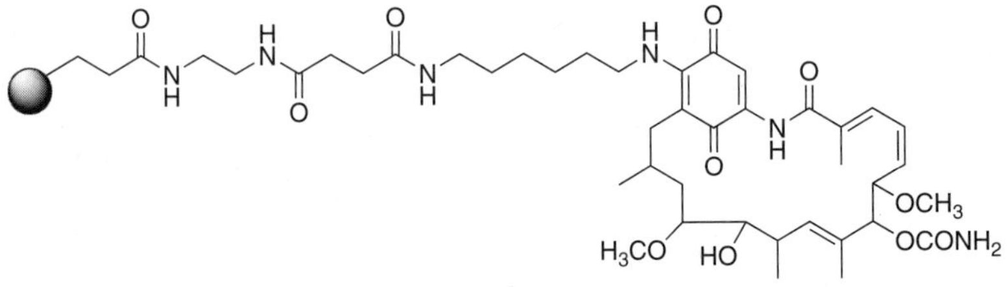

Figure 12-7   Geldanamycin.

affinity resin by immobilizing the diamine derivative of geldanamycin with resin (2) [5]. Immobilized geldanamycin precipitated the HSP90 in a stable and pharmacologically specific manner. Geldanamycin inhibits the Hsp90-mediated conformational maturation/refolding reaction and results in the degradation of HSP90 substrates [6]. X-ray crystallography clearly revealed that geldanamycin binds the N-terminal ATP/ADP binding domain of HSP90 [7] and inhibits ATP/ADP binding [8]. Because HSP90 client proteins include several oncogenic proteins, such as mutated p53, Bcr-Abl, Raf-1, Akt, ErbB2, and HIF−1α, geldanamycin is a promising antitumor compound in a preclinical model system [9]. 17-Allylamino,17-demethoxygeldanamycin (3), a geldanamycin derivative, is now in clinical trial.

REFERENCES

1. DeBoer, C., Meulman, P. A., Wnuk, R. J., Peterson, D. H. (1970). Geldanamycin, a new antibiotic. *J. Antibiot., 23*, 442–447.

2. Sasaki, K., Rinehart, K. L. Jr., Slomp, G., Grostic, M. F., Olson, E. C. (1970). Geldanamycin: I. Structure assignment. *J. Am. Chem. Soc., 92*, 7591–7593.

3. Sasaki, K., Yasuda, H., Onodera, K. (1979). Growth inhibition of virus transformed cells in vitro and antitumor activity in vivo of geldanamycin and its derivatives. *J. Antibiot., 32*, 849–851.

4. Uehara, Y., Hori, M., Takeuchi, T., Umezawa, H. (1986). Phenotypic change from transformed to normal induced by benzoquinonoid ansamycins accompanies inactivation of p60^{src} in rat kidney cells infected with Rous sarcoma virus. *Mol. Cell. Biol., 6*, 2198–2206.

5. Whitesell, L., Mimnaugh, E. G., De Costa, B., Myers, C. E., Neckers, L. M. (1994). Inhibition of heat shock protein HSP90-pp60v-src heteroprotein complex formation by benzoquinone ansamycins: essential role for stress proteins in oncogenic transformation. *Proc. Nat. Acad. Sci. USA, 91*, 8324–8328.

6. Schneider, C., Sepp-Lorenzino, L., Nimmesgern, E., et al. (1996). Pharmacologic shifting of a balance between protein refolding and degradation mediated by Hsp90. *Proc. Natl. Acad. Sci. USA, 93*, 14536–14541.

7. Stebbins, C. E., Russo, A. A., Schneider, C., Rosen, N., Hartl, F. U., Pavletich, N. P. (1997). Crystal structure of an Hsp90-geldanamycin complex: targeting of a protein chaperone by an antitumor agent. *Cell, 89*, 239–250.

8. Grenert, J. P., Sullivan, W. P., Fadden, P., et al. (1997). The amino-terminal domain of heat shock protein 90 (hsp90) that binds geldanamycin is an ATP/ADP switch domain that regulates hsp90 conformation. *J. Biol. Chem., 272*, 23843–23850.

9. Neckers, L. (2002). Hsp90 inhibitors as novel cancer chemotherapeutic agents. *Trends Mol. Med., 8*, S55–S61.

12.8   GERFELIN

Structure   See Figure 12-8.

Discovery   Gerfelin was isolated from culture broth of *Beauveria felina* QN22047 as a compound showing inhibitory activity against human geranylgeranyl diphosphate (GGPP) synthase [1,2]. The structure was determined as **1** [2]. Methylgerfelin (**2**) is a synthetic derivative of **1**.

Biochemical Studies   Gerfelin (**1**) and methylgerfelin (**2**) inhibit the enzymatic activity of human GGPP synthase in a noncompetivive manner against isopentenyl diphosphate (IPP) with K_i values of 19.0 and 2.5 μM, respectively. Kawatani et al. found that gerfelin and methylgerfelin suppress the receptor activator of NF-κB ligand (RANKL)-induced osteoclastogenesis [3].

Identification of Molecular Target   Because gerfelin and methylgerfelin suppress osteoclastogenesis in a GGPP-nonsensitive manner, it was thought that GGPP synthase inhibition by those compounds had little or no involvement in the inhibition of osteoclastogenesis. To identify the target protein, Kawatani et al. prepared methylgerfelin affinity matrix (**3**) by using a photo-cross-linking method [4], and found that methylgerfelin bound directly to glyoxalase I. Gerfelin and methylgerfelin inhibit the enzymatic activity of mouse glyoxalase I in a competitive manner with a K_i value of 0.15 and 0.23 μM, respectively. These results suggest that methylgerfelin inhibits glyoxalase I activity, thereby allowing the accumulation of methylglyoxal, resulting in the inhibition of osteoclastogenesis

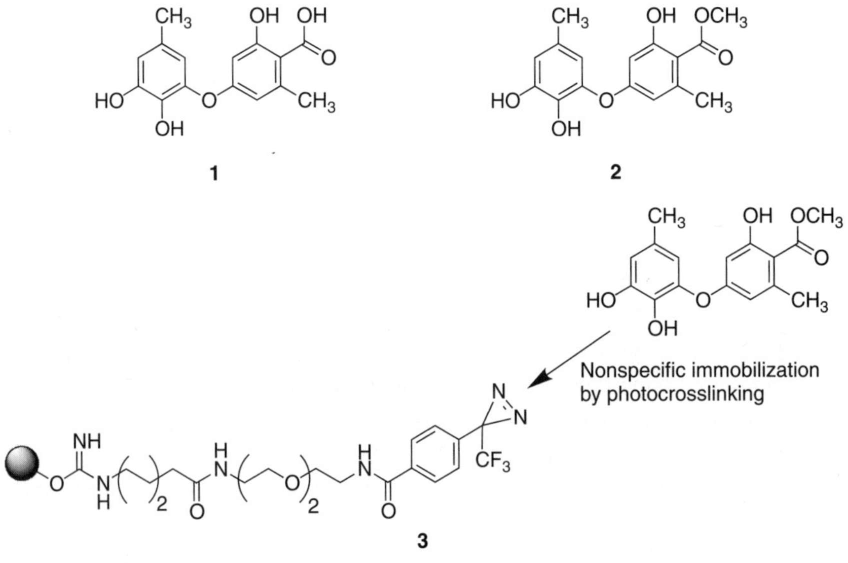

Figure 12-8   Gerfelin.

[3]. X-ray crystallography clearly revealed that methylgerfelin binds the active site of glyoxalase I by coordination bonding via the zinc ion [3].

REFERENCES

1. Zenitani, S., Tashiro, S., Shindo, K., Nagai, K., Suzuki, K., Imoto, M. (2003). Gerfelin, a novel inhibitor of geranylgeranyl diphosphate synthase from *Beauveria felina* QN22047: I. Taxonomy, fermentation, isolation, and biological activities. *J. Antibiot.*, *56*, 617–621.
2. Zenitani, S., Shindo, K., Tashiro, S., et al. (2003). Gerfelin, a novel inhibitor of geranylgeranyl diphosphate synthase from *Beauveria felina* QN22047: II. Structural elucidation. *J. Antibiot.*, *56*, 658–660.
3. Kawatani, M., Okumura, H., Honda, K., et al. (2008). The identification of an osteoclastogenesis inhibitor through the inhibition of glyoxalase I. *Proc. Natl. Acad. Sci. USA*, *105*, 11691–11696.
4. Kanoh, N., Honda, K., Simizu, S., Muroi, M., Osada, H. (2005). Photo-cross-linked small-molecule affinity matrix for facilitating forward and reverse chemical genetics. *Angew. Chem. Int. Ed.*, *44*, 3559–3562.

12.9 MELANOGENIN

Structure   See Figure 12-9.

Discovery and Biochemical Studies   Melanogenin (**1**) was identified as a small molecule that induces pigmentation in melanocytes from a tagged triazine library of 1170 small molecules in a cell-based assay [1]. Melanogenin induced melanin formation in a dose-dependent manner (EC_{50} 2.5 μM) and was more potent than the known stimulator isobutylmethylxanthine. The activity and protein level of tyrosinase, the rate-limiting enzyme in melanogenesis, were up-regulated by melanogenin treatment.

Identification of Molecular Target and Binding Site   Structure–activity relationship analysis using multiple structural derivatives of melanogenin revealed high activities correlating with electronegative fluorinated R_1 and R_2 groups (**2**), with the most significant induction resulting from melanogenin, in which fluorination exists at the *meta* position. To isolate target protein, melanogenin was bound covalently to an affinity matrix via its internal linker (**3**), and the mitochondrial protein, prohibitin, was identified as an intracellular binding target. Studies employing siRNA demonstrate that prohibitin is required for melanogenin to exert its propigmentary effects and reveal an unsuspected functional role for this protein in melanin induction [1]. These results strongly suggested that prohibitin is a pivotal regulator for melanogenesis. Structural models of human prohibitins and two potential melanogenin-binding pockets within the PHB domain in prohibitin were recently proposed [2].

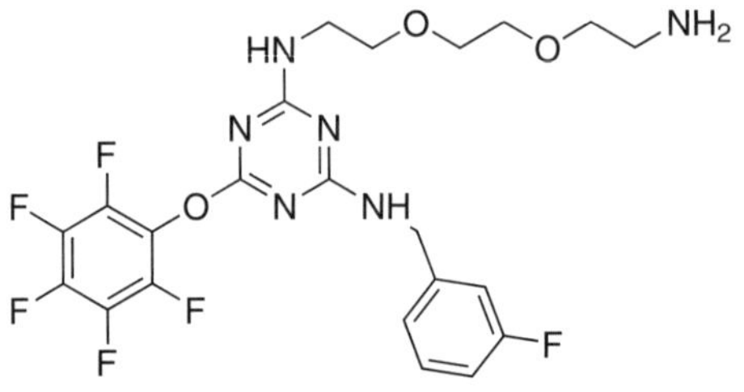

Figure 12-9 Melanogenin.

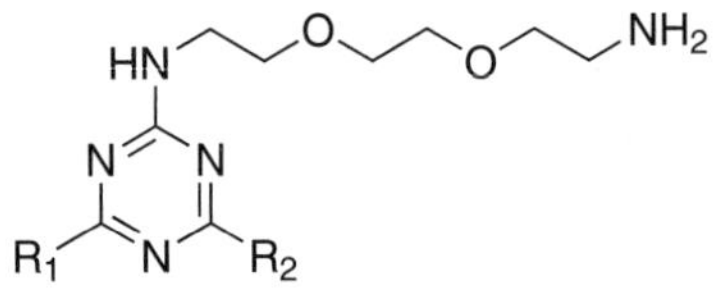

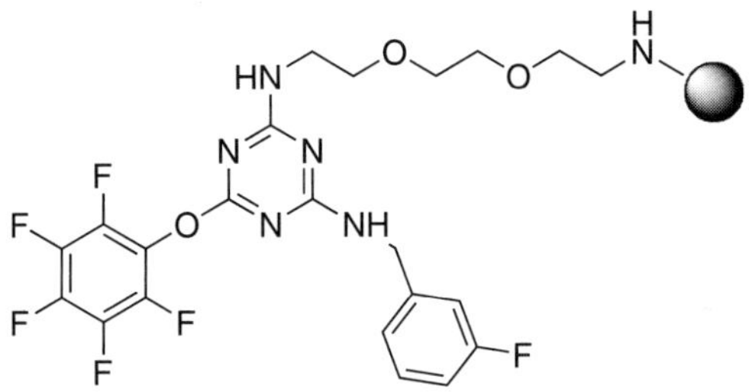

REFERENCES

1. Snyder, J. R., Hall, A., Ni-Komatsu, L., Khersonsky, S. M., Chang, Y. T., Orlow, S. J. (2005). Dissection of melanogenesis with small molecules identifies prohibitin as a regulator. *Chem. Biol.*, *12*, 477–484.
2. Winter, A., Kamarainen, O., Hofmann, A. (2007). Molecular modeling of prohibitin domains. *Proteins*, *68*, 353–362.

12.10   MYRIOCIN (ISP-1)

Structure   See Figure 12-10.

Discovery   Myriocin was isolated as an antifungal compound from *Myriococcum albomyces* [1]. Fujita et al. re-isolated myriocin as a potent immunosuppressant [2]. The stereochemistry was confirmed as **1** [3], and total synthesis was performed by Yoshikawa et al. [4].

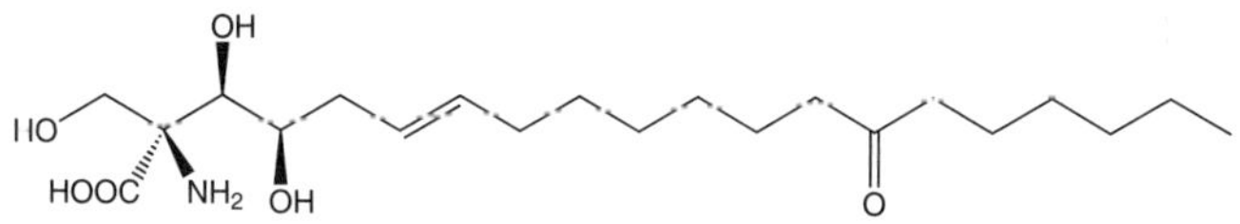

Figure 12-10   Myriocine.

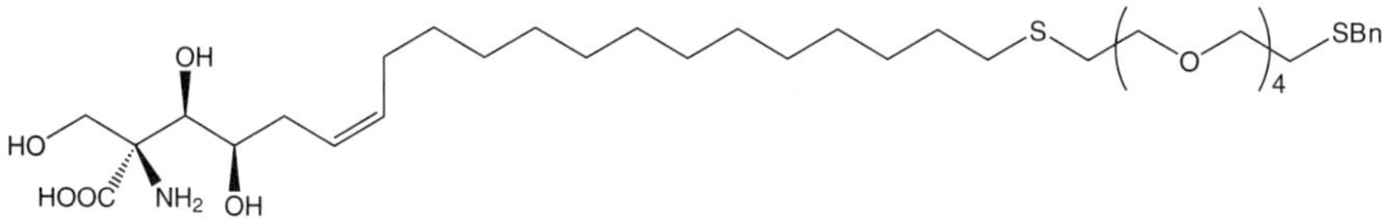

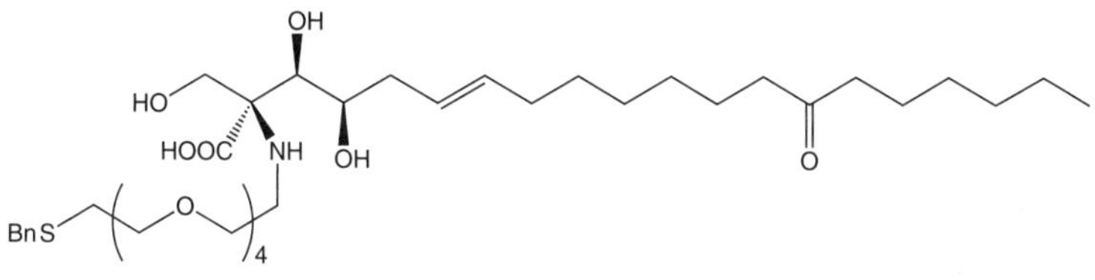

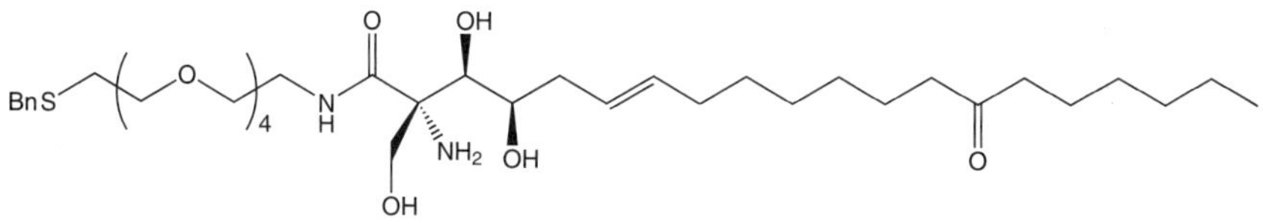

Biochemical Studies   The structure of myriocin is homologous to sphingosine. Miyake et al. found that myriocin inhibits serine palmitoyltransferase (SPT), which catalyzes the first step of sphingolipid biosynthesis [5]. The growth inhibition of CTLL-2 cells induced by myriocin was abolished completely by the addition of sphingosines or sphingosine-1-phosphate, but not by sphingomyelin or glycosphingolipids, suggesting that myriocin suppresses T-cell proliferation by the modulation of sphinogolipid metabolism.

Identification of Molecular Target   To identify the molecular target, Chen et al. synthesized three myriocin derivatives, which bear a pentaethylene glycol linker at (1) the aliphatic extremity of myriocin (C20-linker, **2**), (2) a primary amine (N-linker, **3**), and (3) a carboxyl terminal (C1-linker, **4**) [6]. N-linker and C1-linker myriocin showed the inhibition of SPT activity, but not C20-linker myriocin, suggesting that the aliphatic chain is important for SPT inhibition. To identify the myriocin binding protein, they next synthesized the affinity matrices with N- and C1-linker myriocin. After confirming that N-linker myriocin matrices absorbed the SPT activity from CTLL-2 cell lysate, they purified two specific myriocin-binding proteins with apparent molecular weights of 51 and 55 kDa by using N-linker myriocin matrices. These proteins were identified by mass spectrometory as LCB1 and LCB2, mammalian homologs of two yeast proteins that have been linked genetically to sphingolipid biosynthesis.

REFERENCES

1. Kluepfel, D., Bagli, J., Baker, H., Charest, M. P., Kudelski, A. (1972). Myriocin, a new antifungal antibiotic from *Myriococcum albomyces*. *J. Antibiot.*, *25*, 109–115.
2. Fujita, T., Inoue, K., Yamamoto, S., et al. (1994). Fungal metabolites: II. A potent immunosuppressive activity found in *Isaria sinclairii* metabolite. *J. Antibiot.*, *47*, 208–215.
3. Bagli, J. F., Kluepfel, D., St.-Jacques, M. (1973). Elucidation of structure and stereochemistry of myriocin, a novel antifungal antibiotic. *J. Org. Chem.*, *38*, 1253–1260.
4. Yoshikawa, M., Yokokawa, Y., Okuno, Y., Murakami, N. (1994). Total synthesis of a novel immunosuppressant, myriocin (thermozymocidin, ISP-1), and Z-myriocin. *Chem. Pharm. Bull.*, *42*, 994–996.
5. Miyake, Y., Kozutsumi, Y., Nakamura, S., Fujita, T., Kawasaki, T. (1995). Serine palmitoyltransferase is the primary target of a sphingosine-like immunosuppressant, ISP-1/myriocin. *Biochem. Biophys. Res. Commun.*, *211*, 396–403.
6. Chen, J. K., Lane, W. S., Schreiber, S. L. (1999). The identification of myriocin-binding proteins. *Chem. Biol.*, *6*, 221–235.

12.11   PHOSLACTOMYCIN A

Structure   See Figure 12-11.

Discovery   Phoslactomycins (PLMs, PLM-A **1**) were originally isolated as antifungal antibiotics from the soil bacteria species *Streptomyces* [1,2]. Several related compounds—phosphazomycins [3,4], phospholine [5,6], and leustroducsins [7,8]—were reported as antifungal antibiotics or as inducers of a colony-stimulating factor in bone marrow stromal cells. These compounds commonly contain not only an α, β-unsaturated δ-lactone, but also an amino group, a phosphate ester, and a cyclohexane ring, but have different substituents bound to the cyclohexane ring.

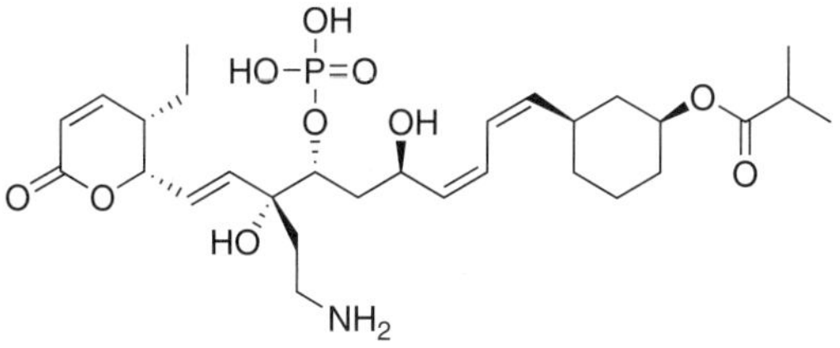

Figure 12-11   Phoslactomycin.

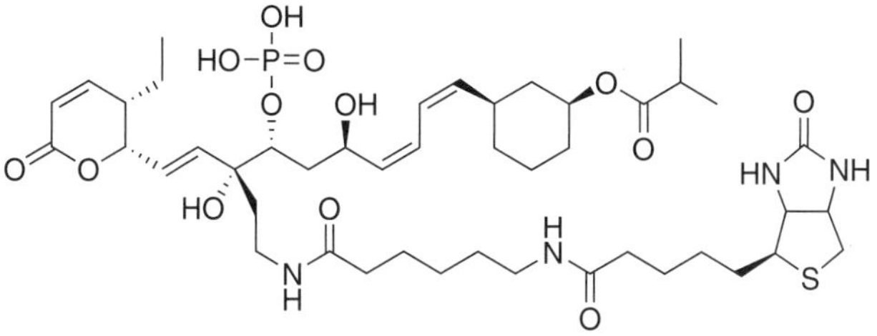

Biochemical Studies   PLMs induced actin filament depolymerization reversibly in murine fibroblast NIH/3T3 cells [9]. Since PLM-F had no effect at all on the polymerization of purified actin in vitro, it is thought that PLMs induce actin depolymerization through an indirect mechanism. An in situ phosphorylation assay revealed that PLM-F treatment stimulated the phosphorylation of intracellular vimentin by the inhibition of phosphatase. An in vitro assay showed that PLMs specifically inhibited protein phosphatase type-2A (PP2A) but not protein phosphatase type-1 (PP1). These results suggest that PLMs are PP2A-specific inhibitors and that PP2A is involved in regulation of the organization of the actin cytoskeleton.

Identification of Molecular Target   Recently, the inhibitory mechanism of PLMs was revealed by Teruya et al. [10]. They synthesized the biotinylated PLM-A (**2**) and investigated the binding proteins in mouse liver extract. They found three proteins to which biotinylated PLM-A binds covalently, and these proteins were identified as PP2A family proteins: PP2Ac (catalytic subunit of PP2A), PP4, and PP6. Furthermore, they clarified the binding site on PP2Ac as Cys269, which is located at the entrance of the catalytic site. Interestingly, this cysteine residue is conserved in PP2A family proteins but not even in PP1, the phosphatase it most resembles. These results strongly suggest that Cys269 is a PP2A-specific gatekeeper residue and that compounds targeting this residue are expected to be PP2A-specific inhibitors.

REFERENCES

1. Fushimi, S., Nishikawa, S., Shimazu, A., Seto, H. (1989). Studies on new phosphate ester antifungal antibiotics phoslactomycins: I. Taxonomy, fermentation, purification and biological activities. *J. Antibiot.*, *42*, 1019–1025.

2. Fushimi, S., Furihata, K., Seto, H. (1989). Studies on new phosphate ester antifungal antibiotics phoslactomycins: II. Structure elucidation of phoslactomycin A to F. *J. Antibiot.*, *42*, 1026–1036.

3. Uramoto, M., Shen, Y.-C., Takizawa, N., Kusakabe, H., Isono, K. (1985). A new antifungal antibiotic, phosphazomycin A. *J. Antibiot.*, *38*, 665–668.

4. Tomiya, T., Uramoto, M., Isono, K. (1990). Isolation and structure of phosphazomycin C. *J. Antibiot.*, *43*, 118–121.

5. Ozasa, T., Suzuki, K., Sasamata, M., et al. (1989). Novel antitumor antibiotic phospholine: I. Production, isolation and characterization. *J. Antibiot.*, *42*, 1331–1338.

6. Ozasa, T., Tanaka, K., Sasamata, M., et al. (1989). Novel antitumor antibiotic phospholine: II. Structure determination. *J. Antibiot.*, *42*, 1339–1343.

7. Kohama, T., Enokita, R., Okazaki, T., et al. (1993). Novel microbial metabolites of the phoslactomycins family induce production of colony-stimulating factors by bone marrow stromal cells: I. Taxonomy, fermentation, and biological properties. *J. Antibiot.*, *46*, 1503–1511.

8. Kohama, T., Nakamura, T., Koinoshita, T., Kaneko, I., Shiraishi, A. (1993). Novel microbial metabolites of the phoslactomycins family induce production of colony-stimulation factors by bone marrow stromal cells: II. Isolation, physico-chemical properties and structure determination. *J. Antibiot.*, *46*, 1512–1519.

9. Usui, T., Marriott, G., Inagaki, M., Swarup, G., Osada, H. (1999). Protein phosphatase 2A inhibitors, phoslactomycins: effects on the cytoskeleton in NIH/3T3 cells. *J. Biochem.*, *125*, 960–965.

10. Teruya, T., Simizu, S., Kanoh, N., Osada, H. (2005). Phoslactomycin targets cysteine-269 of the protein phosphatase 2A catalytic subunit in cells. *FEBS Lett.*, *579*, 2463–2468.

12.12 PIRONETIN

Structure   See Figure 12-12.

Discovery   Pironetin (**1**) was isolated as a plant growth regulator that induces shortening of the plant height of rice [1,2] and as an immunosuppressor that inhibites the blastogenesis of both T and B cells [3].

Biochemical Studies   Pironetin showed antitumor activity against a murine tumor cell line, P388 leukemia, transplanted in mice [4], and induced the

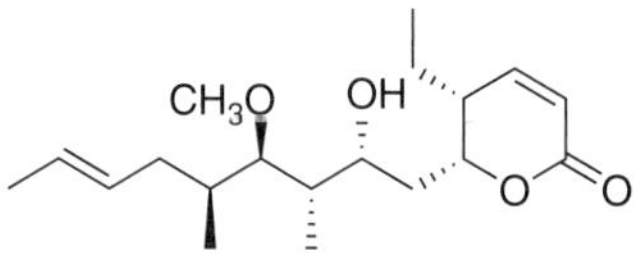

Figure 12-12   Pironetin.

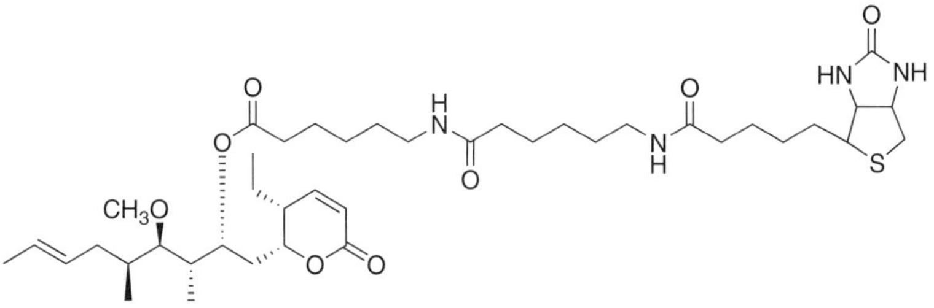

apoptosis via disruption of the cellular microtubule network in situ [5]. The compound inhibited the polymerization of pure tubulin, showing that pironetin is a tubulin inhibitor that inhibits tubulin polymerization by binding to tubulin directly. Binding competition experiments with [³H]colchicine and [³H]vinblastine strongly suggested that pironetin binds on the vinblastine-binding site on β-tubulin.

Identification of Molecular Target   According to the structure–activity relationship analyses, it was speculated that pironetin binds to tubulin covalently by Michael addition via its α,β-unsaturated δ-lactone [6]. The biotinylated pironetin (**2**), which inhibited tubulin assembly both in vitro and in situ, bound covalently with tubulin, and its binding was inhibited by natural pironetin [7]. The finding that inactive pironetin derivatives failed to bind to tubulin in situ suggested that covalent binding is important in inhibiting tubulin polymerization. Partial proteolytic analyses of biotinylated pironetin-bound peptides, followed by systematic alanine scanning of both cysteine and lysine residues, strongly suggested that Lys352 of α-tubulin is the pironetin-binding site. This is surprising, because reactivity of the ε-amino group of lysine is quite low compared with that of the sulfhydryl group of cysteine. In silico analysis of the pironetin–tubulin complex suggested that α helixes 8/10 and β sheet 8 form a cavity that fits pironetin well, and that pironetin binding with Lys352 and Asn258 disrupts the hydrogen network among Glu254, Lys352, and Asn258. The discrepancies in an [³H]vinblastine competition experiment were explained by the structure of vinblastine bound to the tubulin-colchicine : RB3-SLD complex [8]. Gigant et al. revealed that vinblastine binds not only to β-tubulin but also on a hydrophobic groove on the α-tubulin surface that is located at an intermolecular contact in microtubules, and that the pironetin-binding residue Lys352 is part of the vinblastine-binding site. These results strongly suggest that the binding sites of pironetin and vinblastine overlap and that these drugs displace each other.

REFERENCES

1. Kobayashi, S., Tsuchiya, K., Harada, T., et al. (1994). Pironetin, a novel plant growth regulator produced by *Streptomyces* sp. NK10958: I. Taxonomy, production, isolation and preliminary characterization. *J. Antibiot.*, *47*, 697–702.
2. Kobayashi, S., Tsuchiya, K., Kurokawa, T., Nakagawa, T., Shimada, N., Iitaka, Y. (1994). Pironetin, a novel plant growth regulator produced by *Streptomyces* sp. NK10958: II. Structural elucidation. *J. Antibiot.*, *47*, 703–707.
3. Yasui, K., Tamura, Y., Nakatani, T., et al. (1996). Chemical modification of PA-48153C, a novel immunosuppressant isolated from *Streptomyces prunicolor* PA-48153. *J. Antibiot.*, *49*, 173–180.
4. Kondoh, M., Usui, T., Kobayashi, S., et al. (1998). Cell cycle arrest and antitumor activity of pironetin and its derivatives. *Cancer Lett.*, *126*, 29–32.
5. Kondoh, M., Usui, T., Nishikiori, T., Mayumi, T., Osada, H. (1999). Apoptosis induction via microtubule disassembly by an antitumour compound, pironetin. *Biochem. J.*, *340*, 411–416.
6. Watanabe, H., Watanabe, H., Usui, T., Kondoh, M., Osada, H., Kitahara, T. (2000). Synthesis of pironetin and related analogs: studies on structure–activity relationships as tubulin assembly inhibitors. *J. Antibiot.*, *53*, 540–545.
7. Usui, T., Watanabe, H., Nakayama, H., et al. (2004). The anticancer natural product pironetin selectively targets Lys352 of α-tubulin. *Chem. Biol.*, *11*, 799–806.
8. Gigant, B., Wang, C., Ravelli, R. B., et al. (2005). Structural basis for the regulation of tubulin by vinblastine. *Nature*, *435*, 519–522.

12.13 PLADIENOLIDE

Structure   see Figure 12-13.

Discovery   Pladienolides are 12-membered macrolides isolated as an antitumor substance by cell-based hypoxia-induced gene expression assay from *Streptomyces platensis* Mer-11107 [1]. The structures of pladienolide B and D were determined as **1** and **2**, respectively [2,3]. Total synthesis of pladienolide B [4] and biosynthetic gene cluster [5] is also reported.

Biochemical Studies   Pladienolide B is the most potent inhibitor for hypoxia-induced VEGF expression and proliferation of the U251 cancer cell line [6]. COMPARE analyses using a 39-cell line panel showed that pladienolide B has strong antitumor activities in vitro with a novel mechanism of action.

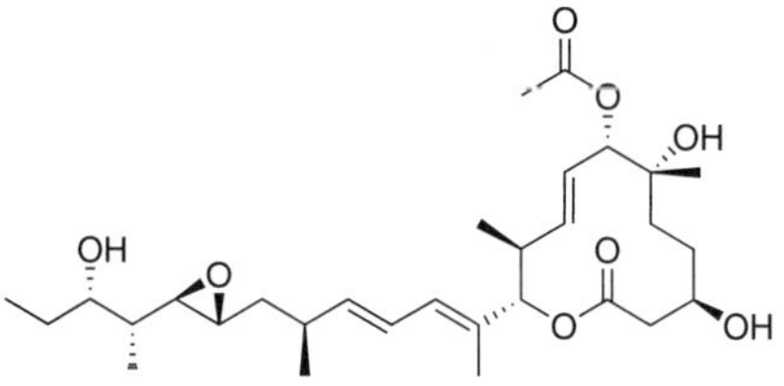

Figure 12-13 Pladienolide.

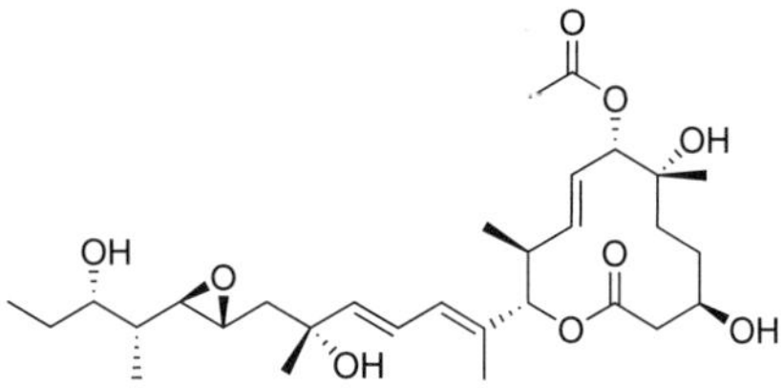

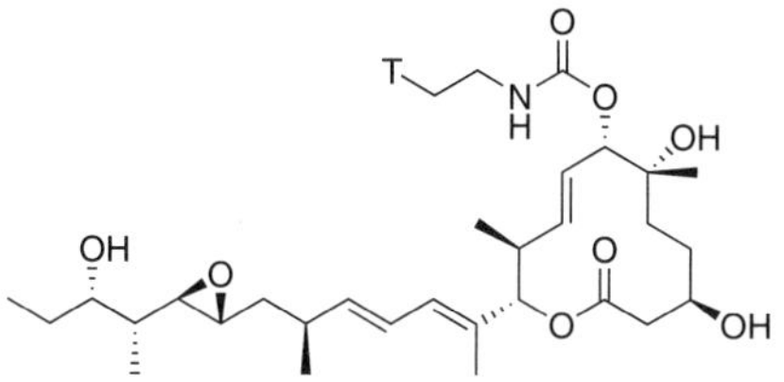

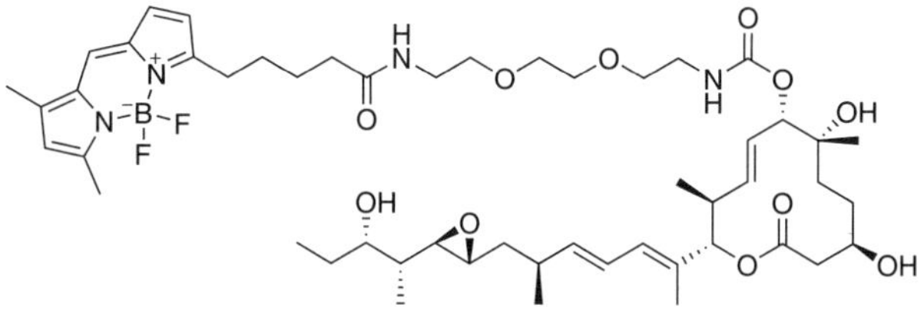

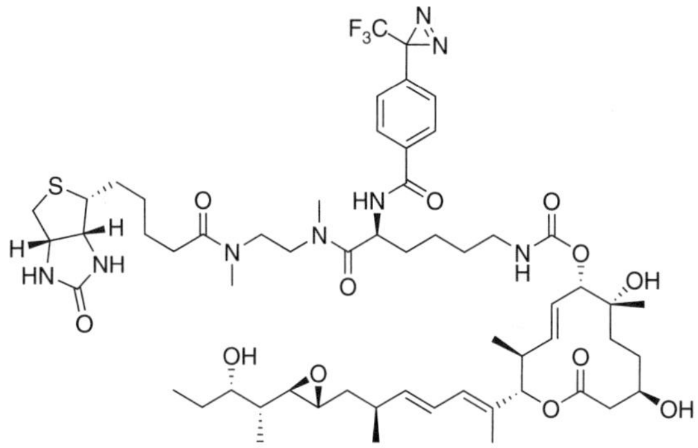

Further, pladienolide B inhibited tumor growth extensively in a xenograft model series.

Identification of Molecular Target To isolate the binding protein of pladienolide, [³H]-labeled (**3**), fluorescence-tagged (**4**) and photoaffinity/biotin (PB)-tagged (**5**) pladienolide B were synthesized and treated with HeLa cells [7]. The [³H] probe was localized specifically in the nuclear fraction with high affinity. The fluorescence probe was concentrated along granular structures in the nuclei. These granules overlapped with the localization of splicing factor SC-35, suggesting that the binding protein of pladienolide B is a splicing- or transcription-related molecule in the nuclear speckle. An immunoprecipitation experiment using several antibodies against nuclear speckle proteins from [^{3}H] probe-treated HeLa cells revealed that pladienolide B–binding protein is in the splicing machinery, U2 snRNP. Finally, SAP130 (SF3b subunit 3), a component of U2 snRNP, was identified as a pladienolide B–binding protein by treatment of immunoprecipitant with the PB probe following ultraviolet irradiation. Since pladienolide B impared splicing in vivo, and the unspliced mRNA accumulation was correlated with the cell growth inhibition against HeLa cells, it is suggested that splicing factor is a potential antitumor drug target.

REFERENCES

1. Sakai, T., Sameshima, T., Matsufuji, M., Kawamura, N., Dobashi, K., Mizui, Y. (2004). Pladienolides, new substances from culture of *Streptomyces platensis* Mer-11107: I. Taxonomy, fermentation, isolation and screening. *J. Antibiot.*, *57*, 173–179.

2. Sakai, T., Asai, N., Okuda, A., Kawamura, N., Mizui, Y. (2004). Pladienolides, new substances from culture of *Streptomyces platensis* Mer-11107: II. Physico-chemical properties and structure elucidation. *J. Antibiot.*, *57*, 180–187.

3. Asai, N., Kotake, Y., Niijima, J., Fukuda, Y., Uehara, T., Sakai, T. (2007). Stereochemistry of pladienolide B. *J. Antibiot.*, *60*, 364–369.

4. Kanada, R. M., Itoh, D., Nagai, M., et al. (2007). Total synthesis of the potent antitumor macrolides pladienolide B and D. *Angew. Chem. Int. Ed. Engl.*, *46*, 4350.

5. Machida, K., Arisawa, A., Takeda, S., et al. (2008). Organization of the biosynthetic gene cluster for the polyketide antitumor macrolide, pladienolide, in *Streptomyces platensis* Mer-11107. *Biosci. Biotechnol. Biochem.*, *72*, 2946–2952.

6. Mizui, Y., Sakai, T., Iwata, M., et al. (2004). Pladienolides, new substances from culture of *Streptomyces platensis* Mer-11107: III. in vitro and in vivo antitumor activities. *J. Antibiot.*, *57*, 188–196.

7. Kotake, Y., Sagane, K., Owa, T., et al. (2007). Splicing factor SF3b as a target of the antitumor natural product pladienolide. *Nat. Chem. Biol.*, *3*, 570–575.

12.14 RADICICOL

Structure   see Figure 12-14.

Discovery   Radicicol was isolated as an antifungal substance from *Monosporuim bonorden* [1]. The molecular formula was revised later as **1** [2].

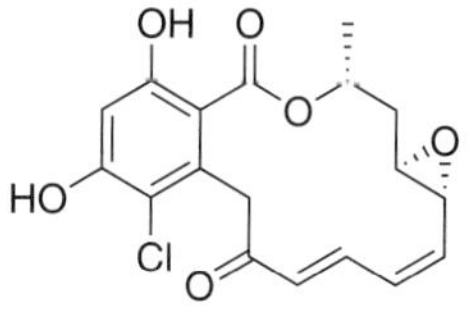

Figure 12-14 Radicicol.

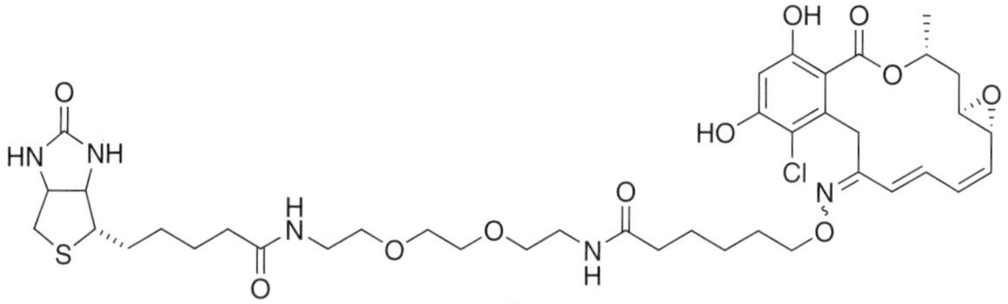

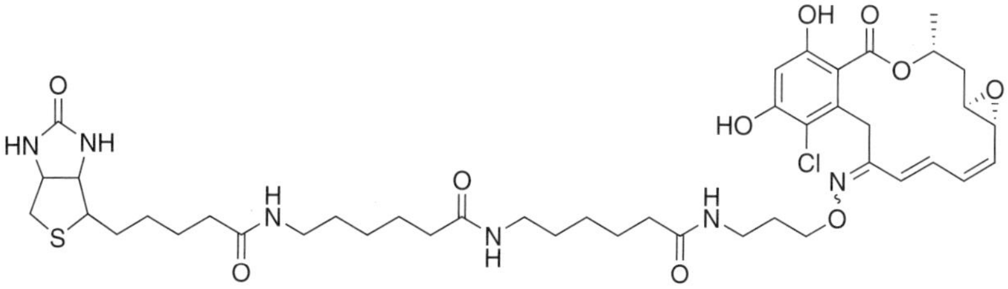

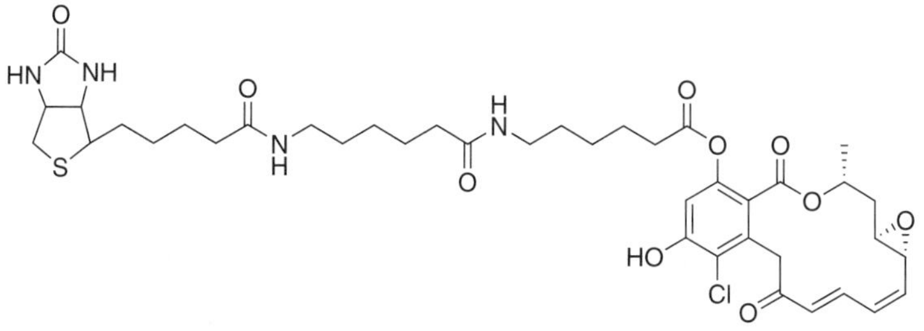

Biochemical Studies Radicicol was re-isolated as the compound that induces the reversal of transformed phenotypes not only of v-*src*-transformed fibroblasts [3] but also *ras*- and *mos*-transformed cells [4,5]. Anti-angiogenic effects were also reported [6].

Identification of Molecular Target The first report of the molecular target of radicicol was made by Sharma et al. [7]. They synthesized a biotinylated radicicol (**2**) and used as a probe to visualize cellular proteins that interact with radicicol. The most prominent cellular protein was determined as HSP90. Radicicol binds on N-terminal ATP/ADP-binding domain of HSP90 [8] and inhibited the inherent ATPase activity [9]. Crystal structure analyses revealed that radicicol plays as nucleotide mimicry [9]. The second report was made by Ki et al. [10]. They also synthesized two functional biotinylated derivatives (**3,4**) and found that derivative **3** binds HSP90, but derivative **4** binds another protein, ATP citrate lyase (ACL), a potential target for hypolipidaemic intervention. These results imply that a different part of radicicol is required for each specific binding and that unmodified radicicol can bind both the proteins. It is likely that introduction of the bulky biotin probe prevents these biotinylated compounds from access to one of these target molecules.

Other Target Proteins Recently it was reported that radiciol also inhibits arheal DNA topoisomerase VI (Topo VI) and human DNA topoisomerase II (Topo II) by a biochemical approach [11,12]. HSP90, Topo VI, and Topo II, which belong to the GHKL superfamily, share a unique structural ATP-binding motif called the Bergerat fold. Structural analysis revealed that the conservation between the radicicol-binding sites of HSP90 and TopoVI is remarkable [13], suggesting that radicicol may be useful in exploring the molecular mechanisms of GHKL family enzymes, notwithstanding that radicicol inhibits the ATPase activity of yeast HSP90 at least 1000-fold more effectively that is seen for archeal TopoVI.

REFERENCES

1. Delmotte, P., Delmotte-Plaquée, J. (1953). A new antifungal substance of fungal origin. *Nature*, *171*, 344.
2. McCapra, F., Scott, A. I., Delmotte, P., Delmotte-Plaquée, J., Bhacca, N. S. (1964). The constitution of monorden, an antibiotic with tranquilising action. *Tetrahedron Lett.*, *5*, 869–875.
3. Kwon, H. J., Yoshida, M., Fukui, Y., Horinouchi, S., Beppu, T. (1992). Potent and specific inhibition of p60^{v-src} protein kinase both in vivo and in vitro by radicicol. *Cancer Res.*, *52*, 6926–6930.
4. Kwon, H. J., Yoshida, M., Muroya, K., et al. (1995). Morphology of ras-transformed cells becomes apparently normal again with tyrosine kinase inhibitors without a decrease in the ras-GTP complex. *J. Biochem.*, *118*, 221–228.
5. Zhao, J. F., Nakano, H., Sharma, S. (1995). Suppression of RAS and MOS transformation by radicicol. *Oncogene*, *11*, 161–173.
6. Oikawa, T., Onozawa, C., Kuranuki, S., et al. (2007). Dipalmitoylation of radicicol results in improved efficacy against tumor growth and angiogenesis in vivo. *Cancer Sci.*, *98*, 219–225.
7. Sharma, S. V., Agatsuma, T., Nakano, H. (1998). Targeting of the protein chaperone, HSP90, by the transformation suppressing agent, radicicol. *Oncogene*, *16*, 2639–2645.

8. Schulte, T. W., Akinaga, S., et al. (1998). Antibiotic radicicol binds to the N-terminal domain of Hsp90 and shares important biologic activities with geldanamycin. *Cell Stress Chaperones*, *3*, 100–108.

9. Roe, S. M., Prodromou, C., O'Brien, R., Ladbury, J. E., Piper, P. W., Pearl, L. H. (1999). Structural basis for inhibition of the Hsp90 molecular chaperone by the antitumor antibiotics radicicol and geldanamycin. *J. Med. Chem.*, *42*, 260–266.

10. Ki, S. W., Ishigami, K., Kitahara, T., Kasahara, K., Yoshida, M., Horinouchi, S. (2000). Radicicol binds and inhibits mammalian ATP citrate lyase. *J. Biol. Chem.*, *275*, 39231–39236.

11. Gadelle, D., Bocs, C., Graille, M., Forterre, P. (2005). Inhibition of archaeal growth and DNA topoisomerase VI activities by the Hsp90 inhibitor radicicol. *Nucleic Acids Res.*, *33*, 2310–2317.

12. Gadelle, D., Graille, M., Forterre, P. (2006). The HSP90 and DNA topoisomerase VI inhibitor radicicol also inhibits human type II DNA topoisomerase. *Biochem. Pharmacol.*, *72*, 1207–1216.

13. Corbett, K. D., Berger, J. M. (2006). Structural basis for topoisomerase VI inhibition by the anti-Hsp90 drug radicicol. *Nucleic Acids Res.*, *34*, 4269–4277.

12.15   SC1 (PLURIPOTIN)

Structure   See Figure 12-15.

Discovery and Development   The core structure of pluripotin was identified from screening of the compound controlling the self-renewal of ES cells, which was derived from heterozygous Oct4 (marker protein of undifferentiated ES cells)-GFP transgenic OG2 mice. From a library of 50,000 discrete heterocycles, a class of 3,4-dihydropyrimido[4,5-*d*]pyrimidines (**1**) was characterized that maintained the undifferentiated phenotype of murine ES cells (mES cells) in a dose-dependent manner. After analyses of structure–activity relationships of a second-generation focused 3,4-dihydropyrimido[4,5-*d*]pyrimidines library, SC1 (pluripotin, **2**) was identified [1].

Biochemical Studies   The self-renewal of mES cells depends largely on LIF (leukemia inhibitory factor) and BMP (bone morphologic protein). However, SC1 maintained an mES ability to self-renew for more than 10 passages in an undifferentiated/pluripotent state and showed relatively low cellular toxicity. The compound reversibly inhibited differentiation of mES cells induced by either FBS or retinoic acid treatment, and after washout SC1, mES cells were selectively induced to differentiate into neural/neuronal, cardiac muscle, and endodermal cells.

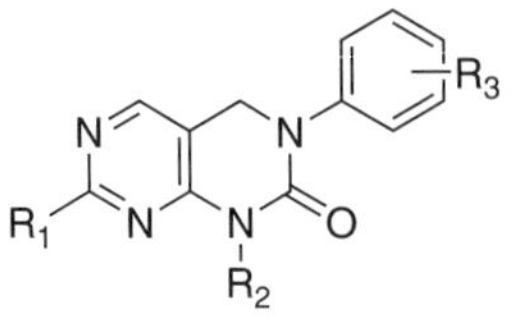

Figure 12-15   SC1.

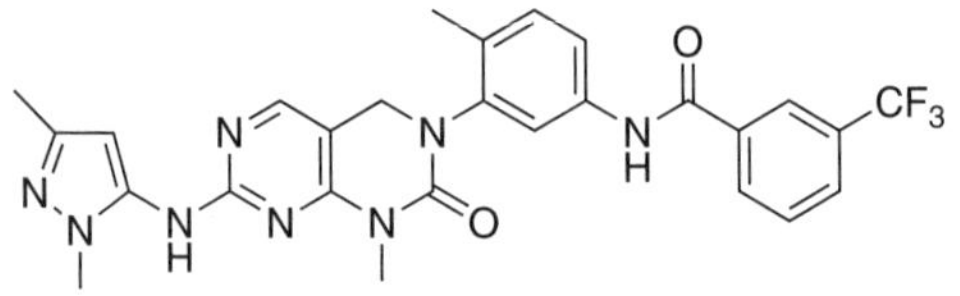

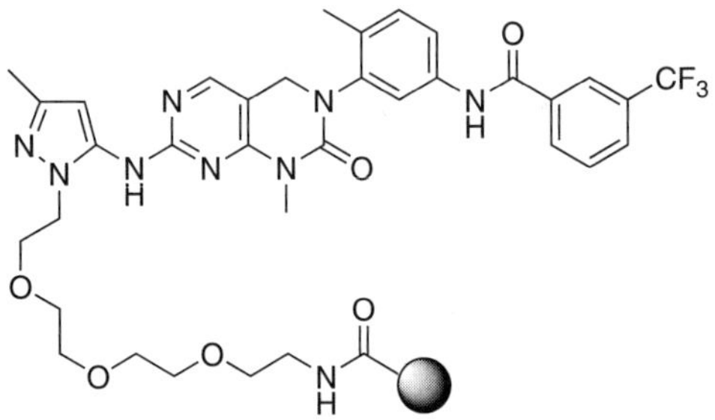

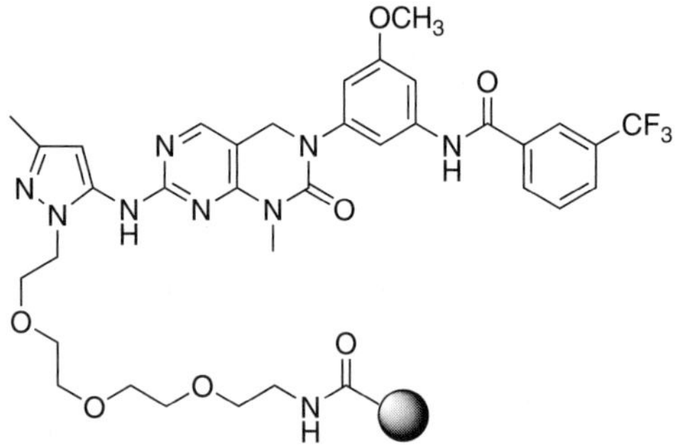

It is known that STAT activation, BMP-signaling-dependent Id1 expression, and canonical Wnt signaling are involved in the self-renewal of mES cells. However, SC1 did not affect these pathways, suggesting that pluripotin may function by a previously uncharacterized mechanism.

Identification of Molecular Target   SC1 and negative control compound was linked through the pyrazole N1 position on an agarose affinity matrix via a poly(ethylene glycol) linker (**3,4**) and identified the two binding proteins, ERK1 and RasGAP. Biochemical analyses showed that pluripotin inhibits ERK1 phosphorylation and increases the level of Ras-GTP. Ras-GTP activates both the ERK and PI3K pathways. The PI3K pathway is required for self-renewal of mES cells, but activation of the ERK pathway promotes differentiation. These results suggested that SC1 maintains the self-renewal of mES by inhibiting two targets; by inhibiting RasGAP, the compound activates both self-renewal and differentiation pathways while blocking ERK-dependent differentiation, which is downstream of Ras. Therefore, SC1 is a dual function small molecule inhibitor, and highlights the value of using small molecules in screens that probe signaling pathways as an alternative to genetic methods. Such a dual function inhibitor could modulate more than one target to achieve a desired biological effect.

REFERENCE

1. Chen, S., Do, J. T., Zhang, Q., et al. (2006). Self-renewal of embryonic stem cells by a small molecule. *Proc. Natl. Acad. Sci. USA*, *103*, 17266–17271.

12.16   SPERGUALIN

Structure   See Figure 12-16.

Discovery   Spergualin was isolated as an antitumor antibiotic from *Bacillus laterosporus* [1], and the structure was determined as **1** [2]. 15-Deoxyspergualin (DSG, **2**) is a synthetic derivative that shows stronger growth inhibition against mouse leukemia L-1210 than spergualin does [3].

Biochemical Studies   DSG inhibited the cell-cycle progression of tumor cell growth [4] and tumorigenic angiogenesis [5]. The compound also shows immuno-suppressive effects [6–9]. DSG has a spermidine and a guanidine moiety in its

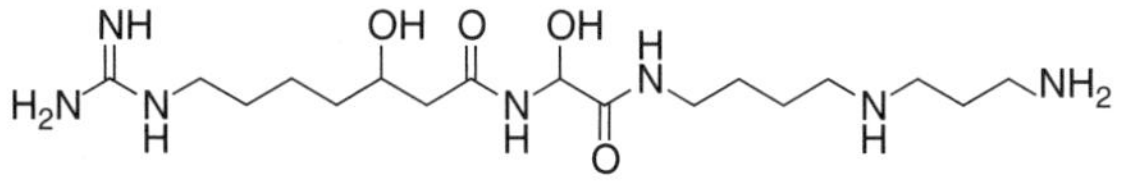

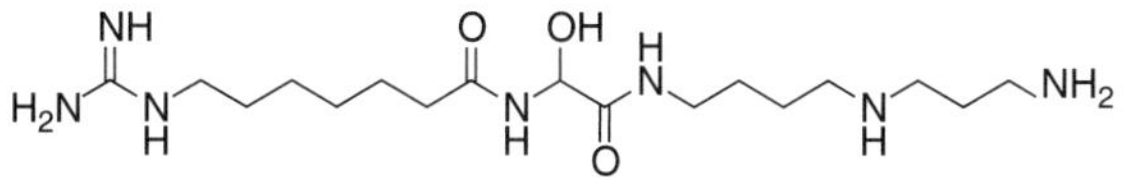

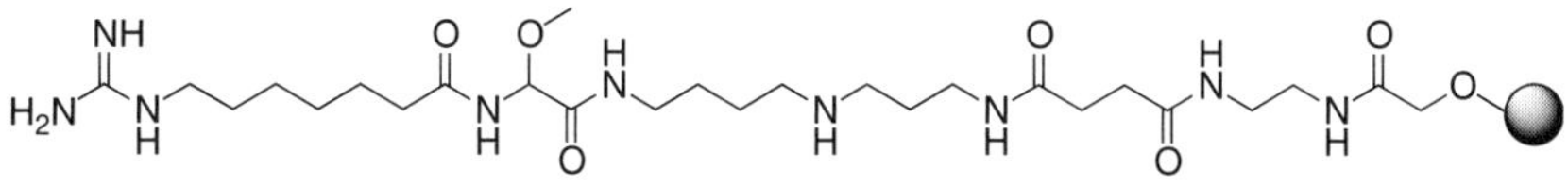

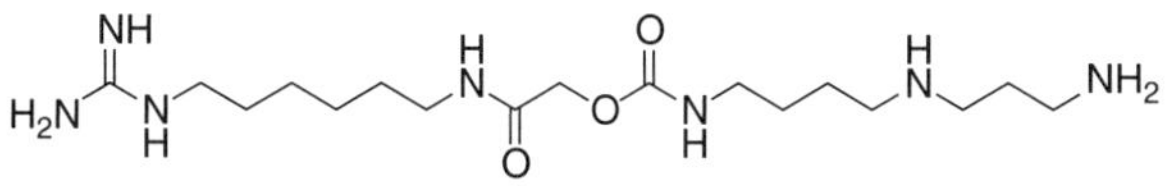

Figure 12-16   Spergualin, DSG, LF08-0299.

structure, and Kawada et al. showed that the spermidine moiety has cell-binding activity and that the guanidine moiety has cytotoxic activity [10].

Identification of Molecular Target   Nadler et al. identified an intracellular-binding protein of DSG by using 11-methoxy-DSG-sepharose (3) as Hsc70, the constitutive or cognate member of the Hsp70 protein family [11]. The members of the Hsp70 family of heat shock proteins are important for many cellular processes, including immune responses, and this finding suggests that heat shock proteins may represent a class of immunosuppressant binding proteins, or immunophilins. The binding site was determined as an EEVD regulatory domain at the extreme C-terminal of Hsc70 by EDC cross-linking of [^{14}C]DSG [12]. However, SDG binding on the EEVD domain did not affect the substrate binding to Hsc70. Furthermore, Tresperimus (4), the other derivative, acts at low concentrations, whereas tissue levels of Hsc70 are almost micromolar, so that only a tiny percentage of Hsc70 activity would be modified [13]. Recently, Kawada et al. reported that DSG inhibits tumor cell growth through the inhibition of protein synthesis and induction of apoptosis by the down-regulation of Akt kinase in a PI3K- and Hsp90-independent manner [14]. They found that phosphatidylcholine (PC) synthesis was inhibited prior to the down-regulation of Akt and suggested that the PC synthesis pathway is one candidate for a real DSG target [15]. To identify the physiological target molecule of DSG, it might be necessary to investigate the physiological phenotypes in detail.

REFERENCES

1. Takeuchi, T., Iinuma, H., Kunimoto, S., et al. (1981). A new antitumor antibiotic, spergualin: isolation and antitumor activity. *J. Antibiot.*, *34*, 1619–1621.

2. Umezawa, H., Kondo, S., Iinuma, H., et al. (1981). Structure of an antitumor antibiotic, spergualin. *J. Antibiot.*, *34*, 1622–1624.

3. Iwasawa, H., Kondo, S., Ikeda, D., Takeuchi, T., Umezawa, H. (1982). Synthesis of (−)-15-deoxyspergualin and (−)-spergualin-15-phosphate. *J. Antibiot.*, *35*, 1665–1669.

4. Nishikawa, K., Shibasaki, C., Takahashi, K., Nakamura, T., Takeuchi, T., Umezawa, H. (1986). Antitumor activity of spergualin, a novel antitumor antibiotic. *J. Antibiot.*, *39*, 1461–1466.

5. Oikawa, T., Shimamura, M., Ashino-Fuse, H., Iwaguchi, T., Ishizuka, M., Takeuchi, T. (1991). Inhibition of angiogenesis by 15-deoxyspergualin. *J. Antibiot.*, *44*, 1033–1035.

6. Umezawa, H., Ishizuka, M., Takeuchi, T., et al. (1985). Suppression of tissue graft rejection by spergualin. *J. Antibiot.*, *38*, 283–284.

7. Nemoto, K., Hayashi, M., Abe, F., Nakamura, T., Ishizuka, M., Umezawa, H. (1987). Immunosuppressive activities of 15-deoxyspergualin in animals. *J. Antibiot.*, *40*, 561–562.

8. Nemoto, K., Hayashi, M., Ito, J., et al. (1987). Effect of spergualin in autoimmune disease mice. *J. Antibiot.*, *40*, 1448–1451.

9. Fujii, H., Takada, T., Nemoto, K., et al. (1990). Deoxyspergualin directly suppresses antibody formation in vivo and in vitro. *J. Antibiot.*, *43*, 213–219.

10. Kawada, M., Someno, T., Inuma, H., Masuda, T., Ishizuka, M., Takeuchi, T. (2000). The long-lasting antiproliferative effect of 15-deoxyspergualin through its spermidine moiety. *J. Antibiot.*, *53*, 705–710.

11. Nadler, S. G., Tepper, M. A., Schacter, B., Mazzucco, C. E. (1992). Interaction of the immunosuppressant deoxyspergualin with a member of the Hsp70 family of heat shock proteins. *Science*, *258*, 484–486.

12. Nadler, S. G., Dischino, D. D., Malacko, A. R., Cleaveland, J. S., Fujihara, S. M., Marquardt, H. (1998). Identification of a binding site on Hsc70 for the immunosuppressant 15-deoxyspergualin. *Biochem. Biophys. Res. Commun.*, *253*, 176–180.

13. Komesli, S., Dumas, C., Dutartre, P. (1999). Analysis of in vivo immunosuppressive and in vitro interaction with constitutive heat shock protein 70 activity of LF08-0299 (*Tresperimus*) and analogues. *Int. J. Immunopharmacol.*, *21*, 349–358.

14. Kawada, M., Masuda, T., Ishizuka, M., Takeuchi, T. (2002). 15-Deoxyspergualin inhibits Akt kinase activation and phosphatidylcholine synthesis. *J. Biol. Chem.*, *277*, 27765–27771.

15. Kawada, M., Ishizuka, M. (2003). Inhibition of CTP: phosphocholine cytidylyltransferase activity by 15-deoxyspergualin. *J. Antibiot.*, *56*, 725–726.

12.17 SPLICEOSTATIN A

Structure   See Figure 12-17.

Discovery   Spliceostatin A (**1**) is a methyl ketal derivative of FR901464 (**2**), which was isolated as an antitumor substance from the culture broth of bacterium *Pseudomonas* sp. No. 2663 [1,2]. The structure of FR901464 was determined to be compound **2** on the basis of spectroscopic and chemical evidence [3]. Total syntheses of FR901464 were reported 4–6.

Biochemical Studies   FR901464 induced G1 and G2/M phase arrest in the cell cycle in M-8 tumor cells [2]. Despite the stimulation of cellular transcription of the SV40 promoter-dependent gene [1], FR901464 suppressed the transcription of some inducible endogenous genes, such as c-myc, E2F-1, p53, and p21 [2]. In the course of investigation of the molecular mechanism of cell-cycle inhibition, Kaida et al. found that FR901464 not only increased the CDK inhibitors p16 and p27, but also produced a C-terminally truncated p27 termed p27* [7]. Because p27* was not generated from p27 cDNA expression, p27* is not the degraded product.

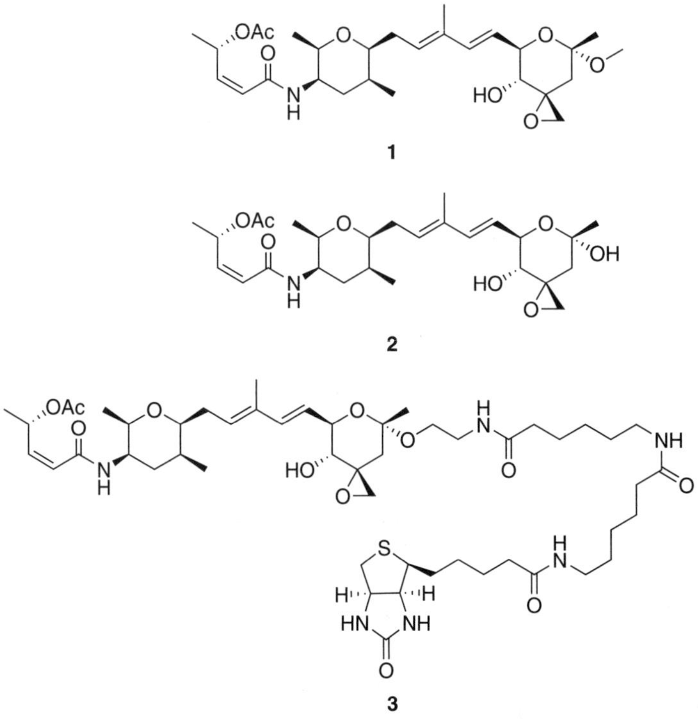

Figure 12-17 Spliceostatin A.

Identification of Molecular Target From SAR studies of FR901464, spliceostatin A was synthesized as the fully active and more stable derivative [8].
Using the biologically active biotinylated probe (biotinylated spliceostatin A,
3), Kaida et al. obtained proteins of molecular size between 130 and 160 kDa
as spliceostatin A-binding proteins [7]. These proteins were determined to be
SAP155, SAP145, and SAP130, all of which are known as components of the
SF3b splicing subcomplex in the U2 snRNP. Spliceostatin A inhibited splicing
not only of p27 in vitro and in vivo, but also of IκBα, β-tubulin, and β-actin
in vivo. Furthermore, spliceostain A treatment resulted in the formation of large
splicing factor–containing speckles in the nucleus, as seen in cell in which
SF3a was knocked down, and leakage of pre-mRNA to the cytoplasm. These
results indicate clearly that spliceostatin A targets SF3b and inhibits splicing of
pre-mRNA, and that the inhibition of pre-mRNA splicing at least during early
steps involving SF3b allows unspliced mRNA leakage and translation. Recently,
it was reported that spliceostatin A also inhibits splicing and nuclear retention of
pre-mRNA in fission yeast, suggesting that the SF3b complex has a conserved
role in pre-mRNA retention [9].

REFERENCES

1. Nakajima, H., Sato, B., Fujita, T., Takase, S., Terano, H., Okuhara, M. (1996). New antitumor substances, FR901463, FR901464 and FR901465: I. Taxonomy, fermentation, isolation, physico-chemical properties and biological activities. *J. Antibiot.*, *49*, 1196.

2. Nakajima, H., Hori, Y., Terano, H., et al. (1996). New antitumor substances, FR901463, FR901464 and FR901465: II. Activities against experimental tumors in mice and mechanism of action. *J. Antibiot.*, *49*, 1204.

3. Nakajima, H., Takase, S., Terano, H., Tanaka, H. (1997). New antitumor substances, FR901463, FR901464 and FR901465: III. Structures of FR901463, FR901464 and FR901465. *J. Antibiot.*, *50*, 96.

4. Thompson, C. F., Jamison, T. F., Jacobsen, E. N. (2001). FR901464: total synthesis, proof of structure, and evaluation of synthetic analogues. *J. Am. Chem. Soc.*, *123*, 9974.

5. Albert, B. J., Sivaramakrishnan, A., Naka, T., Koide, K. (2006). Total synthesis of FR901464, an antitumor agent that regulates the transcription of oncogenes and tumor suppressor genes. *J. Am. Chem. Soc.*, *128*, 2792.

6. Albert, B. J., Sivaramakrishnan, A., Naka, T., Czaicki, N. L., Koide, K. (2007). Total syntheses, fragmentation studies, and antitumor/antiproliferative activities of FR901464 and its low picomolar analogue. *J. Am. Chem. Soc.*, *129*, 48.

7. Kaida, D., Motoyoshi, H., Tashiro, E., et al. (2007). Spliceostatin A targets SF3b and inhibits both splicing and nuclear retention of pre-mRNA. *Nat. Chem. Biol.*, *3*, 576.

8. Motoyoshi, H., Horigome, M., Ishigami, K., et al. (2004). Structure–activity relationship for FR901464: a versatile method for the conversion and preparation of biologically active biotinylated probes. *Biosci. Biotechnol. Biochem.*, *68*, 2178.

9. Lo, C. W., Kaida, D., Nishimura, S., et al. (2007). Inhibition of splicing and nuclear retention of pre-mRNA by spliceostatin A in fission yeast. *Biochem. Biophys. Res. Commun.*, *364*, 573.

12.18   TRAPOXIN B

Structure   See Figure 12-18.

Discovery   Trapoxins A and B (**1** and **2**) were isolated as antitumor agents from the culture broth of *Helicoma ambiens* RF-1023 [1]. The structures were determined by x-ray analysis, mass spectrometry, NMR, and chemical studies [1,2].

Biochemical Studies   These compounds exhibit detransformation activities against v-*sis* oncogene-transformed NIH3T3 cells. Kijima et al. found that

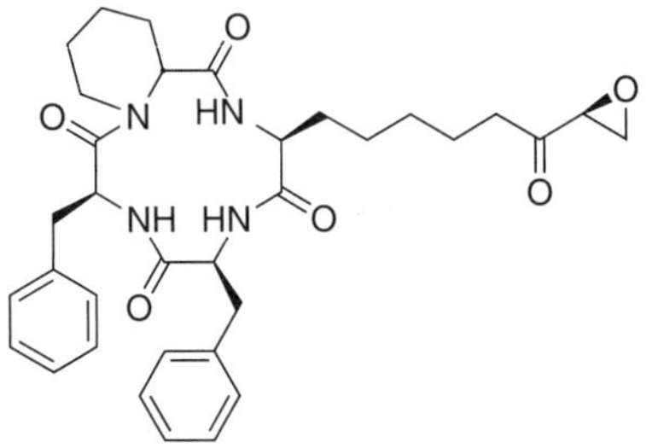

Figure 12-18   Trapoxins.

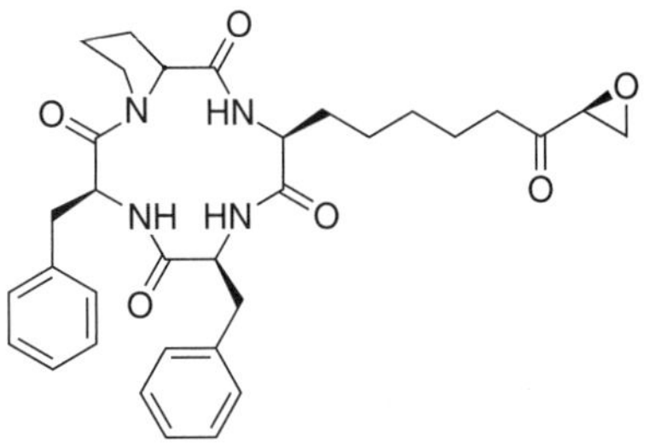

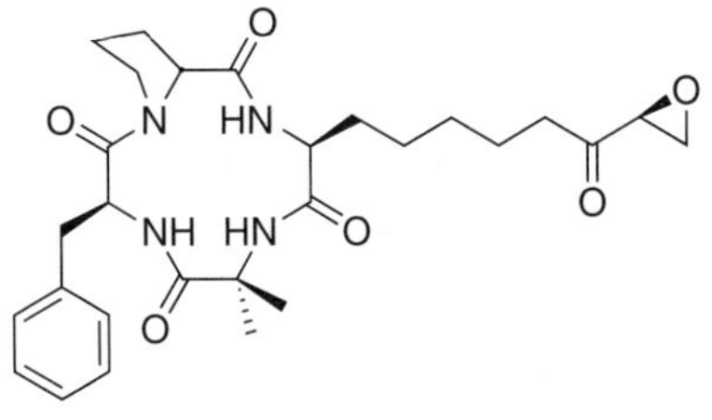

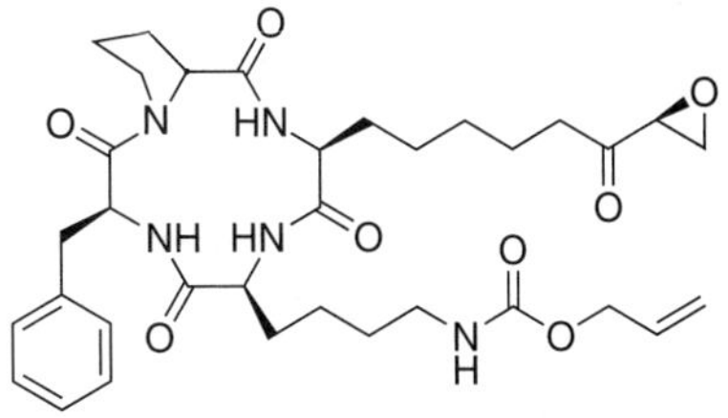

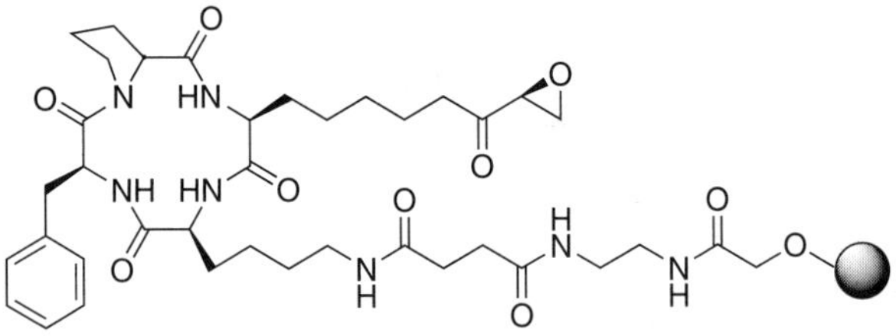

trapoxins caused accumulation of highly acetylated core histones in a variety of mammalian cell lines [3]. In vitro experiments using partially purified mouse histone deacetylase showed that a low concentration of trapoxin irreversibly inhibited deacetylation of acetylated histone molecules. Chemical reduction of an epoxide group in trapoxin completely abolished the inhibitory activity, suggesting that trapoxin binds covalently to the histone deacetylase via the epoxide.

Identification of Molecular Target   From a comparison with chlamydocin (**3**), Taunton et al. speculated that the benzyl side chain was not required for detransformation activity, and synthesized a bioactive trapoxin-like molecule, K-trap (**4**) [4,5]. They succeeded in isolating two nuclear proteins (55 and 50 kDa) that co-purified with histone deacetylase activity by using trapoxin affinity matrix, which was created by cross-linking K-trap with Affi-Gel10 (**5**). Both proteins were identified by peptide microsequencing, and a complementary DNA encoding the histone deacetylase catalytic subunit (p55, HDAC1) was cloned from a human Jurkat T-cell library [4]. After identification of HDAC1, the HDAC family (classes I to III) proteins have been identified and their functions in epigenetic gene regulation and post-translational regulation of nonhistone proteins are investigated using several HDAC inhibitors, including trapoxins [6].

REFERENCES

1. Itazaki, H., Nagashima, K., Sugita, et al. (1990). Isolation and structural elucidation of new cyclotetrapeptides, trapoxins A and B, having detransformation activities as antitumor agents. *J. Antibiot.*, *43*, 1524–1532.
2. Nakai, H., Nagashima, K., Itazaki, H. (1991). Structure of a new cyclotetrapeptide trapoxin A. *Acta Crystallogr. C*, *47*, 1496–1499.
3. Kijima, M., Yoshida, M., Sugita, K., Horinouchi, S., Beppu, T. (1993). Trapoxin, an antitumor cyclic tetrapeptide, is an irreversible inhibitor of mammalian histone deacetylase. *J. Biol. Chem.*, *268*, 22429–22435.
4. Taunton, J., Hassig, C. A., Schreiber, S. L. (1996). A mammalian histone deacetylase related to the yeast transcriptional regulator Rpd3p. *Science*, *272*, 408–411.
5. Taunton, J., Collins, J. L., Schreiber, S. L. (1996). Synthesis of natural and modified trapoxins, useful reagents for exploring histone deacetylase function. *J. Am. Chem. Soc.*, *118*, 10412–10422.
6. Marks, P. A., Miller, T., Richon, V. M. (2003). Histone deacetylases. *Curr. Opin. Pharmacol.*, *3*, 344–351.

12.19 TWS119

Structure   See Figure 12-19.

Discovery and Biochemical Studies   TWS119 (**1**) was discovered as a compound that induces neuronal differentiation in embryonic stem cells by a high-throughput phenotypic cell-based screen of kinase-directed combinatorial libraries [1]. Treatment of a monolayer of mouse P19 EC cells with TWS119 caused cells to differentiate into neuronal lineages with correct neuronal morphology and positive staining for βIII-tubulin, MAP2, and neurofilament-M. Because TWS119 induces neuronal differentiation of embryonic stem cells without embryoid body formation and retinoic acid treatment, it was suggested that the compound acts by a novel mechanism.

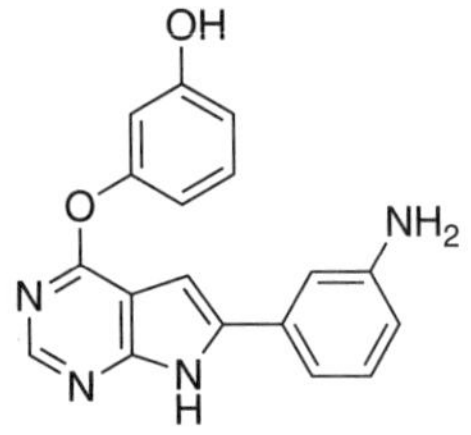

Figure 12-19   TWS119.

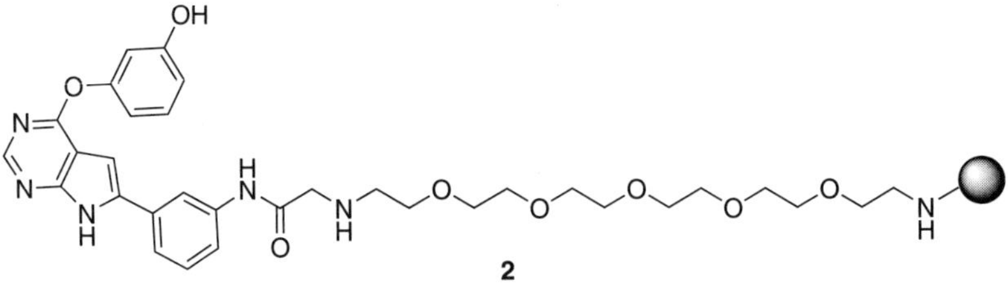

Identification of Molecular Target   To identify the cellular molecular target of TWS119, the compound was linked through the anilino group to an agarose affinity matrix (**2**) based on the structure–activity relationship data. Two proteins (ca. 47 and 49 kDa), which were bound by the affinity matrices in the absence of free TWS119, were detected and identified to be GSK−3β by LC/MS. It is well known that GSK−3β is involved in Wnt signal and down-regulates β-catenin by phosphorylation. TWS119 caused the accumulation of β-catenin and the activation of β-catenin-induced TCF/LEF reporter activity in a dose-dependent manner. These results suggest that TWS119 binds to GSK−3β and activates Wnt signal cascade by stabilizing β-catenin.

REFERENCE

1. Ding, S., Wu, T. Y., Brinker, A., et al. (2003). Synthetic small molecules that control stem cell fate. *Proc. Nat. Acad. Sci. USA*, *100*, 7632–7637.

12.20   WITHAFERIN A

Structure   See Figure 12-20.

Discovery and Biochemical Studies   Withaferin A (WA, **1**) is a highly oxygenated steroidal lactone that is found in the medicinal plant *Withania somnifera* [1,2], a plant widely researched for its anti-inflammatory, cardioactive, and central nervous system effects. WA shows several biological effects, such as antibacterial [3], antitumor [4], anti-angiogenesis [5], and immunosuppressive activities [6], and several target proteins were reported: for example, proteasome [7] and IκB kinase β [8]. However, in this section the molecular target determined by chemical modification is shown below.

Identification of Molecular Target and Binding Site WA was re-isolated as an anticancer agent, showing the unusual pattern of cytoplasmic projections and rapid loss of substrate adhesion [9]. Biotinylated WA (**2**) was synthesized based on the structure–activity relationship data, and the binding protein, which was displaced efficiently by prior addition of unmodified WA, was determined to be annexin II. WA bound to annexin II covalently, and LC/MS-MS analysis revealed that the WA-binding fragment contained Cys133. A binding model of WA bound to Cys133 showed that WA resided in the concave portion of annexin II, which has been described as its actin-binding region. Binding WA to annexin II stimulated the basal F-actin cross-linking activity of annexin II, and WA-mediated disruption of F-actin organization is dependent on the annexin II expression level. Furthermore, WA markedly limits their migratory and invasive capabilities at subcytotoxic concentrations, suggesting that the antitumor and anti-angiogenesis activities of WA might be carried out by stimulation of annexin II activity.

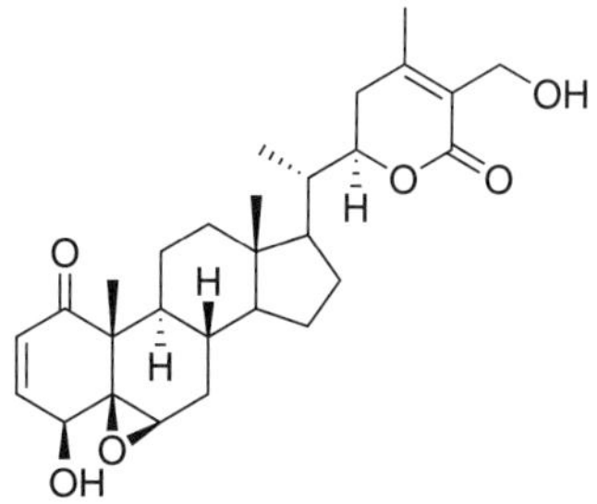

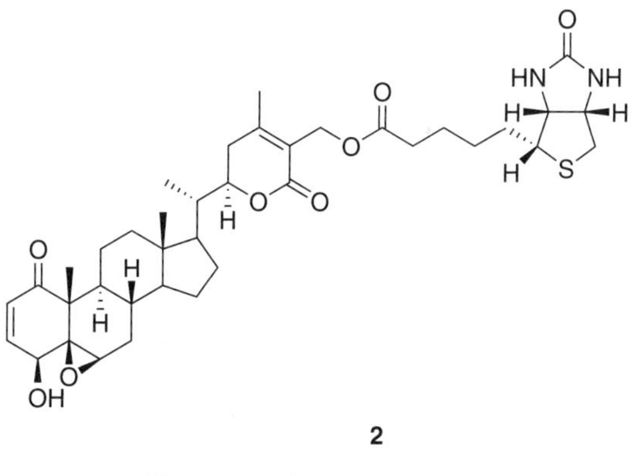

Figure 12-20 Withaferin A.

REFERENCES

1. Lavie, D., Glotter, E., Shvo, Y. (1965). Constituents of *Withania somnifera* Dun: IV. The structure of withaferin A. *J. Chem. Soc.*, 7517–7531.

2. Kupchan, S. M., Doskotch, R. W., Bollinger, P., McPhail, A. T., Sm, G. A., Renauld, J. A. (1965). The isolation and structural elucidation of a novel steroidal tumor inhibitor from *Acnistus arborescens*. *J. Am. Chem. Soc.*, *87*, 5805–5806.

3. Ben-Efraim, S., Yarden, A. (1962). Comments on the activity of some constitutents of *Withania somnifera* Dun: I. The antibacterial activity of five nitrogen-free substances isolated from the leaves. *Antibiot. Chemother. Fortschr. Adv. Prog.*, *12*, 576–582.

4. Shohat, B., Gitter, S., Abraham, A., Lavie, D. (1967). Antitumor activity of withaferin A (NSC-101088). *Cancer Chemother. Rep.*, *51*, 271–276.

5. Mohan, R., Hammers, H. J., Bargagna-Mohan, P., et al. (2004). Withaferin A is a potent inhibitor of angiogenesis. *Angiogenesis*, *7*, 115–122.

6. Shohat, B., Kirson, I., Lavie, D. (1978). Immunosuppressive activity of two plant steroidal lactones withaferin A and withanolide E. *Biomedicine*, *28*, 18–24.

7. Yang, H., Shi, G., Dou, Q. P. (2007). The tumor proteasome is a primary target for the natural anticancer compound withaferin A isolated from "Indian winter cherry." *Mol. Pharmacol.*, *71*, 426–437.

8. Kaileh, M., Vanden Berghe, W., Heyerick, A., et al., (2007). Withaferin A strongly elicits IκB kinase β hyperphosphorylation concomitant with potent inhibition of its kinase activity. *J. Biol. Chem.*, *282*, 4253–4264.

9. Falsey, R. R., Marron, M. T., Gunaherath, G. M., et al. (2006). Actin micro-filament aggregation induced by withaferin A is mediated by annexin II. *Nat. Chem. Biol.*, *2*, 33–38.

INDEX

Protein Targeting with Small Molecules: Chemical Biology Techniques and Applications,
Edited by Hiroyuki Osada
Copyright © 2009 John Wiley & Sons, Inc.